Middle and Upper Ordovician Nautiloid Cephalopods of the Cincinnati Arch Region of Kentucky, Indiana, and Ohio

By ROBERT C. FREY

CONTRIBUTIONS TO THE ORDOVICIAN PALEONTOLOGY OF KENTUCKY AND NEARBY STATES

Edited by JOHN POJETA, JR.

U.S. GEOLOGICAL SURVEY PROFESSIONAL PAPER 1066-P

Prepared in cooperation with the Commonwealth of Kentucky, University of Kentucky, and Kentucky Geological Survey

Stratigraphic distribution, paleoecology, biogeography, and systematic paleontology of 50 species of Ordovician nautiloids from the midcontinent

UNITED STATES GOVERNMENT PRINTING OFFICE, WASHINGTON : 1995

U.S. DEPARTMENT OF THE INTERIOR
BRUCE BABBITT, *Secretary*

U.S. GEOLOGICAL SURVEY
GORDON P. EATON, *Director*

Any use of trade, product, or firm names in this publication is for descriptive purposes only and does not imply endorsement by the U.S. Government.

Library of Congress Cataloging in Publication Data

Frey, Robert C.
Middle and Upper Ordovician nautiloid cephalopods of the Cincinnati arch region of
  Kentucky, Indiana, and Ohio / by Robert C. Frey.
p.   cm. — (U.S. Geological Survey professional paper ; 1066–P) (Contributions to the
  Ordovician paleontology of Kentucky and nearby states)
Includes bibliographical references and index.
Supt. of Docs. no.: I 19.16:1066–P
1. Nautiloidea, Fossil—Kentucky. 2. Nautiloidea, Fossil—Indiana. 3. Nautiloidea,
  Fossil—Ohio. 4. Paleontology—Ordovician. 5. Animals, Fossil—Kentucky.
  6. Animals, Fossil—Indiana. 7. Animals, Fossil—Ohio. I. Title. II. Series.
  III. Series: Contributions to the Ordovician paleontology of Kentucky and nearby states.
QE807.N4F74   1995
564'.52'0977—dc20   94–27353
                                                                                     CIP

For sale by U.S. Geological Survey, Information Services
Box 25286, Federal Center, Denver, CO 80225

CONTENTS

	Page
Abstract	P1
Introduction	2
Acknowledgments	3
Regional Ordovician Paleogeographic Setting	3
Ordovician Lithostratigraphy in the Cincinnati Arch Region	4
Middle Ordovician Champlainian Provincial Series	
Lithostratigraphy in Kentucky	6
High Bridge Group	6
Lexington Limestone	6
Upper Ordovician Cincinnatian Provincial Series	
Lithostratigraphy	8
Edenian Stage	8
Maysvillian Stage	9
Richmondian Stage	9
Distribution and Faunal Affinities of Ordovician Nautiloids in the Cincinnati Arch Region	10
Nautiloid Fauna of the Tyrone Limestone	10
Rocklandian Nautiloid Extinction	12
Nautiloid Fauna of the Lexington Limestone	15

	Page
Distribution and Faunal Affinities of Ordovician Nautiloids in the Cincinnati Arch Region—Continued	
Nautiloid Fauna of the Cincinnatian Provincial Series	P16
Edenian Nautiloid Fauna	16
Maysvillian and Early Richmondian Nautiloid Fauna	17
Nautiloid Fauna of the Leipers Limestone	19
Middle and Late Richmondian Nautiloid Faunas	19
Late Ordovician Extinction Event	20
Nautiloid Taphonomy	21
Nautiloid Biostratinomy	21
Silicified Nautiloids	22
Nonsilicified Nautiloids	22
Nautiloid Classification	23
Systematic Descriptions	24
Systematic Paleontology	27
References Cited	113
Appendix: Locality Register	120
Index	125

ILLUSTRATIONS

[Plates follow index]

PLATE
1. *Cartersoceras* and *Murrayoceras*
2. *Pojetoceras* and *Pleurorthoceras*
3. *Ordogeisonoceras* and *Geisonoceras*
4. *Proteoceras* and *Treptoceras*
5–8. *Treptoceras*
9. *Monomuchites*
10. *Isorthoceras*
11. *Polygrammoceras*?
12. *Polygrammoceras*, *Anaspyroceras*, and *Gorbyoceras*
13. *Richmondoceras*
14. *Cameroceras*
15. *Vaginoceras*
16. *Triendoceras*? and *Actinoceras*
17. *Actinoceras*
18. *Ormoceras*, *Deiroceras*, and *Orthonybyoceras*
19. *Plectoceras*, *Trocholites*, and *Oncoceras*
20. *Oncoceras*, *Beloitoceras*, *Maelonoceras*, *Laphamoceras*, *Allumettoceras*, *Augustoceras*, and *Manitoulinoceras*
21. *Diestoceras* and *Beloitoceras*
22. *Diestoceras* and *Beloitoceras*

CONTENTS

		Page
Figure	1. Map showing Middle Ordovician paleogeography in Eastern North America	P4
	2,3. Charts showing:	
	2. Ordovician lithostratigraphy in the Cincinnati arch region	5
	3. Lexington Limestone lithostratigraphy	7
	4,5. Diagrams showing:	
	4. External features of nautiloids	25
	5. Internal features of nautiloids	26
	6. Drawing showing internal features of *Cartersoceras shideleri*	29
	7. Bivariate plot of measurements made on specimens of *Pojetoceras floweri* and *Orthoceras regularis*	34
	8. Drawing showing internal features of *Pojetoceras floweri*	36
	9–12. Bivariate plot of measurements made on specimens of:	
	9. *Pleurorthoceras clarksvillense* and *P. selkirkense*	38
	10. "*Orthoceras*" *amplicameratum*	41
	11. Middle Ordovician species of *Proteoceras*	44
	12. Cincinnatian species of *Treptoceras*	48
	13. Drawings comparing adapical siphuncle segment shape for specimens of *Treptoceras duseri* and *T. fosteri*	49
	14–16. Bivariate plots of measurements made on specimens of *Treptoceras duseri*, *T. fosteri*, and *T. cincinnatiensis*	50
	17. Drawings comparing external ornament in species of *Monomuchites* from the USGS 6034-CO collection in the Tyrone Limestone	57
	18–19. Bivariate plots of measurements made on specimens of:	
	18. *Isorthoceras*	61
	19. *Richmondoceras brevicameratum* and *Treptoceras fosteri*	72
	20. Drawings comparing cross-sectional features of specimens of *Cameroceras rowenaense*, *C. trentonense*, and *C. inaequabile*	75
	21. Bivariate plot of measurements made on specimens of *Cameroceras rowenaense* and *C. trentonense*	76
	22. Diagram of siphuncular structure in *Triendoceras? davisi*	80
	23. Diagrams showing important morphological features used to distinguish species of *Actinoceras*	83
	24. Bivariate plots of measurements made on specimens of *Ormoceras ferecentricum*, *O. allumettense*, and *O. cannonense*	87
	25. Drawing reconstructing the complete shell of *Plectoceras* cf. *P. carletonense* from the USGS 6034-CO collection in the Tyrone Limestone	93
	26. Drawings of the apertures of species of *Diestoceras*	107

TABLES

		Page
Table	1. Distribution of species of nautiloids in the Tyrone Limestone, the Lexington Limestone, and the Clays Ferry Formation in the Blue Grass region of Kentucky	P11
	2. Comparison of the 6034-CO nautiloid fauna from the Tyrone Limestone with nautiloid taxa from coeval strata exposed in Eastern and Central North America	12
	3. List of Middle Ordovician (Blackriveran and Rocklandian) nautiloid genera from Eastern North America	14
	4. Nautiloid species from the Strodes Creek Member of the Lexington Limestone, north-central Kentucky	17
	5. Distribution of species of nautiloids from the Cincinnatian Provincial Series in the Cincinnati arch region	18
	6–11. Comparisons of the morphological features of:	
	6. Baltoceratids *Cartersoceras* and *Murrayoceras*	30
	7. Cincinnatian species of *Treptoceras*	48
	8. Species of *Monomuchites* from the Middle Ordovician Rocklandian Stage of Eastern North America	56
	9. Species of *Cameroceras* from the Middle Ordovician of New York and the Upper Ordovician of the Cincinnati arch region of Kentucky, Indiana, and Ohio	76
	10. Middle Ordovician (Blackriveran and Rocklandian) species of *Actinoceras* from Eastern North America	84
	11. Cincinnatian species of *Diestoceras*	106

CONTRIBUTIONS TO THE ORDOVICIAN PALEONTOLOGY OF KENTUCKY AND NEARBY STATES

MIDDLE AND UPPER ORDOVICIAN NAUTILOID CEPHALOPODS OF THE CINCINNATI ARCH REGION OF KENTUCKY, INDIANA, AND OHIO

By ROBERT C. FREY

ABSTRACT

This paper describes the taxonomy, phylogenetic relations, biostratigraphy, biogeography, and paleoecology of 50 species and 30 genera of Middle and Upper Ordovician (Rocklandian to Richmondian) nautiloid cephalopods from the Cincinnati arch region of Kentucky, Indiana, and Ohio. The study centers on collections of silicified material made by the U.S. Geological Survey and the Kentucky Geological Survey during the geologic mapping program of the State of Kentucky. This material, primarily of Middle Ordovician age, is augmented by additional nonsilicified specimens from the Upper Ordovician Cincinnatian Provincial Series in Kentucky, Indiana, and Ohio.

Silicifed collections from the Ordovician of Kentucky yielded a number of more or less complete, uncrushed, nautiloid specimens that commonly preserve the details of the shell's external ornament. Key internal features, such as connecting rings and septal necks, are often lacking, complicating systematic placement of some orthocerid taxa. Critical cameral and endosiphuncular deposits are, however, well preserved in some actinocerid and endocerid taxa. Well-preserved nautiloids from the Cincinnatian Provincial Series consist primarily of calcite-replaced internal molds from clay shale units in Indiana and Ohio, and from the Leipers Limestone in south-central Kentucky. Additional well-preserved, but fragmentary nautiloids also occur in the Hitz Limestone Member of the Saluda Formation (Richmondian) in southeastern Indiana and in equivalent rocks in adjacent portions of northern Kentucky.

Nautiloids from the Tyrone Limestone (Middle Ordovician, Rocklandian) in central Kentucky consist of 18 species belonging to 14 genera, representing 6 nautiloid orders. This fauna is characterized by an abundance of small longicones, namely, the baltoceratids *Cartersoceras* and *Murrayoceras* and the orthocerids *Pojetoceras* and *Proteoceras*, plus the large actinocerid *Actinoceras*, the large endocerid *Vaginoceras*, the annulated orthocerid *Monomuchites*, and a number of oncocerids: species of *Oncoceras*, *Beloitoceras*, *Maelonoceras*, and *Laphamoceras*. The coiled tarphycerid *Plectoceras* is also present, but typically is represented by fragmentary specimens.

Nautiloid faunas in the overlying Lexington Limestone (Middle Ordovician, Kirkfieldian and Shermanian) in central Kentucky are less diverse, being most abundant in collections from low-energy, deeper-water facies in the Logana Member and Grier Limestone Member.

Manuscript approved for publication May 31, 1994.

Faunas consist of abundant small longicones, primarily the orthocerid *Isorthoceras*, and lesser numbers of the orthocerids *Gorbyoceras* and *Polygrammoceras*, the endocerid *Cameroceras*, and the oncocerids *Oncoceras* and *Allumettoceras*. Nautiloid faunas increase in diversity toward the top of the Middle Ordovician section in northern Kentucky. *Isorthoceras* occurs in association with the orthocerids *Ordogeisonoceras* and *Treptoceras* and with the small coiled tarphycerid *Trocholites* in the Point Pleasant Tongue of the Clays Ferry Formation.

A low-diversity nautiloid fauna consisting of the small orthocerid *Treptoceras* and the endocerid *Triendoceras*? occurs in the basal portions of the succeeding Cincinnatian Provincial Series in the Kentucky-Indiana-Ohio area (Kope Formation, Edenian, and Maysvillian). Abundant well-preserved specimens of several species of the orthocerid *Treptoceras* and the large endocerid *Cameroceras* occur in younger Maysvillian to early Richmondian blocky claystone facies within the Fairview, Grant Lake, and Bull Fork Formations. A diverse nautiloid fauna characterized by the abundance of the large discosorid *Faberoceras* and the slender oncocerid *Augustoceras*, as well as the ubiquitous *Treptoceras* and *Cameroceras* occurs in the Leipers Limestone (Maysvillian) in the valley of the Cumberland River in south-central Kentucky.

Younger Richmondian nautiloid faunas in the upper part of the Cincinnatian Provincial Series in the region consist of a diverse set of immigrant taxa, more characteristic of older or equivalent carbonate facies exposed in the Western United States, Canada, and northern Greenland. These faunas are species of the orthocerids *Gorbyoceras* and *Pleurorthoceras*, the ascocerid *Schuchertoceras*, the actinocerids *Armenoceras* and *Lambeoceras*, the coiled tarphycerid *Charactoceras*, and numerous oncocerids. The latter include *Beloitoceras*, *Neumatoceras*, *Manitoulinoceras*, *Oonoceras*, and *Diestoceras*. These taxa are common to facies within the Saluda, Whitewater, upper part of the Bull Fork, and Drakes Formations in Indiana, southwestern Ohio, and north-central Kentucky.

Several families within the order Orthocerida are redefined, and the diagnostic morphologic features of the endocerids *Cameroceras* and *Vaginoceras* are clarified. Three new genera, all members of the order Orthocerida, are described. These are *Pojetoceras*, type species *P. floweri*, new species, from the Tyrone Limestone in Kentucky; *Ordogeisonoceras*, type species *O. amplicameratum* (Hall), from the Point Pleasant Tongue of the Clays Ferry Formation in northern Kentucky; and *Richmondoceras*, type species *R. brevicameratum* n. sp., from the Saluda Formation in Indiana and the upper part of the Bull Fork Formation in Ohio. Additional new species described here are *Cartersoceras popei*,

Monomuchites obliquum, *M. annularis*, and *Oncoceras major*, all from the Tyrone Formation in central Kentucky; *Cameroceras rowenaense* from the Leipers Limestone in south-central Kentucky; and *Triendoceras? davisi* from the Kope Formation in northern Kentucky.

Nautiloids from Middle and Upper Ordovician strata in the Cincinnati arch region can be divided into four major time-bounded faunas, separated by regional environmental and (or) biological events. A major extinction of nautiloid taxa coinciding with a widespread sea-level regression and a catastrophic volcanic event separates the diverse, tropical "Black River" fauna (Chazyan to Rocklandian) from the species-poor "Trenton" fauna characteristic of the Lexington Limestone (Kirkfieldian and Shermanian). A regional transgressive event coupled with an increase in clastic sedimentation in the area led to the development of the "Cincinnatian" fauna characteristic of Edenian to early Richmondian strata across the area. This fauna is superseded by the "Richmondian" fauna in the middle and late Richmondian, following another regional transgressive event and a shift to more carbonate-rich lithofacies. This fauna is an immigrant fauna that was derived from older Edenian and Maysvillian faunas characteristic of carbonate platform facies in western and arctic portions of North America. This "Richmondian" fauna was terminated by the Late Ordovician mass extinction event, which is associated with the withdrawal of epeiric seas from the Cincinnati arch region.

INTRODUCTION

This study is one of a series of papers documenting the Ordovician fossils of the Cincinnati arch region of Kentucky, Indiana, and Ohio. Previous chapters in this series are the introductory chapter, Chapter A, by Pojeta (1979); papers on the articulate brachiopods, Neuman (1967), Chapter B (Alberstadt, 1979), Chapter C (Howe, 1979), Chapter L (Pope, 1982), and Chapter M (Walker, 1982); trepostome and cystoporate bryozoans, Chapter I (Karklins, 1984); corals, Chapter N (Elias, 1983); symmetrical univalved mollusks, Chapter O (Wahlman, 1992); trilobites, Ross (1967), Chapter D (Ross, 1979); ostracodes, Chapter H (Warshauer and Berdan, 1982); and Chapter J (Berdan, 1984); echinoderms, Chapter K (Parsley, 1981); edrioasteroids, Chapter E (Bell, 1979); asteroids, Chapter F (Branstrator, 1979); and a discussion of conodont faunas and their biostratigraphy in the region, Chapter G (Sweet, 1979).

This chapter describes the nautiloid cephalopod faunas of the Middle Ordovician (Rocklandian) Tyrone Limestone and the overlying Middle Ordovician (Kirkfieldian and Shermanian) Lexington Limestone from the Blue Grass region of central and north-central Kentucky. Additional nautiloids from the Upper Ordovician Cincinnatian Provincial Series in northern Kentucky, the Cumberland region of south-central Kentucky, and portions of Indiana and southwestern Ohio are also described. An index map of the area of this report is presented by Pojeta (1979, fig. 3).

With a few notable exceptions, nautiloid cephalopods from the Ordovician section exposed in the study area, especially faunas from the Middle Ordovician Tyrone Limestone and Lexington Limestone, have not been previously extensively studied in a comprehensive manner. S.A. Miller (1875) described and illustrated a number of species of "*Orthoceras*" and "*Cyrtoceras*," primarily from the Cincinnatian Provincial Series exposed in the Cincinnati, Ohio, area. Hall and Whitfield (1875) described the additional species, *Orthoceras carleyi*, *O. duseri*, *O. turbidum*, and *Gomphoceras eos.*, from Upper Ordovician strata exposed in southwestern Ohio. James (1886) published a compendium of the described Ordovician cephalopod taxa from the Cincinnati area, consisting of brief descriptions and notations on the occurrence of these taxa within these rocks. Miller and Faber (1894) described *Orthoceras ludlowense* and *O. albersi* from the Middle Ordovician rocks of north-central Kentucky and *Gomphoceras indianense* from Cincinnatian strata in Indiana. Cumings (1908) presented descriptions and illustrations of a number of the more common species of nautiloids that had been described from Upper Ordovician strata in Indiana over the previous half century. Foerste (1910) identified the endocerid *Suecoceras inaequabile* (Miller) from the Waynesville Formation in Ohio, described *Orthoceras hammelli* from the Hitz Limestone Member of the Saluda Formation in Indiana, and *Orthoceras bilineatum-frankfortense* and *Orthoceras (Loxoceras) milleri* from what he called the Trenton Limestone of the Blue Grass region of Kentucky. In a second paper, Foerste (1912) described *Orthoceras tyronensis* from the Tyrone Limestone in central Kentucky. Bassler (1915) included Ordovician nautiloids from the region in his index of Ordovician and Silurian fossils, borrowing heavily from a list of taxa and their occurrences provided by Nickles (1902). Foerste (1914a) described *Orthoceras rogersensis* from the "Cynthiana Formation" exposed at Rogers Gap in north-central Kentucky. Foerste published a number of papers through the 1920's and early 1930's, illustrating and redescribing a number of nautiloids from the Ordovician strata exposed in the area. The same author (Foerste, 1928b) erected the new genus *Troedssonoceras* and figured and described under this name a specimen from the Leipers Formation of south-central Kentucky. Foerste (1929b) figured the apical end of an endocerid from Richmondian rocks in Indiana and described and illustrated *Trocholites faberi* from the "Cynthiana Limestone" of north-central Kentucky.

The first modern systematic study of nautiloids from the region was the study by Foerste and Teichert (1930) describing actinocerid cephalopods from Kentucky and adjacent portions of Tennessee and Missouri. This paper provided detailed descriptions and photographic plates of six species of *Actinoceras* and one species of *Ormoceras* from the Tyrone Limestone in central Kentucky, one species each of *Actinoceras* and *Deiroceras* from the Curdsville Limestone Member of the Lexington Limestone in the same area, and *Ormoceras? covingtonense* from the "Maysville Group" (Upper Ordovician) of northern Kentucky. Flower (1942) described an unusual nautiloid fauna from a lens within the Middle Ordovician "Cynthiana Limestone" exposed at the Poindexter Quarry in north-central Kentucky. This paper was followed by his classic monograph on nautiloid cephalopods from the Cincinnati area (Flower, 1946). This volume reviewed nautiloid systematics, the history of nautiloid study in the region, the distribution of nautiloids throughout the Ordovician strata exposed in the Kentucky-Indiana-Ohio area, and provided descriptions and photographic plates of numerous new and previously described nautiloid taxa. Flower's study concentrated on species from younger Cincinnatian rocks in the Cincinnati area, but omitted discussions of the abundant orthocerid and endocerid nautiloids present in these strata, deferring description and study of these forms to a later time.

Since the publication of Flower's monograph, very little has been published concerning the paleontology of nautiloids from the Ordovician of the Kentucky-Indiana-Ohio area. Flower (1955c) described trace fossils he ascribed to nautiloid activity from the "McMillan Formation" at Cincinnati, Ohio, and later (Flower, 1957b) described the paleoecology of several nautiloid assemblages from the Cincinnatian Provincial Series in southwestern Ohio. Aronoff (1979) documented the morphological distinctiveness of the actinoceroid *Orthonybyoceras* Shimizu and Obata, type species *O. covingtonense* from the Upper Ordovician of northern Kentucky, and the orthocerid *Treptoceras* Flower, type species *T. duseri*, from equivalent strata in southwest Ohio. Frey (1981) noted the presence of the aberrant orthocerid nautiloid *Narthecoceras* in Upper Ordovician rocks in southwestern Ohio and discussed its paleobiogeographic significance. Frey (1988, 1989) described the paleoecology of a well-preserved nautiloid fauna from the *Treptoceras duseri* shale unit in southwestern Ohio, illustrating a number of the taxa present.

ACKNOWLEDGMENTS

I thank John Pojeta of the U.S. Geological Survey (USGS) for his encouragement, support, and editorial assistance during the course of this study. Also, I express my appreciation to W.A. Ausich, S.M. Bergstrom, and the staff of the Department of Geological Sciences at Ohio State University for use of library and laboratory facilities. Special thanks go to C. Teichert for his comments concerning nautiloid systematics. J.A. Catalani, R.E. Crick, and W.C. Swadley critically read an earlier draft of this paper and offered many helpful suggestions that substantially improved its quality. M. Balanc developed and printed many of the photographs used here. Thanks also go to W. Berry, University of California at Berkeley (UCal); J. Thompson at the U.S. National Museum (USNM); F. Escrivan of the Museum of Comparative Zoology (MCZ) at Harvard; N. Landman and M. Hinkley of the American Museum of Natural History (AMNH); J. Marak of the Limper Museum of Geology at Miami University (MU); and D. Meyer of the University of Cincinnati (UC) for the loan of specimens critical to the completion of this study. I also wish to acknowledge my debt of gratitude to members of USGS and Kentucky Geological Survey field parties, who collected much of the Kentucky silicified material in the 1960's, and to the late R.H. Flower and R.A. Davis of the College of Mt. St. Joseph, who initially assembled and organized the bulk of the collection of nautiloids described here.

REGIONAL ORDOVICIAN PALEOGEOGRAPHIC SETTING

The Ordovician strata exposed in the Cincinnati arch region of Kentucky, Indiana, and Ohio were deposited on the southeastern portion of a cratonic platform that made up much of the Laurentian continent in the early Paleozoic. Modern paleogeographic reconstructions for the Ordovician (Scotese and McKerrow, 1990) place the Kentucky-Indiana-Ohio area in tropical or subtropical latitudes, 20°–30°S of the Ordovician equator and within the "carbonate belt" (Witzke, 1990). For much of the Ordovician, this region was inundated by a shallow epicontinental seaway, its lithofacies, fossils, and sedimentary structures all indicating intertidal to subtidal marine conditions and maximum water depths of 82–98 ft (25–30 m).

The central Kentucky Blue Grass region was the site of nearly continuous, very shallow water, carbonate platform deposition throughout much of Ordovician time (Cressman, 1973; Cressman and Noger, 1976). Adjacent portions of the northern Kentucky-Indiana-Ohio area formed the shallow, northward-ramping margin of what has been termed the Lexington Platform by Mitchell and Bergstrom (1991) and gently sloped toward deeper water basins to the north and west (Cressman, 1973) (fig. 1). Throughout the Ordovician, a series of volcanic island-arcs collided with the eastern margin of the Laurentian continent, culminating in the Middle Ordovician Taconic orogeny. This resulted in the formation of a rising landmass that produced a clastic source area along the eastern margin of the continent (Bird and Dewey, 1970; Ettensohn, 1991). Vast quantities of detrital sediments were eroded from this Taconic landmass and were deposited as the Queenston delta, which spread clastic sediments north and west into adjacent epeiric sea-shelf areas (Denison, 1976). These clastic sediments, mostly fine silt and clay, began to appear in the Kentucky-Indiana-Ohio area in the late Middle Ordovician (Kirkfieldian) and

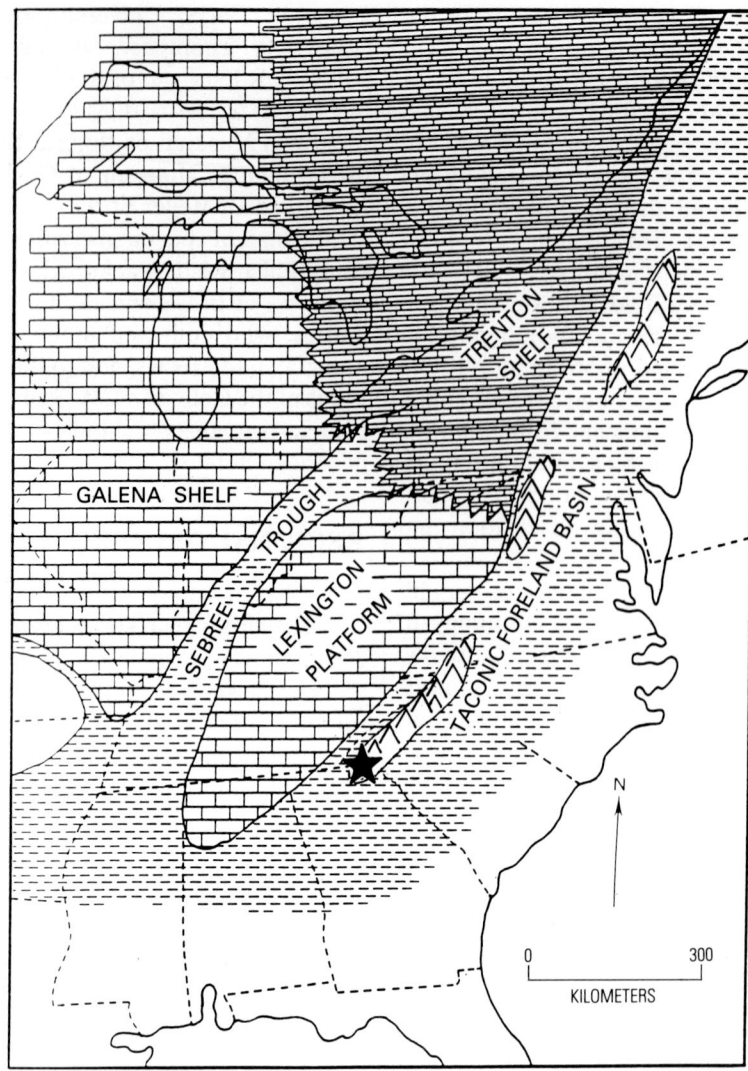

FIGURE 1.—Late Middle Ordovician (Kirkfieldian and Shermanian) paleogeography in Eastern North America (Laurentia). Volcanic island arcs occur in the Taconic foreland basin. Star symbol indicates probable source area for Deicke Bentonite bed (modified from Mitchell and Bergstrom, 1991).

increased throughout the succeeding Late Ordovician. This sediment influx resulted in the deposition of increasingly clastic lithofacies in the region, from the pure carbonates of the Middle Ordovician High Bridge Group to the dominantly mudstone facies at the top of the Upper Ordovician Cincinnatian Provincial Series.

In addition to the effects of tectonism in the east, sedimentation in the Cincinnati arch area in the Ordovician was affected by sea-level fluctuations that resulted in the development of at least three post-Tyrone Limestone transgressive-regressive cycles, at least one cycle during the deposition of the Middle Ordovician Lexington Limestone (Cressman, 1973) and two or three during the deposition of the Upper Ordovician Cincinnatian Provincial Series (Hay, 1981; Tobin, 1982; Shrake and others, 1988).

ORDOVICIAN LITHOSTRATIGRAPHY IN THE CINCINNATI ARCH REGION

Ordovician rocks in the East-Central United States are exposed at the surface along a series of north-south-trending structural highs that extend across portions of central Tennessee, central Kentucky, western Ohio, and eastern Indiana. This north-south trend roughly parallels the trend of the Appalachian basin to the east. Wahlman (1992) used the term "Cincinnati arch" for this entire

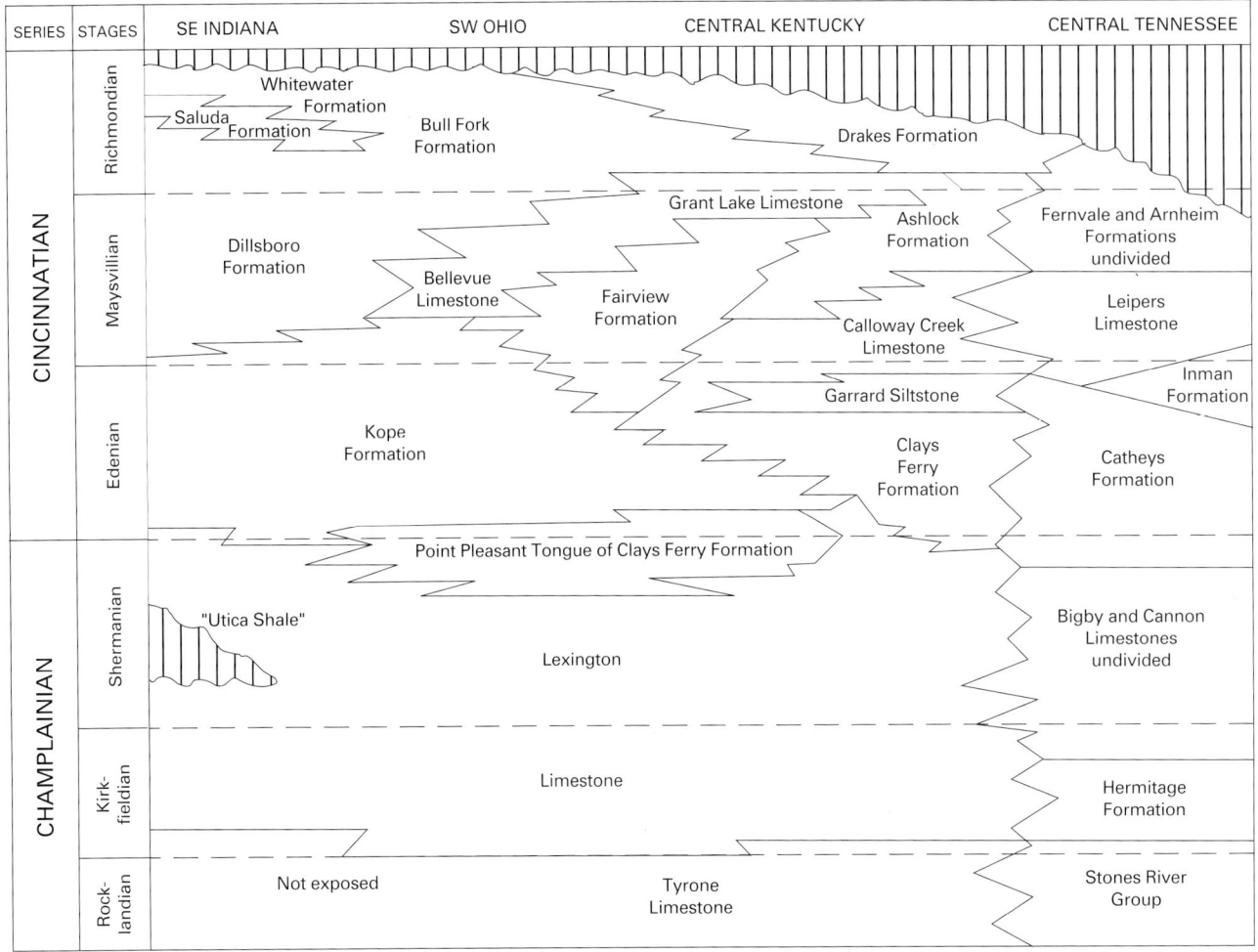

FIGURE 2.—Lithostratigraphic relations of Middle and Upper Ordovician stratigraphic units in the Cincinnati arch region of Kentucky, Indiana, Ohio, and Tennessee (modified from Wahlman, 1992; after Sweet, 1979).

structural trend. He noted that the arch consisted of five primary structural elements. These are, from north to south, the Kankakee arch of Indiana and the Findlay arch of Ohio, which merge near Cincinnati; the Jessamine dome in central Kentucky; the Cumberland sag of the Kentucky-Tennessee border; and the Nashville dome of central Tennessee. The nautiloid faunas described here come from the Ordovician strata exposed in the Cumberland sag in south-central Kentucky, in the vicinity of the Jessamine dome in central Kentucky, and northward from the tristate region of northern Kentucky, southeastern Indiana, and southwestern Ohio.

Specifically, nautiloid faunas described here were collected from the Tyrone Limestone of the Middle Ordovician High Bridge Group and the overlying Middle-Upper Ordovician Lexington Limestone, both exposed in the Inner Blue Grass region of central Kentucky. Additional nautiloid specimens are described from the Upper Ordovician Cincinnatian Provincial Series exposed in northern Kentucky and adjacent portions of Indiana and Ohio. These consist of nautiloids from the Clays Ferry and Kope Formations in northern Kentucky; the Leipers Limestone in south-central Kentucky; the Fairview, Grant Lake, and Bull Fork Formations in southwestern Ohio; and the Dillsboro Formation, the Saluda Formation, and the Hitz Limestone Member of the Saluda Formation in southeastern Indiana (fig. 2). Brief descriptions of most of these lithostratigraphic units are found in the work by Pojeta (1979). More detailed descriptions of the Tyrone Limestone are found in Cressman and Noger (1976), descriptions of the various members of the Lexington Limestone in Cressman (1973), and descriptions of a majority of the Upper Ordovician lithostratigraphic units are in Weir and others (1984). Wahlman (1992) also gave an extensive account of the regional Ordovician lithostratigraphy and nomenclatural history of these stratigraphic units. The present work has only a broad overview of Ordovician lithostratigraphy and depositional environments across the northern Cincinnati arch region during the time of deposition of the Tyrone Limestone, the Lexington Limestone, and the Cincinnatian Provincial Series. An expanded description of the Tyrone Limestone is provided, as this formation was not extensively discussed by either Pojeta (1979) or Wahlman (1992), and it is the source of a large collection of nautiloids that is described here for the first time.

MIDDLE ORDOVICIAN CHAMPLAINIAN PROVINCIAL SERIES LITHOSTRATIGRAPHY IN KENTUCKY

HIGH BRIDGE GROUP

The High Bridge Group consists of 590–750 feet (ft) (180–229 meters (m)) of high-calcite limestone and dolostone and is exposed in the Inner Blue Grass region of Kentucky, primarily in the gorge of the Kentucky River and tributaries southwest of Lexington. The High Bridge Group, Middle Ordovician (Blackriveran and Rocklandian) in age, consists of the oldest strata exposed in Kentucky. In ascending order, it consists of the shallow-marine, burrow-mottled Camp Nelson Limestone (Blackriveran); the intertidal dolostone and cryptalgal limestones of the Oregon Formation (Blackriveran); and the complex intertidal and shallow subtidal marine facies of the Tyrone Limestone (Blackriveran-Rocklandian). In this study, only Tyrone Limestone taxa were examined.

Tyrone Limestone.—In central Kentucky, the Tyrone Limestone is 55–155 ft (17–47 m) thick, thickening at the expense of the underlying dolostone and dense limestone of the Oregon Formation. Cressman and Noger (1976) and Kuhnhenn and others (1981) describe the formation as consisting of basal argillaceous dolomitic micrite (5–50 ft thick), followed by up to 50 ft of interbedded micrite, biopelmicrite, and biopelsparite, and an upper 40-ft interval consisting of interbedded argillaceous dolomitic micrite, cherty burrowed micrite, and biopelmicrite. The formation thins to the north and east within the central Kentucky area. The Tyrone Limestone is conformable with the underlying Oregon Formation but has a disconformable contact with the overlying Lexington Limestone.

Two distinct altered volcanic ash beds occur in the upper half of the formation and can be correlated with similar K-bentonite beds in the upper Mississippi valley, central Tennessee, and the southern Appalachians (Huff and Kolata, 1990; Bergstrom and others, 1991). The Pencil Cave bentonite of drillers occurs about 21 ft (7 m) below the top of the formation and can be traced throughout the outcrop area (Cressman and Noger, 1976). The Mud Cave bentonite of driller, when present, marks a disconformity between the Tyrone and the overlying Lexington Limestone. The Mud Cave bentonite is locally eroded away at the contact with the Lexington Limestone, particularly in the area north of the Kentucky River in central Kentucky (Cressman, 1973, fig. 7).

Cressman and Noger (1976) described the various lithofacies within the Tyrone Limestone as being representative of alternating supratidal islands (mud-cracked argillaceous dolostone), intertidal mudflats (burrowed micrite with ostracodes), and shallow, subtidal marine conditions (fossiliferous micrite and biopelsparite). The latter facies contains a fauna that consists of the coral *Tetradium*, bryozoans, gastropods, polyplacophorans, nautiloids, and pelecypods. Subtidal marine facies thicken to the south from the type locality at Tyrone, in Woodford County, Ky.

Ross and others (1982) assigned a Middle Ordovician (Blackriveran and Rocklandian) age to the Tyrone Limestone in the Lexington region of central Kentucky. These authors correlate the Tyrone Limestone with the lower half of the Carters Limestone in central Tennessee; the Ottosee Shale, Mocassin Limestone, and Eggleston Formation in eastern Tennessee; the upper half of the Platteville Formation and the basal portion of the Decorah Formation in Iowa, Missouri, Wisconsin, and Minnesota and the Platteville Group, Decorah "Subgroup" in Illinois; and the Watertown, Selby, and Napanee Limestones in New York.

All of the Tyrone nautiloids described here are from USGS collection 6034-CO. This silicified collection is from a hard, white to gray, crystalline fossiliferous limestone unit, 3 ft thick, exposed 25 ft (8 m) below the Pencil Cave bentonite. This unit is well exposed in roadcuts along Watts Mill Road, 0.8 mi (1.3 km) south of Sulfur Well, Jessamine County, Ky. (Little Hickman 7.5 min. quadrangle, Little Hickman A section). Petrographically, this limestone would be classified as a biopelsparrudite (classification of Folk, 1962) or a fossiliferous grainstone (classification of Dunham, 1962). The unit contains poorly sorted whole and fragmentary silicified shells of nautiloids, gastropods, solitary corals, trepostome bryozoans, and pelecypods, plus intraclasts consisting of subrounded fragments of micrite, all in a matrix of sand-sized peloidal and skeletal grains.

LEXINGTON LIMESTONE

The Lexington Limestone, Middle Ordovician and Late Ordovician (Kirkfieldian to Edenian) in age, is well exposed in the Inner Blue Grass region of central Kentucky (Sweet, 1979; Wahlman, 1992). The formation consists of approximately 300 ft (98 m) of thin-bedded fossiliferous limestone, calcarenite, and calcilutite, and lesser amounts of clay shale and siltstone. The formation (Pojeta, 1979, p. A10) has been divided into 12 named members and several named beds (fig. 3). Limestone facies have more siliciclastic sediments than those of the

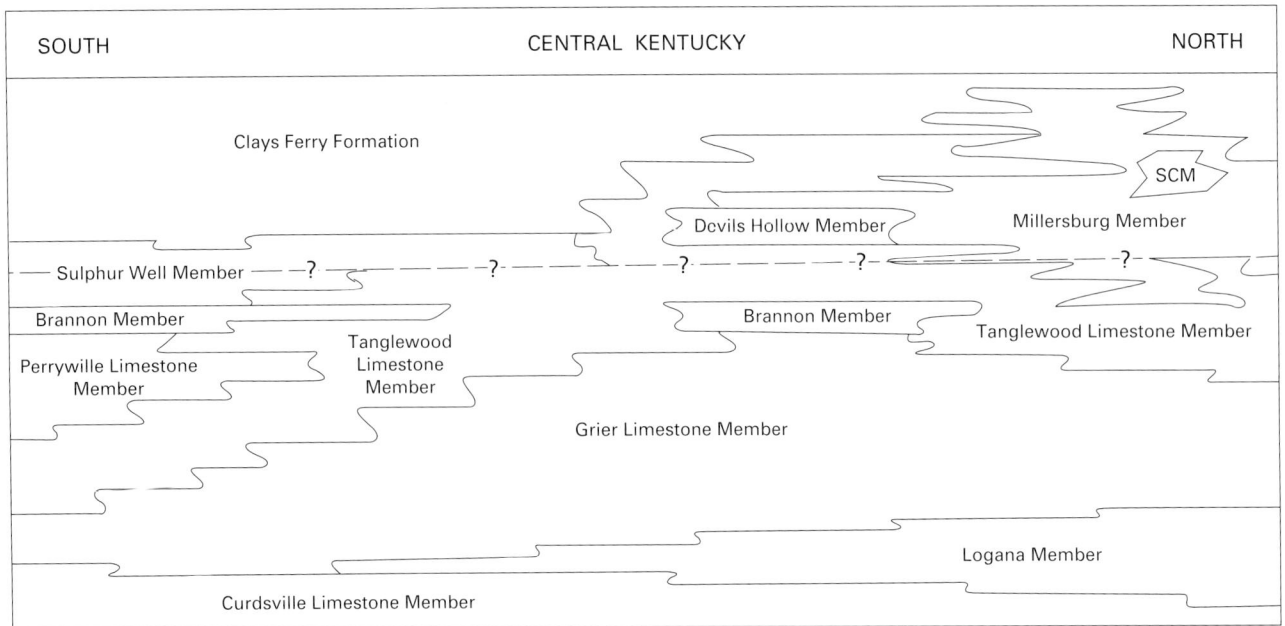

FIGURE 3.—Facies relations among the lithostratigraphic units of the Lexington Limestone and Clays Ferry Formation in central Kentucky. Taken from Wahlman (1992) who modified the figure from Cressman (1973) and an unpublished Cressman figure. Cressman (written commun., January 1994) added the modifications shown herein: (1) the position of the Strodes Creek Member (SCM) of the Lexington Limestone and (2) the approximate Middle Ordovician-Upper Ordovician boundary shown by the dashed line bearing question marks.

underlying Tyrone Limestone, reflecting an increase in the influx of sediments from the rising tectonic belts to the east. Many of these carbonates also differ texturally from Tyrone lithologies in being coarser grained shelly packstone and grainstone in contrast to the fine-grained, dense micritic limestone, dolostone, and cryptalgal limestone of the Tyrone Limestone. Beginning with deposition of the Lexington Limestone, the abundance and diversity of shelly benthic filter-feeding faunas increases, especially brachiopods, bryozoans, pelecypods, and crinoids. This is in contrast to the typically lower diversity *Tetradium*-ostracode-mollusk biota characteristic of marine facies within the Tyrone Limestone. Lithofacies and faunas collectively indicate a change from a restricted, low-energy carbonate platform that consisted of a mosaic of subtidal marine coralline thickets, shallow-marine lagoons, and intertidal mudflats (Cressman and Noger, 1976; Kuhnhenn and others, 1981) to a more open-shelf, higher energy, subtidal marine carbonate ramp complex that dipped to the north (Cressman, 1973).

Coeval with the carbonate ramp environments of the Lexington Limestone in Kentucky were similar carbonate shelf environments to the south in central Tennessee (Nashville Group; Wilson, 1949) and to the northeast into New York (Trenton Group; Titus and Cameron, 1976; Titus, 1982). Immediately to the northwest of Kentucky was the Sebree trough, a deeper water linear trough that extended from central Ohio southwestward through southern Indiana, western Kentucky, and western Tennessee (fig. 1). This trough was the site of graptolitic shale deposition (Cressman, 1973; Bergstrom and Mitchell, 1990). A shallow-water carbonate shelf flanked this trough to the northwest and was continuous from Iowa in the west (Galena Formation), northeastward into Ontario and Quebec (Simcoe and Ottawa Groups).

In Kentucky the basal member of the Lexington Limestone, the Curdsville Limestone Member (Kirkfieldian), records the initial marine transgression over the eroded surface of the Tyrone Limestone. The Curdsville Limestone Member consists of 20–40 ft of cross-bedded basal grainstone beds that grade up into lower-energy fossiliferous packstone and grainstone beds having a diverse fauna of brachiopods, bryozoans, pelecypods, trilobites, and crinoids. The Curdsville Limestone Member is overlain conformably by the interbedded, thin, planar, argillaceous limestone and shale of the Logana Member (Kirkfieldian), up to 50 ft thick, thickening north across central Kentucky, or the nodular Grier Limestone Member (fig. 3). Cressman (1973) indicated that the Logana Member represented the height of the initial marine transgression, deposited under quiet-water conditions at water depths below wave base and containing a more sparse fauna of dalmanellid brachiopods, bryozoans, pelecypods, univalved mollusks, and nautiloids. The Logana

Member interfingers with and is overlain by the Grier Limestone Member (Kirkfieldian and Shermanian), consisting of 100–180 ft of thin, irregularly bedded, often nodular, very fossiliferous limestone. These limestones contain a diverse marine fauna of brachiopods, bryozoans, and crinoids in coarser grained lithologies and ostracodes, univalved mollusks, and pelecypods in finer grained, more argillaceous lithologies. Cressman (1973) inferred that the Grier Limestone Member was deposited under shallow-water (less than 50 ft deep), well-aerated, but normally low energy subtidal marine conditions.

In latest Middle Ordovician and earliest Late Ordovician times (Shermanian-Edenian), there was subsidence in the southern half of the Lexington platform and the formation of a west-northwest-trending bank (Tanglewood bank of Hrabar and others, 1971) across the north-central portion of the platform (Cressman, 1973, pls. 9c–d). Interbedded, thin argillaceous limestone and shale of the Brannon Member were deposited in the subsiding basin to the south. High-energy grainstone facies were deposited in the vicinity of the bank (Tanglewood Limestone Member). Nodular, very fossiliferous limestones of the Millersburg Member were deposited to the east (Cressman, 1973, pls. 9c–d). These extensively bioturbated limestones were deposited in shallow marine subtidal environments in water depths of less than 45 ft. Between the Tanglewood Limestone Member and Millersburg Member in the east-central Blue Grass region there occurs the Strodes Creek Member of the Lexington Limestone. This unit consists of up to 30 ft of "bouldery" gray, micrograined limestone and characteristically abundant, large stromatoporoids. Black and Cuppels (1973) inferred low-energy, quiet-water, mud-bottom marine conditions for the Strodes Creek Member and development of local stromatoporoid biostromes. These biostromes provided a habitat for a diverse fauna of orthid and rhynchonellid brachiopods, bryozoans, the coral *Favistina*, pelecypods, univalved mollusks, nautiloids, trilobites, and ostracodes. Other named members and named beds of the Lexington Limestone are discussed by Cressman (1973) and Pojeta (1979).

Above the Lexington Limestone, thin-bedded, planar packstones and clay shales of the Clays Ferry Formation were deposited at the northward-sloping northern edge of the platform in deeper water, low-energy, mud-bottom environments. Water depths in these environments were estimated by Cressman (1973) to have exceeded 80 ft (24 m), below fair-weather wave base, but within the zone of storm-current reworking. These facies extend north into Ohio as the Point Pleasant Tongue of the Clays Ferry Formation (Swadley and others, 1975). The Point Pleasant Tongue consists of 100 ft of interbedded fossiliferous limestone and shale that underlies the Kope Formation in the Cincinnati, Ohio, area.

UPPER ORDOVICIAN CINCINNATIAN PROVINCIAL SERIES LITHOSTRATIGRAPHY

The Upper Ordovician (Edenian to Richmondian) Cincinnatian Provincial Series continues the trend initiated in the underlying Lexington Limestone, namely lithofacies indicating the increasing influx of fine-grained clastic sediments into the northern Cincinnati arch region from the Taconic landmass to the east. These lithofacies, collectively as much as 985 ft (323 m) thick, consist primarily of interbedded clay shale, fossiliferous wackestone and packstone, having less amounts of siltstone, dolomitic mudstone, and dolomitic limestone. Marine faunas are diverse and abundant in most of these facies, dominated by brachiopods, trepostome bryozoans, and crinoids in skeletal facies and by mollusks and trilobites in mudstone facies (Frey, 1987b).

The tristate area surrounding Cincinnati, Ohio (Kentucky-Indiana-Ohio), exposes the type section for the Upper Ordovician Cincinnatian Provincial Series of North America. The Cincinnatian Provincial Series is divided into three stages. These are, in ascending order, the Edenian, the Maysvillian, and the Richmondian Stages. The following overview of Upper Ordovician lithostratigraphy in the region is organized by stage and presented in ascending order. The complex facies relations of these formations across the Cincinnati arch region are illustrated in figure 2.

EDENIAN STAGE

During Edenian time, transgressive conditions were prevalent across much of the northern Cincinnati arch region, inundating much of the Lexington platform under shallow seas having water depths of at least 80 ft (24 m). These low-energy, muddy-bottom environments were the site of the deposition of the widespread clay shale and thin limestone of the Clays Ferry Formation, which reach a maximum thickness of 250 ft in the southwestern Outer Blue Grass region (Weir and others, 1984, pl. 9). To the north, the Clays Ferry Formation interdigitates with the more clastic-rich facies of the Kope Formation, thinning northward due to the thickening wedge of mudstone and shale of the Kope (up to 280 ft thick in southwestern Ohio). Biotas in both formations are most common in the thin wackestone or packstone beds and consist of small dalmanellid and plectambonitid brachiopods, bryozoans, small pelecypods, gastropods, monoplacophorans, nautiloids, trilobites, ostracodes, and crinoids. These low-energy, mud-bottom environments were periodically reworked by storm events, commonly resulting in sets of storm-formed "tempestite" beds consisting of fining-up packets of packstone/grainstone-siltstone-shale (Hay and others, 1981; Meyer and others, 1981; Tobin, 1982;

Kepferle and others, 1987; Shrake and others, 1988). In the northern Kentucky–southwestern Ohio area there is an increase in limestone beds upsection, indicating a shallowing of conditions upward toward the partially equivalent and overlying more calcareous, shallow-water Fairview Formation.

MAYSVILLIAN STAGE

Shallower water marine conditions returned to the Lexington platform area in central Kentucky in Maysvillian time, resulting in the deposition of the thin-bedded, locally nodular, fossiliferous limestone of the Calloway Creek Limestone across much of the central Kentucky region. The even-bedded fossiliferous limestone and shale of the Fairview Formation were deposited at the same time in somewhat deeper water environments at the northern edge of the platform in the northern Kentucky-southwestern Ohio area. These latter strata, 40–130 ft thick, represent transitional environments between the deep-water, transgressive conditions associated with the underlying Kope Formation and the very shallow water, locally shoaling conditions, inferred for basal portions of the overlying Grant Lake Limestone. Preserved biotas in these carbonate facies are dominated by diverse brachiopods and trepostome bryozoans, evidence of the development of dense bryozoan thickets that supported large populations of "nestling" brachiopod species. Associated with these thicket facies are lesser numbers of mollusks, trilobites, ostracodes, and crinoids.

By late Maysvillian time, restricted, very shallow water, low-energy environments were prevalent across the south-central portion of the Lexington platform, resulting in the deposition of the sparsely fossiliferous calcitic and dolomitic mudstone and micrograined limestone of the Ashlock Formation (Weir and others, 1984). More fossiliferous, open-marine facies of the Grant Lake Limestone were deposited to the north in the Maysville and Covington, Ky., areas. This formation, consisting of up to 160 ft of irregularly bedded, poorly sorted fossiliferous limestone, extends northward into southwestern Ohio, where it contains more shale as a result of the somewhat deeper water, lower energy, bottom conditions inferred for this area at this time (Schumacher and others, 1991).

Coeval with the above-stated lithostratigraphic units are the irregularly bedded, nodular, fossiliferous limestone and thin shale of the Leipers Limestone, exposed in the Cumberland sag region along the Kentucky-Tennessee State line. This formation, up to 180 ft thick in this area, overlies the even-bedded limestone of the Catheys Formation and, in south-central Kentucky, is overlain by the dolostone and mudstone of the Cumberland Formation (Weir and others, 1984). Flower (1946) described the occurrence of two distinct coralline biostromes in the lower half of the Leipers Limestone. The inferred environment of deposition was a shallow-marine, mixed carbonate and clastic bottom that becomes lower energy with time, the lithofacies becoming more fine-grained upsection. This normal marine environment had a diverse fauna of brachiopods, bryozoans, the coral *Tetradium*, gastropods, nautiloids, pelecypods, ostracodes, and trilobites.

RICHMONDIAN STAGE

Shallow epeiric seas again transgressed across the northern portion of the Cincinnati arch region in early Richmondian time, leading to the deposition of the interbedded clay shale and thin fossiliferous limestone of the Bull Fork Formation (10–240 ft thick) across northern Kentucky and southwestern Ohio. The lithologically and faunally similar upper Dillsboro Formation was deposited at the same time in adjacent portions of southeastern Indiana (Sweet, 1979). These interbedded shales and limestones were deposited in deeper water, lower-energy environments, compared with the underlying, primarily carbonate facies of the Grant Lake Limestone. Water depths were below fair-weather wave base but within the zone of storm-current reworking, similar to those inferred for the Clays Ferry and Kope Formations (Weir and others, 1984). Preserved faunas contains diverse, abundant brachiopods, bryozoans, corals, stromatoporoids, gastropods, nautiloids, pelecypods, ostracodes, trilobites, and crinoids.

Very shallow water, restricted marine and intertidal mudflat conditions continued to occur across central Kentucky, resulting in the deposition of the dolomitic mudstone and micrograined limestone of the Ashlock Formation and the overlying Rowland Member of the Drakes Formation. Limestone beds in these units commonly contain mud cracks (Weir and others, 1984) and are only sparsely fossiliferous having local occurrences of brachiopods, bryozoans, gastropods, and ostracodes.

In later Richmondian time, regressive conditions spread throughout the northern Cincinnati arch region, the central Kentucky area being the site of complex facies consisting of alternating thin, shallow-marine and intertidal mudflat lithofacies. These are the coralliferous, thin-bedded limestone and shale of the Bardstown Member of the Drakes Formation, 20–65 ft thick, exposed along the western edge of the Outer Blue Grass region, and the maroon and green dolomitic mudstones of the Preachersville Member of the Drakes, 100 ft thick in the southeastern Blue Grass region and thinning to the northwest (Weir and others, 1984, pl. 7). The latter member is inferred to represent restricted shallow-marine and intertidal mudflat environments that prograded into the northern

Cincinnati arch region from the southeast, indicative of the distal edge of the Queenston delta.

The Saluda Formation, consisting of a broad lens of sparsely fossiliferous dolomitic limestone and calcitic dolostone, was deposited at the same time along a north-south trend through southeastern Indiana and northern Kentucky. The formation (recognized as the Saluda Dolomite Member of the Drakes Formation in Kentucky), averages 52 ft in thickness between Madison, Ind. and Jefferson County, Ky. Thickness decreases north of Madison, where the formation wedges out due to the irregularly bedded, very fossiliferous limestones of the Whitewater Formation, which thickens to 90 ft northward at Richmond, Ind. The Saluda Dolomite Member thins southward to a thickness of 20 ft in Marion County, Ky. In the Madison, Ind.–Jefferson County, Ky., area, the Saluda Formation or Saluda Dolomite Member is separated from overlying Silurian limestones by the Hitz Limestone Member of the Saluda, or the Hitz Limestone Bed of the Saluda Dolomite Member for Indiana or Kentucky, respectively, consisting of up to 15 ft of interbedded, thin, fossiliferous limestone, shale, and dolomitic limestone (Kepferle, 1976).

Hatfield (1968) described the Saluda Formation as being deposited under very shallow-water, possibly hypersaline, lagoonal conditions that developed behind an initially northward-advancing and then southward-retreating coral-sponge bank. The Whitewater Formation and the Hitz Limestone represent the return of more normal marine shallow water conditions to the area prior to the development of the regional erosional surface that separates these Ordovician strata from the overlying Silurian carbonates. Along the east flank of the Cincinnati arch in southwestern Ohio and adjacent portions of north-central Kentucky, this erosional surface truncates the upper portions of the Bull Fork Formation or the Preachersville Member of the Drakes Formation.

DISTRIBUTION AND FAUNAL AFFINITIES OF ORDOVICIAN NAUTILOIDS IN THE CINCINNATI ARCH REGION

Study of the Middle and Upper Ordovician nautiloids of the Cincinnati arch region of Kentucky, Indiana, and Ohio has led to the recognition of four major time-bounded Ordovician nautiloid faunas in this region. These are characteristic of four time intervals: Blackriveran to Rocklandian, Kirkfieldian to early Edenian, middle Edenian to early Richmondian, and middle Richmondian to late Richmondian. The Edenian to Richmondian nautiloid fauna, in turn, can be further subdivided into three distinctive subfaunas characteristic of specific time intervals and (or) paleogeographic regions within the Cincinnatian Provincial Series in the tristate area of northern Kentucky, southeastern Indiana, and southwestern Ohio.

The four major nautiloid faunas—"Black River" (found in the Tyrone Limestone and equivalent units), "Trenton" (found in the Lexington Limestone and equivalent units), "Cincinnatian" (found in the Kope Formation through part of the Bull Fork Formation and equivalent units), and "Richmondian" (found in part of the Bull Fork Formation through the Saluda Formation and equivalent units)—are separated by three regional sedimentological-paleontological events. A regional unconformity and a major extinction of nautiloid taxa separate the nautiloid faunas of the Tyrone and Lexington Limestone. A marine transgression coupled with an influx of clastic sediments from the rising tectonic areas in eastern Laurentia led to the more gradual turnover in nautiloid faunas from the Lexington Limestone to the Edenian Clays Ferry and Kope Formations. Within the Cincinnatian Provincial Series, the early Richmondian transgression coincides with the migration into the Kentucky-Indiana-Ohio area of the "Richmondian" nautiloid fauna derived from taxa more characteristic of coeval or slightly older Edenian and Maysvillian faunas in the Western United States, Canada, the Arctic archipelago, and Greenland (Flower, 1946, 1976; Frey, 1981). Flower (1942, 1946) also indicated an earlier incursion of these "Arctic" nautiloid taxa into the area, associated with the latest Shermanian to Edenian transgressive episode (Clays Ferry and Kope Formations). These Upper Ordovician nautiloid faunas were terminated by the Late Ordovician mass extinction event (Sheehan, 1988; Brenchley, 1989) associated with the withdrawal of epeiric seas from the Cincinnati arch region.

NAUTILOID FAUNA OF THE TYRONE LIMESTONE

The nautiloid fauna from the 6034-CO collection in the Tyrone Limestone consists of at least 18 species, belonging to 14 genera, representing 6 nautiloid orders. A complete listing of the nautiloid species identified from this collection is provided in table 1. The collection is dominated numerically by the small longiconic baltoceratids *Cartersoceras shideleri* Flower (pl. 1, figs. 1–6), *C. popei* new species (n. sp.) (pl. 1, figs. 7–10), and *Murrayoceras* cf. *M. murrayi* (Billings) (pl. 1, figs. 12–17), and the small longiconic orthocerids *Pojetoceras floweri* n. sp. (pl. 2, figs. 1–9) and *Proteoceras tyronensis* (Foerste) (pl. 4, figs. 1–7). Small, typically fragmentary specimens of oncocerid nautiloids are second in abundance to the small longiconic taxa. Many of these specimens, however, are too incomplete to identify to the genus. Identified oncocerids from the 6034-CO collection are *Oncoceras major* n. sp. (pl. 19, figs. 11–15), *Beloitoceras* cf. *B. huronense* (Billings) (pl. 20, figs. 3P6), *Laphamoceras* cf.

TABLE 1.—Distribution of species of nautiloid cephalopods in the Tyrone Limestone of the High Bridge Group, Members of the Lexington Limestone, and the Clays Ferry Formation, all Middle Ordovician from the Blue Grass region of central Kentucky

[Data include only those nautiloid species described in this paper, based on U.S. Geological Survey silicified collections and on museum collections. Middle Ordovician stratigraphic units from which no nautiloid species were recovered are omitted. TY, Tyrone Limestone; CF, Clays Ferry Formation; Members of the Lexington Limestone are designated: CL, Curdsville Limestone; LO, Logana; GL, Grier Limestone; ML, Millersburg; SC, Strodes Creek. X indicates presence of species]

Species	TY	CL	LO	GL	ML	SC	CF	Location
Ellesmerocerida								
Cartersoceras popei	X							6034-CO
Cartersoceras shideleri	X							6034-CO
Murrayoceras cf. *M. murrayi*	X							6034-CO
Orthocerida								
Pojetoceras floweri	X							6034-CO
Anaspyroceras cylindricum		X						Curdsville, Ky.
Ordogeisonoceras amplicameratum							X	Ludlow, Ky.
Proteoceras tyronensis	X							6034-CO
Isorthoceras albersi			X	X	X		X	5067-CO, 5073-CO, 5092-CO
Isorthoceras rogersensis							X	Rogers Gap, Ky.
Gorbyoceras cf. *G. tetreauense*			X					5073-CO, 5092-CO, 7791-CO
Monomuchites annularis	X							6034-CO
Monomuchites obliquum	X							6034-CO
Monomuchites cf. *M. costalis*	X							6034-CO
Polygrammoceras sp A.				X				5067-CO
Polygrammoceras? sp. B	X							6034-CO
Polygrammoceras? cf sp. P. A						X		7310-CO
Endocerida								
Cameroceras cf. *C. trentonense*				X				4073-CO
Vaginoceras sp. A	X							6034-CO
Vaginoceras sp. B		X						5072-CO
Actinocerida								
Actinoceras altopontense	X							6034-CO, High Bridge, Ky.
Actinoceras kentuckiense	X							6034-CO; High Bridge, Ky.
Actinoceras curdsvillense		X						Curdsville, Ky.
Ormoceras ferecentricum	X							6034-CO; High Bridge, Ky.
Deiroceras curdsvillense		X						5100-CO; Curdsville, N.Y.
Tarphycerida								
Plectoceras cf. *P. carletonense*	X							6034-CO
Trocholites faberi							X	Sadieville, Ky.
Oncocerida								
Oncoceras major	X							6034-CO
Oncoceras sp.			X					4879-C0, 4883-CO
Beloitoceras cf. *B. huronense*	X							6034-CO
Maelonoceras cf. *M. praematurum*	X							6034-CO
Laphamoceras cf. *L. scofieldi*	X							6034-CO
Allumettoceras cf. *A. tenerum*			X					4865-CO, 6419-CO

TABLE 2.—Comparison of the nautiloid fauna from the 6034-CO collection in the Tyrone Limestone in Jessamine County, Ky. with those described from equivalent strata exposed in Eastern and Central North America

[Data from (1) Catalani (1987), (2) Steele and Sinclair (1971), (3) Wilson (1961), (4) Foerste (1932, 1933), (5) Flower (1984) plus manuscript notes and plates, and (6) Hall (1847)]

Tyrone Taxa Kentucky	Platteville Minnesota Wisconsin[1,4]	"Black River" St. Joseph Isl. Ontario[4]	"Leray-Rockland" Ottawa Lowland Ontario-Quebec[2,3]	Black River Group Black River valley New York[6]
Ellesmerocerida				
Cartersoceras	X	X	X	X
Murrayoceras	X	X	X	X
Orthocerida				
Pojetoceras	X		X	
Polygrammoceras?				
Monomuchites	X	X	X	X
Proteoceras	X		X	X
Endocerida				
Vaginoceras	X		X	X
Actinocerida				
Actinoceras	X	X	X	X
Ormoceras		X	X	X
Tarphycerida				
Plectoceras	X		X	X
Oncocerida				
Oncoceras	X	X	X	X
Beloitoceras	X	X	X	
Laphamoceras	X		X	
Maelonoceras				
Species in common	11/14	8/14	12/14	8/14

L. scofieldi Foerste (pl. 20, figs. 13–18), and *Maelonoceras* cf. *M. praematurum* (Billings) (pl. 20, figs. 7–12). Lesser numbers of the large annulated orthocerid *Monomuchites annularis* n. sp. (pl. 9, figs. 1–3, 11–15), the actinocerids *Actinoceras altopontense* Foerste and Teichert (pl. 17, figs. 6–10), *A. kentuckiense* Foerste and Teichert (pl. 17, figs. 1–5), and *Ormoceras ferecentricum* Foerste and Teichert (pl. 18, figs. 1–7), and the large endocerid *Vaginoceras* sp. A (pl. 15, figs. 1–4, 8–12), are also present in the 6034-CO collection. The tarphycerid *Plectoceras* cf. *P. carletonense* Foerste occurs in this collection but is known only from fragmentary specimens (pl. 19, figs. 1–3).

Sloan (1987) placed the Blackriveran-Rocklandian boundary at the Deicke Bentonite bed (Willman and Kolata, 1978). This bed is regarded as being equivalent to the Pencil Cave bentonite of drillers in central Kentucky (Huff and Kolata, 1990). This would make the nautiloid fauna from the 6034-CO collection Blackriveran in age, as it occurs 25 ft (8 m) below the Pencil Cave bentonite of drillers at the Little Hickman section. Sweet (1984), however, placed the Blackriveran-Rocklandian boundary at the base of the Platteville Formation in Minnesota, well below the Deicke Bentonite bed, indicating a probable Rocklandian age, for the 6034-CO fauna. Until these differences in opinion are resolved, the specific age of the diverse Tyrone nautiloid fauna in the Tyrone is uncertain, but it can be placed with confidence in the latest Blackriveran-Rocklandian interval.

The nautiloid fauna from the Tyrone Limestone is similar to nautiloid faunas described from coeval strata at other localities in Eastern North America. This "Black River" nautiloid fauna is characterized by the association of species of the actinocerids *Actinoceras*, *Gonioceras*, and *Ormoceras*; the endocerid *Vaginoceras*; the baltoceratids *Cartersoceras* and *Murrayoceras*; the orthocerids *Proteoceras* and *Monomuchites*; the oncocerids *Oncoceras*, *Beloitoceras*, *Laphamoceras*, and *Zitteloceras*; and the tarphycerids *Avilionella*, *Barrandeoceras*, and *Plectoceras*. This diverse fauna is characteristic of shallow-water carbonate facies deposited in cratonic shelf seas in eastern Laurentia in the post-Whiterockian Middle Ordovician. Elements of this fauna made their initial appearance in Chazyan strata in the Lake Champlain region of New York and Canada (Ruedemann, 1906; Foerste, 1938; Flower, 1955b). The fauna became more diverse and more widespread in succeeding Blackriveran and Rocklandian times. Seventy-six species of nautiloids, belonging to 39 genera and 8 orders, have been reported from the Blackriveran to early Rocklandian rocks in Wisconsin, Minnesota, and Illinois (Catalani, 1987). Wilson (1961) listed 75 species, belonging to 29 genera and 6 orders, from the coeval "Leray-Rockland beds" exposed in the Ottawa Lowland region of Ontario and Quebec. The Tyrone Limestone nautiloid fauna described here is compared with these faunas plus those from the "Black River Formation" exposed on St. Joseph Island in Lake Huron (Foerste, 1932 and 1933) and the Black River Group in New York (Hall, 1847) in table 2.

ROCKLANDIAN NAUTILOID EXTINCTION

This diverse and widespread "Black River" nautiloid fauna thrived in carbonate shelf environments in eastern Laurentia, including Kentucky, for approximately 4 million years (m.y.). This nautiloid fauna was rather abruptly terminated (in the geologic sense) in the late Blackriveran or Rocklandian by an extinction event that may have taken place over a time span of less than a million years (455–454 Ma; Sloan, 1987). The 6034-CO collection in the Tyrone Limestone comes from a limestone bed 25 ft below the Pencil Cave bentonite of drillers and 46 ft below the Tyrone Limestone–Lexington Limestone contact. There are no data on nautiloid faunas from this intervening 46 ft interval in the Tyrone Limestone. With the possible exception of *Vaginoceras* sp., none of the 18 species identified from this Tyrone Limestone collection occur in the overlying basal Curdsville Limestone Member of the Lexington Limestone (Kirkfieldian). Five out of 14 of the genera present in the Tyrone Limestone (36 percent) became extinct prior to the deposition of the Curdsville Limestone Member. In contrast to the Tyrone Limestone fauna, the identified nautiloid fauna from the Curdsville Limestone Member consists of only four species, belonging to four genera and three orders.

Catalani (1987) listed 76 species of nautiloids, belonging to 39 genera and 8 orders from the various formations of the Platteville Group in Illinois or the Platteville Formation in Wisconsin. He listed only 13 species, belonging to 10 genera and 4 orders, from the overlying Decorah Shale (Rocklandian and Kirkfieldian) in the same area. Only four Platteville species occur in the overlying Decorah Shale, indicating extinction of 95 percent of the Platteville species in the latest Blackriveran to early Rocklandian. Eight out of the 39 listed Platteville genera (21 percent) were extinct prior to the deposition of the Decorah Shale.

Wilson (1961) listed 75 species belonging to 29 genera and 6 orders from the "Leray-Rockland beds" exposed in the Ottawa region of Ontario and Quebec. Only nine species, belonging to eight genera and four orders, are listed for the overlying Hull beds (Kirkfieldian). Wilson listed seven species that crossed over from the "Leray-Rockland beds" into the Hull beds, indicating that 91 percent of the Leray-Rockland species became extinct prior to the deposition of the Hull. Seven out of the 29 genera (24 percent) listed for the "Leray-Rockland beds" do not occur in the overlying Hull beds.

Hall (1847), Young (1943), Flower (1952, 1957a), and Cameron and Mangion (1977) have collectively listed 18 species of nautiloids, belonging to 14 genera and 6 orders from the Watertown Limestone (Blackriveran) in New York. Titus and Cameron (1976) listed only three species of nautiloids, representing two genera and two orders, from the overlying Napanee Limestone (Rocklandian). The same authors listed three nautiloid species from the overlying Kings Falls Limestone (Kirkfieldian). None of the Watertown Limestone nautiloid species occur in the overlying Napanee or Kings Falls Limestones. Five out of 14 genera (36 percent) were extinct regionally by the time of the initial deposition of the Napanee Limestone.

All of these examples documented from the literature indicate a major turnover in nautiloid taxa in latest Blackriveran to early Rocklandian time. The interval of time during which this regional extinction event occurred appears to have been geologically short. This extinction is significant not only for the large number of species that became extinct, but also for the large number of genera that became extinct. A search of the literature has indicated that the "Black River" nautiloid fauna consisted of at least 54 genera belonging to 8 orders. Twenty-six genera (48 percent) were extinct prior to the deposition of Kirkfieldian strata in eastern Laurentia (table 3). Succeeding Kirkfieldian nautiloid faunas throughout eastern Laurentia were depauperate in terms of taxonomic diversity, especially in comparison with the older Black River fauna.

Some Blackriveran genera apparently found refugia in similar tropical carbonate platform environments in central Tennessee and in western and northern portions of Laurentia (Foerste and Teichert, 1930; Flower, 1957a). Some of these taxa became significant components of younger Shermanian to Richmondian nautiloid faunas in these regions. These were actinocerids like *Actinoceras* and *Ormoceras*, true *Endoceras*, and the oncocerid *Beloitoceras*. Other genera that were evidently terminated throughout their geographic range are the baltoceratids *Cartersoceras* and *Murrayoceras*, the orthocerids *Monomuchites* and *Proteoceras*, the tarphycerids *Avilionella*, *Barrandeoceras*, and *Plectoceras*, and the oncocerid *Maelonoceras*. The endocerid *Vaginoceras* occurred in the Kirkfieldian Curdsville Limestone Member in Kentucky and then became extinct. Similarly, the actinocerid *Gonioceras* survived into the lower members of the Decorah Shale in the upper Mississippi valley region and then went extinct. Taxa like *Anaspyroceras*, *Gorbyoceras*, *Deiroceras*, *Oncoceras*, and *Zittelloceras* continue into the Kirkfieldian and Shermanian in eastern Laurentia, apparently unaffected by this extinction.

Determining the specific cause(s) of this turnover in nautiloid taxa is difficult. There is a widespread regional unconformity at the base of the Kirkfieldian across much of Eastern North America. This unconformity has been recognized in Kentucky (Cressman, 1973), central Tennessee (Wilson, 1949; Wahlman, 1992), the upper Mississippi valley region (Sloan, 1987; Kolata, 1987), the Upper Peninsula of Michigan (Votaw, 1980), and the Adirondack uplift region of New York (Cameron and Mangion, 1977). This unconformity is apparently absent in the Ottawa Lowlands region of Ontario and Quebec, and no significant break in sedimentation is noted within the "Ottawa Megagroup" (Wilson, 1961; Cameron and Mangion, 1977). Although no distinct unconformity was recognized at this interval in southern Ontario, Liberty (1969) has indicated shoaling facies corresponding to this time interval within the Bobcaygeon Formation.

The widespread nature of this unconformity indicates a major marine regression from these cratonic shelf areas at this time. This regression may have been the result of collisional events along the eastern margin of

TABLE 3.—Middle Ordovician (Blackriveran and Rocklandian) nautiloid genera from Eastern North America (Laurentia)

[Data from Foerste (1932, 1933), Flower (1952), Flower and Teichert (1957), Wilson (1961), Catalani (1987), and Stait (1988). IL, Illinois; KY, Kentucky; LAB, Labrador; MN, Minnesota; MO, Missouri; NFL, Newfoundland; NY, New York; ONT, Ontario; PA, Pennsylvania; QUE, Quebec; TN, Tennessee; WI, Wisconsin. *, genera that went extinct at the end of the Rocklandian]

Genus	Distribution
Order Ellesmerocerida	
1. *Cyrtocerina*	WI, MN, ONT
2. *Cartersoceras**	KY, TN, ONT
3. *Murrayoceras**	KY, NY, ONT
Order Orthocerida	
4. *Anaspyroceras*	NY, ONT, QUE
5. *Metaspyroceras*	WI, MN
6. *Pojetoceras*	KY, ONT
7. *Proteoceras**	KY, ONT, NFL
8. *Monomuchites**	KY, WI, MN, NY, ONT
9. *Gorbyoceras*	QUE, NFL
10. *Whitfieldoceras*	WI, ONT
Endocerida	
11. *Endoceras*	NY, ONT, NFL
12. *Vaginoceras*	KY, NY, ONT, QUE, NFL
Actinocerida	
13. *Actinoceras*	KY, TN, MO, WI, MN, PA, NY, ONT, QUE, NFL
14. *Gonioceras**	TN, WI, MN, IL, NY, QUE, NFL
15. *Deiroceras*	MN, NY, ONT, QUE
16. *Ormoceras*	KY, TN, NY, ONT, QUE
Ascocerida	
17. *Hebertoceras*(?)**	IL
18. *Redpathoceras**	QUE
Tarphycerida	
19. *Avilionella**	TN, WI
20. *Barrandeoceras**	TN, ONT
21. *Bodeiceras**	WI
22. *Centrocyrtoceras*	TN, ONT, QUE
23. *Chidleyenoceras**	LAB, NFL
24. *Paquettoceras**	QUE
25. *Plectoceras**	KY, TN, WI, IL, NY, ONT, QUE
26. *Rhynchorthoceras**	NFL
Oncocerida	
27. *Beloitoceras*	KY, WI, MN, IL, ONT
28. *Oncoceras*	KY, ONT, QUE
29. *Richardsonoceras*	WI, ONT, QUE
30. *Laphamoceras*	KY, MN
31. *Zitteloceras*	WI, ONT, QUE, NFL

TABLE 3.—Middle Ordovician (Blackriveran and Rocklandian) nautiloid genera from Eastern North America (Laurentia)—Continued

Genus	Distribution
Oncocerida—Continued	
32. *Maelonoceras**	KY, ONT
33. *Cyrtorizoceras*	WI, MN
34. *Dunleithoceras**	IL
35. *Ehlersoceras**	ONT
36. *Kentlandoceras**	IND, WI, ONT
37. *Loganoceras**	ONT, QUE
38. *Rizosceras*	ONT
39. *Romingeroceras**	ONT
40. *Scofieldoceras**	MN
41. *Allumettoceras*	MN, ONT, QUE
42. *Tripteroceras*	WI, MN, NY, ONT, QUE
43. *Fayettoceras*(?)	WI
44. *Manitoulinoceras*(?)	ONT
45. *Staufferoceras*	MN
46. *Valcouroceras*(?)*	WI
47. *Actinomorpha**	WI, MN
48. *Danoceras*	ONT
49. *Diestoceras*	ONT
Discosorida	
50. *Reedsoceras*	WI
51. *Ruedemannoceras**	TN
52. *Simardoceras**	QUE
53. *Sinclairoceras**	QUE
54. *Teichertoceras*	QUE

There were 26/54 genera (48 percent) extinct by the end of the Rocklandian.

Laurentia (Kay, 1951; Neuman, 1955; Bird and Dewey, 1970; Ettensohn, 1991), due to sea-level fluctuations associated with continental glaciation in North Africa being initiated at this time (McKerrow, 1979) or due to a combination of these events.

There is also evidence of a change in lithofacies from nearly pure carbonate platform facies in the Blackriveran and Rocklandian to mixed carbonate and clastic facies in the Kirkfieldian and Shermanian. This appears to be true for much of Eastern North America (Wilson, 1949; Liberty, 1969; Cressman, 1973; Titus and Cameron, 1976; Cameron and Mangion, 1977; Sloan, 1987). As indicated above, this increase in clastic sedimentation is the result of successive collisional events between volcanic arc belts and the eastern margin of Laurentia, leading to the development of a siliciclastic source area (Taconic landmass) to the east (Bird and Dewey, 1970; Ettensohn, 1991) and uplift along the Transcontinental arch in the west (Witze, 1980; Sloan, 1987). Sediments eroding off of these rising landmasses spread into adjacent cratonic shelf areas.

Due to this sediment influx, Kirkfieldian and younger marine environments were more mud-rich, necessitating the development of new marine faunas adapted to these mud-bottom conditions. These faunal changes were evident to many of the paleontologists who have studied these faunas over the past 150 years, leading to the traditional differentiation of Middle Ordovician Mohawkian fossil biotas into either "Black River" or "Trenton" faunas (Hall, 1847; Winchell and Schuchert, 1895; Raymond, 1903; Kay, 1937; Young, 1943; Cooper, 1956). This change in benthic faunas would have, in turn, impacted the nautiloid faunas feeding on these organisms, possibly leading to the turnover in nautiloid faunas that occurred at this time.

Huff and Kolata (1990) traced the Deicke Bentonite Bed from southern Minnesota through eastern Missouri, across Kentucky and Tennessee, and into the valley and ridge of the southern Appalachians. Equivalent bentonite beds in Kentucky and Tennessee have also been termed the Pencil Cave bentonite of drillers or T-3 bentonite. As indicated above, the Pencil Cave occurs in the upper 50 ft of the Tyrone Limestone in central Kentucky, 25 ft above the biosparrudite that is the source of the 6034-CO collection of Tyrone nautiloids. Huff and Kolata (1990) describe the Deicke bed as ranging from 5 cm to 1.5 m thick and believe it represents the accumulation of air-fall volcanic ash on a shallow-marine platform during Rocklandian time. These authors noted that this bentonite bed thickens toward the southeast, indicating a volcanic source area that was probably east of the Tennessee-Georgia State line (fig. 1). They suggested that Southern Hemisphere trade winds carried the ash northwest across the craton. The Deicke was considered to be a single-graded bed and the result of a solitary eruptive event. Taking into account postdepositional compaction, Huff and Kolata estimated that the original volume of ash deposited was 1,122 cubic kilometers (km^3). They indicated that the Deicke Bentonite Bed and the overlying Millbrig Bentonite Bed (Willman and Kolata, 1978) are among the largest air-fall pyroclastic deposits reported from the geologic record. By way of comparison, the 1815 eruption of Tambora in Indonesia, one of the most violent eruptions recorded during historical times, had a volume of 175 km^3 (Self and others, 1984).

Sloan (1987) describes the Deicke volcanic eruption as the cause of a major provincial extinction event in the upper Mississippi valley region. At the species level, he lists conodont extinction at 10 percent, brachiopod extinction at 39 percent, gastropod extinction at 80 percent, trilobite extinction at 90 percent, and echinoderm extinction at 100 percent. As indicated above, nautiloid extinction at or near this horizon in this region was 95 percent based on the data presented by Catalani (1987). Sloan (oral commun., 1992) described these extinctions as varying variable from class to class but averaging 65 percent at the species level. He indicated that this extinction event marked the "Black River"–"Trenton" faunal boundary. Sloan also pointed out that the level of generic extinction was, however, much lower. Based on Catalani's (1987) list of taxa, 21 percent of the nautiloid genera in the Platteville Formation (or Group) became extinct at this time. Citing Kunk and Sutter (1984), Kolata and others (1986), and Samson (1986), Sloan gave the Deicke volcanic event an age of 454.2 Ma.

Several models proposed recently to explain the terminal Cretaceous extinction event (Sharpton and Ward, 1990) have postulated that large volumes of dust in the atmosphere generated by a bolide impact or by a large volcanic event would have had a catastrophic effect on both terrestrial and marine ecosystems. Deleterious conditions associated with these events are abrupt climatic cooling, the emission of large volumes of toxic gases, acidic rainfalls, and disruption of food chains. Huff and Kolata (1990) have suggested that the Middle Ordovician Deicke and Millbrig volcanic events deposited ash rapidly over broad areas and ejected large amounts of volcanic ash to high levels in the atmosphere. Consequently, these occurrences would both be expected to result in conspicuous extinction events.

Flower (1946, 1976) considered "Black River" nautiloid faunas (Blackriveran and Rocklandian) in Eastern North America to be tropical faunas compared with middle "Trenton" nautiloid faunas (Kirkfieldian and Shermanian), which he termed cooler water, temperate faunas. If Flower is correct, then this would give some support to the idea that climatic cooling resulting from either the Deicke or the Millbrig volcanic events contributed to the Rocklandian turnover in nautiloid faunas in Laurentia. Huff and others (1992), in discussing the biological effects of the widespread Millbrig ash fall, citing the work of Sharpf (1990) in Kentucky, indicated that this volcanic event had a minimal impact on benthic faunas. Sharpf suggested that the turnover from "Black River" to "Trenton" faunas was more likely the result of environmental change associated with the transgression early during the deposition of the Trenton Group. More detailed study of faunas on either side of the Deicke and Millbrig Bentonite Beds in Kentucky and elsewhere is needed to better understand the causes of this extinction event.

NAUTILOID FAUNA OF THE LEXINGTON LIMESTONE

As indicated above, the later Middle Ordovician "Trenton" nautiloid fauna in Kentucky and elsewhere in

Eastern North America is depauperate in terms of diversity in comparison with pre-Kirkfieldian Middle Ordovician "Black River" fauna (table 1). Kirkfieldian nautiloids, including Lexington Limestone taxa, are associated with mixed carbonate-clastic, shallow-marine shelf facies across Eastern North America. Faunas similar to the Lexington Limestone nautiloid fauna occur in the Galena Formation in Wisconsin or Galena Group in Minnesota and Illinois (Catalani, 1987), the Trenton Group in New York (Hall, 1847; Titus and Cameron, 1976); the Verulam and Lindsay Formations in southern Ontario (Liberty, 1969); and the Hull and Sherman Fall Formations in the Ottawa region of Ontario and Quebec (Wilson, 1961).

The nautiloid fauna collected from basal portions of the Curdsville Limestone Member of the Lexington Limestone in the central Blue Grass region is distinct from that in overlying Lexington Limestone strata in Kentucky in its relict "Black River" appearance. Blackriveran holdovers are *Actinoceras curdsvillense* Foerste and Teichert (pl. 17, fig. 11), *Deiroceras curdsvillense* Foerste and Teichert (pl. 18, figs. 8, 9), *Vaginoceras* sp. B (pl. 15, figs. 5–7), and *Anaspyroceras* cf. *A. cylindricum* (Foerste) (pl. 12, fig. 5). None of these taxa are present in collections from overlying members of the Lexington Limestone in Kentucky (table 1). Foerste (1910) also described "*Orthoceras*" *bilineatum-frankfortense* from "the cherty limestone beneath the Logana Bed." It is not known whether this limestone bed is part of the Curdsville Limestone Member or forms the base of the overlying Logana Member. This species appears to be assignable to the genus *Gorbyoceras* Shimizu and Obata.

Nautiloids in the Logana Member are abundant but of low diversity, the silicified collections from central Kentucky containing only three recognizable species. Faunas are dominated numerically by the small longiconic pseudorthocerid *Isorthoceras albersi* (Miller and Faber) (pl. 10, figs. 1–9), and contain lesser numbers of the annulated longiconic orthocerid *Gorbyoceras* cf. *G. tetreauense* Wilson (pl. 12, fig. 3) and the depressed orthoconic oncocerid *Allumettoceras* cf. *A. tenerum* (Billings) (pl. 20, figs. 19–22). This nautiloid fauna, especially the abundance of *Isorthoceras albersi*, is characteristic of the remainder of the Lexington Limestone.

Nautiloid diversity increases in the overlying Grier Limestone Member (Shermanian), where there are numerous specimens of *Isorthoceras albersi* (pl. 10, figs. 1–9) associated with the longiconic orthocerid *Polygrammoceras* sp. A (pl. 12, figs. 1, 2), the endocerid *Cameroceras* cf. *C. trentonense* Conrad (pl. 14, figs. 3, 4), and the oncocerid *Oncoceras* sp. (pl. 20, figs. 1, 2). The Grier Limestone Member nautiloid fauna is most similar to the nautiloid fauna described by Hall (1847) and Titus and Cameron (1976) from the Trenton Group in New York.

Nautiloids from collections in the upper Middle-lower Upper Ordovician Millersburg Member in central Kentucky consist only of the orthocerid *Isorthoceras albersi*. Nautiloids in the Strodes Creek Member in the vicinity of Winchester, Ky., consist of well-preserved calcite-replaced internal molds of the large longiconic orthocone *Polygrammoceras*? cf. sp. A (pl. 11, figs. 1–3) and fragmentary specimens of a small, slender cyrtocone similar to the discosorid *Faberoceras sonnenbergi* Flower.

Flower (1942) described a diverse nautiloid assemblage from a fossiliferous lens exposed at the Poindexter Quarry at Cynthiana, Ky. This fauna contains six species of the slender cyrtoconic oncocerid *Oonoceras* and lesser numbers of the oncocerids *Oncoceras* and *Rizosceras*, and rare specimens of the longiconic orthocerid *Isorthoceras* and the actinocerids *Deiroceras* and *Orthonybyoceras*. Flower also described the actinocerids *Armenoceras vaupeli* and *Troedssonoceras obscuroliratum*, the oncocerid *Danoceras cynthianense*, and the discosorids *Faberoceras ooceriforme*, *F. saffordi*, *F. sonnenbergi*, and *Reedsoceras mcfarlani*, all from the "Cynthiana Limestone" at Cynthiana, Ky. Comparison of nautiloid material from Winchester and Cynthiana indicates similar preservation and lithologies. Black and Cuppels (1973) indicate that the Strodes Creek Member extends northward from Winchester to Cynthiana. This evidence suggests that all of these nautiloid taxa may come from the Strodes Creek Member. If so, the Strodes Creek Member contains the most diverse nautiloid fauna in the Lexington Limestone exposed in the Blue Grass region of Kentucky (table 4). This nautiloid fauna consists of largely endemic taxa whose closest affinities are with nautiloids from the Catheys Formation in central Tennessee (Bassler, 1932) and nautiloid faunas from the Leipers Limestone (Late Ordovician, Edenian and Maysvillian) in the Cumberland valley in south-central Kentucky (Flower, 1946).

NAUTILOID FAUNA OF THE CINCINNATIAN PROVINCIAL SERIES

EDENIAN NAUTILOID FAUNA

As indicated above, the central Kentucky region and adjacent portions of northern Kentucky, southeastern Indiana, and southwestern Ohio were inundated by a marine transgression in the latest Shermanian and Edenian. The deeper water marine environments generated by this transgression were the sites of the deposition of the clay shale, siltstone, and thin fossiliferous limestone of the Clays Ferry and Kope Formations throughout this area. Cressman (1973) and Weir and others (1984) indicate that both of these stratigraphic units were deposited

TABLE 4.—Nautiloid species from the Strodes Creek Member of the Lexington Limestone, north-central Kentucky

[Data for species from the Cynthiana, Ky., area from Flower (1946); data for species from the Winchester, Ky., area from collections made by USGS field parties and the writer's field work in the area. PO, Poindexter Quarry; CY, Cynthiana, Ky., area; WIN, Winchester, Ky., area; 7310-CO, USGS collection locality number]

Species	Occurrence
Order Orthocerida	
1. *Isorthoceras praenuntium*	PO
2. *Polygrammoceras*? sp.	WIN (7310-CO)
Order Endocerida	
3. *Cameroceras* sp.	CY
Order Actinocerida	
4. *Armenoceras vaupeli*	PO
5. *Deiroceras perisiphonatum*	PO
6. *Troedssonoceras obscuroliratum*	CY
7. *Orthonybyoceras*? *perseptatus*	PO
Order Oncocerida	
8. *Oncoceras carlsoni*	PO
9. *O. fossatum*	CY
10. *O.* sp.	CY
11. *Oonoceras acutum*	PO
12. *O. gracilicurvatum*	PO
13. *O. multicameratum*	PO, WIN (7328-CO)
14. *O. planiseptum*	PO
15. *O. suborthoforme*	PO
16. *O. triangulatum*	PO
17. *O.*? *brevidomum*	PO
18. *Rizosceras graciliforme*	PO
19. *R. conicum*	PO
20. *Danoceras cynthianaense*	CY
21. *Diestoceras* sp.	PO
Order Discosorida	
22. *Faberoceras ooceriforme*	CY
23. *F. saffordi*	CY, WIN (7331-CO)
24. *F. sonnenbergi*	CY, WIN (7310-CO)

at water depths at or below 80 ft (24 m), below fair-weather wave base, but still within the zone of storm-current reworking.

Nautiloids are locally common in the interbedded clay shale and thin limestone of the Clays Ferry Formation in north-central Kentucky and the Point Pleasant Tongue of the Clays Ferry Formation in the greater Cincinnati region of northern Kentucky and southwestern Ohio. Characteristic species are the small longiconic orthocerids *Isorthoceras albersi* (pl. 10, figs. 1–9) and *I. rogersensis* (Foerste) (pl. 10, figs. 10-12); *Treptoceras* cf. *T. duseri* (Hall and Whitfield); the large longiconic orthocerid *Ordogeisonoceras amplicameratum* (Hall) (pl. 3, figs. 1–3); an unidentified species of the endocerid *Cameroceras*, the coiled tarphycerid *Trocholites faberi* Foerste, and the oncocerid *Oncoceras covingtonense* Flower.

Nautiloid faunas in the more clay-rich Kope Formation are dominated by common, but typically incomplete specimens of the longiconic orthocerid *Treptoceras transversum* (Miller) (pl. 8, figs. 4–9) and the large endocerid *Triendoceras*? *davisi* n. sp. (table 5). The latter species is typically represented in these strata by only the matrix-filled molds of the siphuncle tube (pl. 16, figs. 1–5). Less common in the Kope Formation are specimens of *Ordogeisonoceras amplicameratum* (pl. 3, figs. 4, 5, 8) and *Trocholites faberi* (pl. 19, figs. 6–10). Flower (1946) also lists rare specimens of the oncocerids *Augustoceras*? sp. and *Diestoceras*? *edense* Flower and the discosorids *Faberoceras magister* (Miller), *Westonoceras ventricosum* Flower, and *W.*? *ortoni* (Meek) from the Kope Formation at Cincinnati.

Clays Ferry and Kope nautiloid faunas are most similar to the coeval Utica Shale fauna in New York (Hall, 1847; Ruedemann, 1926) and the fauna of the Whitby Formation in southern Ontario (Liberty, 1969). Associated with the regional transgressive event represented by the deposition of these formations in the Cincinnati arch region is a brief incursion of the so-called "Arctic" fauna into the northern Kentucky-southwestern Ohio region (Flower, 1946, 1976). This "Arctic" nautiloid fauna has been described from late Middle Ordovician (Shermanian) to Late Ordovician (Richmondian) tropical platform facies in New Mexico (Flower, 1970), Wyoming (Foerste, 1935; Miller and Carrier, 1942), Manitoba (Foerste, 1929b), the Hudson Bay region (Nelson, 1963), the Canadian Arctic Archipelago (Miller and others, 1954), and northern Greenland (Troedsson, 1926). Edenian elements of this fauna in the Cincinnati area are species of the oncocerid *Diestoceras* and the discosorid *Westonoceras*. This incursion of these "Arctic" nautiloid taxa into the Cincinnati arch region coincides with the fuller development of this fauna to the north in the Cobourg Limestone in Ontario, Quebec, and adjacent portions of New York (Flower, 1952; Wilson, 1961; Liberty, 1969).

MAYSVILLIAN AND EARLY RICHMONDIAN NAUTILOID FAUNA

Nautiloids are abundant and typically well preserved in a number of blocky claystone beds in the Fairview Formation, Grant Lake Limestone, and Bull Fork Formation (Maysvillian-early Richmondian) in northern Kentucky, southeastern Indiana, and southwestern Ohio. Nautiloid

TABLE 5.—Distribution of species of nautiloid cephalopods in the Upper Ordovician Cincinnatian Provincial Series in the Cincinnati arch region to Kentucky, Indiana, and Ohio

[Data include only those nautiloid species described in this paper, primarily nonsilicified material from museum collections. Upper Ordovician stratigraphic units from which no nautiloid species are described are omitted. CF, Clays Ferry Formation; KO, Kope Formation; LL, Leipers Limestone; FA, Fairview Formation; GL, Grant Lake Limestone; BF, Bull Fork Formation; DR, Drakes Formation; SA, Saluda Formation; WW, Whitewater Formation; HL, Hitz Limestone Member. X indicates presence of species. For locality information, see appendix]

Species	CF	KO	LL	FA	GL	BF	DR	SA	WW	HL	Location
Orthocerida											
Pleurorthoceras clarksvillense						X					IND-1, OH-7, OH-15
Ordogeisonoceras amplicameratum		X									Cincinnati, Ohio
Treptoceras byrnesi				X							OH-4
Treptoceras cincinnatiensis			X		X	X					KY-3, OH-6 to 9, 6127-CO
Treptoceras duseri	X			X	X	X					KY-3, OH-2, 3, 5–9
Treptoceras fosteri						X					IND-1, 2; OH-6 to 9
Treptoceras transversum		X		X							KY-1, 2
Gorbyoceras gorbyi								X			IND-4
Gorbyoceras hammelli								X		X	IND-4 to 6
Richmondoceras brevicameratum					X		X				Osgood, Ind.; OH-10
Endocerida											
Cameroceras rowenaense			X								KY-4
Cameroceras? sp.						X					KY-5
Triendoceras? davisi		X									KY-1, 2
Actinocerida											
Orthonybyoceras dyeri			X	X	X						Covington, Ky., Cincinnati, Ohio
Tarphycerida											
Trocholites faberi		X									KY-1
Oncocerida											
Beloitoceras amoenum						X			X		IND-7, OH-13
Beloitoceras bucheri							X				IND-4
Augustoceras shideleri			X								KY-4
Manitoulinoceras gyroforme						X					OH-12
Diestoceras indianense								X			IND-4
Diestoceras shideleri								X			IND-4
Diestoceras eos					X						OH-10

faunas in these claystone beds are usually of low diversity, dominated by an abundance of several species of the longiconic orthocerid *Treptoceras* Flower. These species are *T. duseri* (Hall and Whitfield) (pl. 4, figs. 8–12; pl. 5, figs. 1–6), *T. fosteri* (Miller) (pl. 5, figs. 7–9; pl. 6, figs. 1–3, 5, 7, 8), and *T. cincinnatiensis* (Miller) (pl. 7, figs. 1–8; pl. 8, figs. 1–3). Species of *Treptoceras* typically make up 75 percent of the nautiloids present in these strata, usually associated with lesser numbers of the endocerid *Cameroceras inaequabile* (Miller), the actinocerid *Orthonybyoceras dyeri* (Miller) (pl. 18, figs. 10–17), and rare specimens of the oncocerids *Oncoceras* and *Manitoulinoceras* (Frey, 1988, 1989). This is the typical "Cincinnatian" fauna that is characteristic of shallow-marine strata of this age across the northern Cincinnati arch region.

Nautiloids are less common in brachiopod-bryozoan packstone facies of the Calloway Creek Limestone, the Fairview Formation, and the Grant Lake Limestone in the Kentucky-Indiana-Ohio region. Specimens are often fragmentary and consist of poorly preserved internal molds. Faunas in these limestones are also dominated by species of *Treptoceras* and *Cameroceras*, appearing with less common specimens of the actinocerids *Orthonybyoceras dyeri* and *Troedssonoceras turbidum* (Hall and

Whitfield) and rare specimens of the oncocerids *Beloitoceras* and *Oncoceras* (Flower, 1946).

Nautiloid taxa in Maysvillian and early Richmondian strata in the Cincinnati region are also characteristic of the "Lorraine" fauna of Foerste (1914b). This fauna is well developed in coeval clastic marine facies of the Pulaski Formation in New York (Hall, 1847; Ruedemann, 1926) and in the Georgian Bay Formation in southern Ontario (Parks and Fritz, 1922; Foerste, 1924a; Liberty, 1969).

NAUTILOID FAUNA OF THE LEIPERS LIMESTONE

The Leipers Limestone (Edenian and Maysvillian) exposed in the Cumberland River valley in south-central Kentucky has a distinctive, diverse nautiloid fauna of 17 species belonging to 9 genera. These nautiloid taxa consist of four species of the large discosorid *Faberoceras* and three species of the slender cyrtoconic oncocerid *Augustoceras*, associated with the oncocerids *Danoceras* and *Oncoceras*, as well as the ubiquitous orthocerid *Treptoceras* and the endocerid *Cameroceras* (Flower, 1946). The latter genus is represented by *C. rowenaense* n. sp. (pl. 14, figs. 1, 2). The most abundant nautiloid in collections available to the writer is the oncocerid *Augustoceras shideleri* Flower (pl. 20, figs. 23–27).

Flower (1946) indicated that nautiloids were most abundant in two *Tetradium* biostromal beds in the lower half of the formation. The Leipers Limestone nautiloid fauna is endemic to the outcrop belt in south-central Kentucky, and some of the characteristic genera also occur in the Catheys Formation and the Leipers Limestone in adjacent portions of central Tennessee (Bassler, 1932). Many of the Leipers Limestone nautiloid taxa also occur in similar facies in the older Strodes Creek Member of the Lexington Limestone (Shermanian and Edenian) in east-central Kentucky.

MIDDLE AND LATE RICHMONDIAN NAUTILOID FAUNAS

Younger Richmondian strata in the northern Cincinnati arch region of Kentucky, Indiana, and Ohio indicate the influx of many new nautiloid taxa into the region (Flower, 1946, 1976; Frey, 1981). As indicated above, these immigrant taxa have their origin in the distinctive "Arctic" nautiloid fauna characteristic of late Middle Ordovician to Late Ordovician tropical carbonate platform facies in western and northern portions of North America (Foerste, 1929a). The incursion of these taxa into the Kentucky-Indiana-Ohio area coincides with an early Richmondian transgression across the area and the subsequent shallowing-up cycle in which there is a shift from primarily clastic to carbonate sedimentation (fig. 2 and table 5).

The "Richmondian" fauna from the northern Cincinnati arch region consists of smaller, local versions of the giant forms characteristic of this "Arctic" nautiloid fauna. These Ohio valley species also differ from the typical "Arctic" fauna characteristic of Wyoming, Manitoba, and the Hudson Bay region in the absence of many important taxa. These are the actinocerids *Actinoceras*, *Kochoceras*, and *Huronia*; the orthocerid *Ephippiorthoceras*; the tarphycerids *Aspidoceras*, *Digenoceras*, and *Wilsonoceras*; and the discosorids *Cyrtogomphoceras*, *Westonoceras*, and *Winnipegoceras*. The "Richmondian" fauna in the Cincinnati arch region is more similar to nautiloid faunas described from the more carbonate-rich facies of the Maquoketa Shale (Late Ordovician, Maysvillian and Richmondian) in northeast Iowa and Minnesota (Foerste, 1936a). The Ohio valley "Richmondian" fauna is also similar to coeval nautiloid faunas described from the Vaureal Formation on Anticosti Island, Quebec (Foerste, 1928a) and the Whitehead Formation at Perce, Quebec (Foerste, 1936b).

The first incursions of this "Arctic" fauna into the Kentucky-Indiana-Ohio area during Richmondian times occur within the "Waynesville" biofacies (in the sense of Wahlman, 1992, p. 19) in the Bull Fork Formation (early Richmondian) in southwestern Ohio. Early immigrants are the ascocerid *Schuchertoceras obscurum* Flower and the annulated orthocerid *Gorbyoceras curvatum* Flower, both in the *Treptoceras duseri* shale unit (Frey, 1988, 1989). The longiconic orthocerid *Pleurorthoceras clarkesvillense* (Foerste) (pl. 2, figs. 10–15), the aberrant "flatfish" actinocerid *Lambeoceras richmondense* (Foerste), and the large oncocerid *Diestoceras waynesvillense* Flower occur slightly higher up (30–40 ft) within the formation in southwestern Ohio (Flower, 1946).

The "Arctic" nature of the "Richmondian" fauna in the tristate area reaches its fullest development, however, in the Bardstown Member of the Drakes Formation in northwest Kentucky, in the overlying Saluda Dolomite Member in northwest Kentucky and adjacent portions of southeastern Indiana, and in the Whitewater Formation in southeastern Indiana. A compilation of data from field work done for this report in Indiana and Ohio and a list of taxa presented by Flower (1946) indicate the presence of 65 described nautiloid species in these strata, belonging to 24 genera and 7 orders. Only about 20 of these species, however, are abundant enough to be commonly encountered in these strata. A number of late Richmondian nautiloid taxa from the Saluda Formation, its Hitz Limestone Member, and the overlying Whitewater Formation in southeastern Indiana are described and illustrated here. These taxa are the large longiconic orthocerid *Richmondoceras brevicameratum* n. sp. (pl. 13, figs. 1–4), the annulated orthocerids *Gorbyoceras gorbyi* (Miller) (pl. 12, figs. 12, 13) and *G. hammelli* (Foerste) (pl. 12, figs. 4,

6, 7, 10, 11), and the oncocerids *Beloitoceras amoenum* (Miller) (pl. 22, fig. 2), *B. bucheri* Flower (pl. 22, figs. 4, 5), *Diestoceras shideleri* (Foerste) (pl. 21, figs. 1, 2, 5; pl. 22, figs. 1, 3) and *D. indianense* (pl. 22, figs. 6–9).

The top of the Upper Ordovician rocks in southwestern Ohio consists primarily of blue clay shale and thin fossiliferous limestone west of the Dayton, Ohio, area and maroon and pale green mudstone interspersed with minor thin siltstone and limestone southeast of Dayton. In southwestern Ohio the blue shale and limestone facies is referred to as either the Bull Fork or Whitewater Formation (Weir and others, 1984, p. 81–90; Shrake and others, 1988), and the maroon and green mudstone facies is placed within the Preachersville Member of the Drakes Formation. The contact between these units is gradational and diachronous across southwestern Ohio, with the Preachersville Member thickening to the southeast (Sweet, 1979).

These formations in southwestern Ohio contain a sparse nautiloid fauna consisting of the ubiquitous *Treptoceras* and *Cameroceras* plus the large longiconic orthocerid *Richmondoceras brevicameratum* (pl. 13, figs. 5–8), rare specimens of the actinocerid *Lambeoceras richmondense* (Foerste), and the oncocerids *Beloitoceras amoenum* Miller (pl. 21, fig. 3), *Manitoulinoceras gyroforme* Flower (pl. 20, figs. 28, 29), and *Diestoceras eos* (Hall and Whitfield) (pl. 21, fig. 4). All of these taxa were collected from the blue shale and thin limestone lithology (=uppermost Bull Fork Formation).

The writer has collected a small, incomplete oncocerid (cf. *Beloitoceras*) from the maroon mudstone facies in Adams County, Ohio. Flower (1957b) also listed *Treptoceras* associated with trilobites and eurypterids from this lithology exposed at Manchester, Adams County, Ohio. Nautiloids have not been described from the same facies in adjacent portions of northern Kentucky.

LATE ORDOVICIAN EXTINCTION EVENT

One of the largest extinction events in the Phanerozoic fossil record occurred at the Ordovician-Silurian boundary. Sepkoski (1982) has estimated that 22 percent of all marine invertebrate families became extinct during this event, second in magnitude only to the extinction at the Permian-Triassic boundary. Brenchley (1989) summarized the effects of this extinction on the major invertebrate groups, noting that the Late Ordovician extinction sharply reduced the number of genera and species among trilobites, nautiloids, corals, brachiopods, crinoids, graptolites, conodonts, and acritarchs. Both sessile filter feeders and vagile benthos were affected. His figures indicate that graptolite genera were reduced nearly 100 percent, corals and trilobites suffered 75 percent extinction at the genus level, crinoid genera suffered 70 percent extinction, and brachiopod genera were reduced by 25 percent.

House (1988) indicated approximately 100 nautiloid genera in the Upper Ordovician and a loss of nearly 55 percent of these genera by the end of the period. The writer's tabulations of Upper Ordovician nautiloid taxa from the published literature indicate that Richmondian nautiloid faunas in Laurentia consisted of 48 genera belonging to 8 orders. Succeeding Gamachian faunas, known primarily from Anticosti Island, Quebec (Foerste, 1928a), are represented by 15 genera belonging to 6 orders. The nautiloid order Ellesmerocerida went extinct at this time, and the order Endocerida was reduced to several poorly known species of *Cameroceras* and *Endoceras* and the problematic genus *Humeoceras*. Only 13 genera survived from the Ordovician into the Silurian, indicating an extinction of 73 percent of Richmondian nautiloid genera by the end of the Gamachian. These Silurian survivors are elements of the Ordovician "Arctic" fauna that persisted into the Silurian in carbonate basin areas in the Hudson Bay region of Canada (Foerste and Savage, 1927; Flower, 1968b) and in the Anticosti Island region of Quebec (Foerste, 1928a).

In the northern Cincinnati arch region, nautiloids declined from a high of 65 species, belonging to 24 genera, in the late Richmondian Saluda and Whitewater Formations to 7 species, belonging to 6 genera in the youngest Ordovician rocks immediately below the Brassfield Limestone (Silurian-Llandoverian) in southeastern Indiana. In southwestern Ohio, nautiloid diversity drops from a high of eight species belonging to seven genera in the upper part of the Bull Fork Formation to a low of two species belonging to two genera in the Preachersville Member of the Drakes Formation, within 10 ft of the Ordovician-Silurian contact. While Brassfield Limestone nautiloids are poorly known and in need of study, the possible exception being the orthocerid *Treptoceras*, none of the Richmondian genera occur in the overlying Silurian. The Brassfield Limestone nautiloid fauna in the Kentucky-Indiana-Ohio region (Foerste, 1893) is a mix of Ordovician survivors and direct descendants (orthocerids *Treptoceras* Flower and *Euorthoceras* Foerste, actinocerid *Ormoceras* Saeman, and the discosorid *Glyptodendron* Claypole) plus a number of newly evolved, distinctly Silurian taxa (actinocerid *Elrodoceras* Foerste and the discosorid *Discosorus* Hall).

The mass extinction event of the Late Ordovician is coincident with a worldwide regressive episode that has been associated with continental glaciation centered in African and South American portions of Gondwanaland (Berry and Boucot, 1973; Denison, 1976; Brenchley and Newall, 1980; Brenchley, 1984). This extinction has been attributed to climatic cooling (Stanley, 1984a, 1984b), a loss of shallow-marine shelf area (Sheehan, 1988), and a combination of these factors (cooling, regression, changes in ocean chemistry, and paleogeography), all associated with the growth and decay of the Gondwanan ice sheets (Brenchley, 1989). A lack of supporting geochemical evidence across the Ordovician-Silurian boundary makes an extraterrestrial impact unlikely as the cause of this extinction (Orth and others, 1986).

Sedimentological and paleontological evidence suggests that the late Richmondian (=Ashgillian) extinction in Laurentia was primarily the result of the regression of shallow epicontinental seas off extensive cratonic shelf areas, eliminating marine habitats and the "perched faunas" that were dependent on these environments (Sheehan, 1988), the so-called "species-area effect" (MacArthur and Wilson, 1963). Evidence of this Late Ordovician regression is well documented in the northern Cincinnati arch area.

In the Kentucky-Indiana-Ohio region, this Late Ordovician regression is represented by the Richmondian shallowing-up cycle, consisting of the basal deeper water marine shelf shale and fossiliferous limestone of the Bull Fork Formation grading up into shallow water carbonates (Saluda and Whitewater Formations) in western portions of the Cincinnati arch region and intertidal mudstone and mud-cracked siltstone (Preachersville Member of the Drakes Formation) along the east flank of the arch (fig. 2). Uppermost Ordovician strata (Gamachian) are not exposed anywhere in the Cincinnati arch region. Contrary to the observations of Gray and Boucot (1972), at most localities studied for this report in southeastern Indiana and southwestern Ohio, a disconformity marks the Ordovician-Silurian boundary. This distinctive erosional surface is diachronous across the region, being older to the southeast and younger in the western portions of the Cincinnati arch region (Sweet, 1979) (fig. 2).

Uppermost Ordovician nautiloid faunas in the Cincinnati arch region are dominated by elements of the "Arctic" fauna, characterized by Flower (1946, 1976) as being a tropical fauna. Nautiloids from the Ordovician rocks, immediately underlying the Brassfield Limestone (Llandoverian, C1–C2) in Decatur County, Ind. (Fluegeman and Pope, 1983) have "Arctic" affinities. These taxa consist of the endocerid *Endoceras*, the oncocerids *Beloitoceras* and *Oonoceras*, and large tarphycerids tentatively assigned to *Charactoceras* Foerste and *Deckeroceras* Foerste. Ordovician nautiloid taxa surviving into the Early Silurian in Laurentia are also largely members of this "Arctic" fauna (Foerste and Savage, 1927; Foerste, 1928a; Flower, 1968b). These taxa are associated with carbonate platform facies in the Hudson Bay region and mixed carbonates and shale at Anticosti Island, Quebec. Succeeding Llandoverian nautiloid faunas across Laurentia are also primarily tropical taxa associated with carbonate platform facies. This "*Discosorus-Huronia* fauna" consists of a mix of Ordovician "Arctic" relict forms and new actinocerid and discosorid taxa, the latter two orders being typically most abundant in tropical carbonate facies (Flower and Teichert, 1957; Flower, 1957b, 1968b). None of these occurrences seem to support the idea that climatic cooling associated with the Late Ordovician glaciation was a major factor affecting the extinction of these nautiloid taxa in Laurentia in the late Richmondian, as a majority of the surviving taxa were stenothermic tropical forms.

NAUTILOID TAPHONOMY

NAUTILOID BIOSTRATINOMY

Nautiloid remains occur in a number of the lithofacies that make up the Ordovician strata exposed in Kentucky and adjacent States. Nautiloid preservation in these rocks is a function of several factors: original depositional environment; nautiloid shell strength, morphology, and size; and effects of a variety of postdepositional processes. The latter are biological processes like bioturbation, mechanical processes like storm reworking of bottom sediments, and chemical processes such as dissolution and replacement of the original aragonitic shell.

Nautiloids are typically restricted to subtidal marine facies representative of the environments occupied by the living cephalopod. Reports of extensive postmortem transport of *Nautilus* shells in the Indo-Pacific region (Stenzel, 1964; Toriyama and others, 1965; Teichert, 1970), coupled with descriptions of a number of current-affected, shallow-water, or strand-line nautiloid assemblages (Miller and Furnish, 1937; Reyment, 1958, 1968; Zinsmeister, 1987), have led to the general assumption that many or all fossil nautiloid accumulations were deposited after some unknown period or distance of transport from the habitat of the living nautiloid animal. Many Paleozoic nautiloids possessed phragmocones that were counterweighted against positive buoyancy and probably would have rapidly filled with water and sank after death (Crick, 1988, 1990; Frey, 1989). Studies of *Nautilus* (Chamberlain and others, 1981) have indicated that extensive postmortem drift of *Nautilus* is a relatively rare event. Ordinarily, upon the death of the animal, the shells sink to the sea floor in the area in which the animals lived. Studies of the ontogeny of living *Nautilus* (Saunders and Spinosa, 1979; Landman and others, 1983) indicate that these cephalopods lack a planktonic larval stage capable of being dispersed by oceanic currents. Additionally, the distribution of adult *Nautilus* is constrained by stretches of deep water in excess of 800 m (2,625 ft) as well as areas of open ocean (Ward, 1987). These studies of living *Nautilus* coupled with investigations of the biogeography of fossil nautiloids (Flower, 1957b, 1976; Crick, 1980, 1990; Frey, 1989) indicate that nautiloids were not part of the nekton, but were nektobenthic, that is, part of the shallow-shelf benthos and capable of dispersal only along continuous shelf areas or over shallow stretches of open ocean. The facies distribution of the nautiloid taxa described here also supports this point of view.

SILICIFIED NAUTILOIDS

Many of the nautiloids described here, especially specimens from the Tyrone Limestone and the Lexington Limestone, are silicified. These specimens were obtained from bulk rock collections (36,000 pounds (lbs) of limestone) made by USGS field parties in the 1960's and 1970's. Details of this acid-etching program can be found in the works by Pojeta (1979) and Pope (1982). Silicified nautiloid material presents both problems and opportunities in terms of nautiloid study. As indicated by Flower (1965), nautiloids are often common in massive-bedded limestone facies, but they are difficult to study because of the complexity of extracting large nautiloid shells from these dense limestones. Usually, only fragments of such specimens can be obtained. Lower Paleozoic nautiloid species are often based on single fragments or a number of fragments representing different parts of one or more specimens.

Silicified material from the Ordovician of Kentucky yielded a number of more or less complete, uncrushed specimens, commonly with details of the shell's exterior ornament preserved. Unfortunately, key internal features, especially septal necks and connecting rings, typically are not preserved, making family- and genus-level identifications difficult. These features are especially critical in the systematic treatment of morphologically conservative longiconic groups, namely, the baltoceratids, orthocerids, and actinocerids. These taxa make up the bulk of the nautiloids from Ordovician strata in Kentucky. Surprisingly, cameral and endosiphuncular deposits are commonly preserved in these specimens (pls. 1, 4, 17) and can, if sufficiently developed, give some indication of siphuncle segment shape and septal neck type. Coarse silicification typically destroys key internal features of the siphuncle used to differentiate endocerid taxa. Fortunately, in the case of the Tyrone Limestone material, silicification was such that these important features were preserved (pl. 15).

Silicified nautiloid material from the 6034-CO collection in the Tyrone Limestone consisted of many unidentifiable fragments and fewer sections of phragmocone up to 35 cm in length. Baltoceratids and orthocerids in this collection consist primarily of adapical portions of phragmocones (less than 15 cm in length) that lack preserved body chambers (pls. 1, 2, 4, 9). Endocerids and actinocerids consist of larger portions of the phragmocone (15–35 cm in length) that preserve the large siphuncle and attached camerae (pls. 15, 17). Oncocerids are preserved as adapical portions of the phragmocone, rarely having shells that have body chambers intact (pl. 19, figs. 11–15; pl. 20, figs. 7–12). Tarphycerids are preserved as the inner whorls of the coiled phragmocone (pl. 19, figs. 1–3) and fragments of the body chamber. The fragmental nature of this assemblage as well as the enclosing biopelsparite lithology attest to deposition in a shallow-water, wave- or current-affected environment. However, the poor sorting of this nautiloid material suggests only periodic current reworking of these shells, reworking by episodic storms rather than the constant milling action of "normal" wave and current activity in a shoaling environment.

The basal Curdsville Limestone Member of the Lexington Limestone yielded less well preserved, coarsely silicified, fragmentary phragmocones and external molds (pl. 17, fig. 11; pl. 18, figs. 8, 9). In contrast, nautiloids from the overlying Logana Member and Grier Limestone Member of the Lexington Limestone and from the Clays Ferry Formation consist of finely silicified fragmentary and complete phragmocones that have the external ornament well preserved (pl. 10, figs. 4–7; pl. 12, figs. 1, 2; pl. 20, figs. 1, 2, 19–22). The more complete nature of this material is probably a function of the lower energy, quiet-bottom environments indicated for these stratigraphic units.

NONSILICIFIED NAUTILOIDS

Nautiloid specimens from more clastic-rich lithofacies of the Lexington Limestone and the overlying Cincinnatian Provincial Series in the Kentucky-Indiana-Ohio area usually are poorly preserved, matrix-filled internal molds or better preserved, calcite-replaced internal molds. Shell exteriors are rarely encountered, usually as external or composite molds or, less commonly, as the calcite-replaced outer shell. Nautiloids in shalier, mud-rich facies are commonly complete and consist of the phragmocone and the body chamber but are usually crushed and flattened by compaction of the enclosing sediment. These molds typically fail to preserve the critical internal structures of the siphuncle. Nautiloid material from the various limestone lithologies in Cincinnatian strata often consists of fragmentary material that is generally less deformed than specimens from shalier facies. Preservation of internal features is dependent on whether or not the septa and siphuncle are replaced by calcite. Silicification is uncommon in Cincinnatian rocks and, if present, is commonly coarse and incomplete. Replacement of nautiloid shells by sparry calcite is far more common and results in the best preserved nautiloids collected from these Upper Ordovician rocks. However, in some specimens, especially endocerids, the coarse sparry calcite has obliterated the critical internal structure of the siphuncle.

Claystone facies and, less commonly, thin packstone beds in the Kope, Clays Ferry, Fairview, Grant Lake, and Bull Fork Formations locally yield abundant well-preserved nautiloids. This material typically consists of complete, calcite-replaced phragmocones with body chambers more or less intact (pl. 3, figs. 1–5; pl. 4, figs. 8–12; pl. 5; pl. 6; pl. 7). Some of these also preserve the outer shell, occasionally with the original color pattern of the shell (pl. 6, fig. 3). Similar calcite-replaced material is also common in the dense, dark-gray fossiliferous wackestone of the Leipers Limestone in south-central Kentucky (pl. 14, figs. 1, 2; pl. 20, figs. 23–27). Less well preserved, but commonly complete, matrix-filled internal molds are locally common in the muddy wackestone facies of the Mount Auburn Member of the Grant Lake Formation (Schumacher and others, 1991) in the Cincinnati area; in local facies within the Bull Fork Formation in southwestern Ohio (pl. 2, figs. 10–15; pl. 6, figs. 4, 8; pl. 13, figs. 5–8); and in dense, micritic limestone beds within the Saluda Formation in southeastern Indiana (pl. 12, figs. 12, 13; pl. 13, figs. 1, 3, 4; pl. 21, figs. 1, 2). Spar-replaced, well-preserved fragmentary nautiloid material is abundant in the thin, platy, silty limestone of the Hitz Limestone Member at the top of the Saluda Formation in southeastern Indiana and northern Kentucky (pl. 12, figs. 4, 6, 7, 10, 11).

NAUTILOID CLASSIFICATION

The classification used here for the higher level nautiloid taxa is that of Teichert (1988). Teichert's classification divides the nautiloid cephalopods into four subclasses: the Orthoceratoidea, consisting of the orders Ellesmerocerida, Orthocerida, and Ascocerida, plus a number of short-lived Cambrian orders (Chen and Teichert, 1983); the Endoceratoidea, consisting of the order Endocerida and the problematic Interjectocerida; the Actinoceratoidea, consisting of the order Actinocerida; and the Nautiloidea, consisting of the orders Tarphycerida, Oncocerida, Discosorida, and Nautilida. As emphasized by Crick (1988) and by Wade (1988), this classification stresses the importance of buoyancy regulation as the central problem that fueled the evolution of the group. Morphological differentiation of these taxa centers on the development of various buoyancy-regulating devices. Crick (1988) noted that the Paleozoic evolutionary history of the cephalopods is "a series of experiments in an attempt to find the most energy-efficient compromise between buoyancy regulation, stability, orientation, locomotor design, and environment." Crick also listed five major solutions to the buoyancy control problem that were devised by these nautiloids. These are (1) short orthoconic and cyrtoconic longicones with water ballast (Ellesmerocerida and the various Cambrian orders); (2) large orthoconic and cyrtoconic longicones with calcareous ballast (Ellesmerocerida, Orthocerida, Endocerida, Actinocerida); (3) orthoconic and cyrtoconic brevicones with water ballast (Oncocerida and Discosorida); (4) planispiral coiling (Tarphycerida, Nautilida); and (5) shell truncation (Ascocerida and rare Orthocerida).

Family- and genus-level systematics largely follow Teichert and others (1964) with the exception of the Orthocerida and the Tarphycerida. The familial descriptions for the orthocerid Orthoceratidae, Michelinoceratidae, and Geisonoceratidae are emended to better define these groups. Descriptions of a number of genera within these families were also revised to better define these taxa. The classification of the Tarphycerida used here is that of Flower (1984); however, the Barrandeocerida is abandoned and its families placed in the Tarphycerida. New genera described are the longiconic orthocerids *Pojetoceras*, *Ordogeisonoceras*, and *Richmondoceras*. Familial and generic differentiation of taxa stresses the morphology of the siphuncle and the septa, especially siphuncle segment position, segment shape, segment diameter, endosiphuncular deposits, and nature of the septal necks.

The species problem in nautiloid systematics.—Species-level nautiloid taxa tend to be, on the whole, poorly defined. Most described Paleozoic species are based on fragmentary, often poorly preserved material. Studies of the few well-preserved, complete fossil nautiloids known to the writer have indicated that pronounced ontogenetic changes in the details of the shell morphology are typical of these nautiloid taxa. Species based on fragmentary specimens often have been found to represent separate ontogenetic stages of the same animal. This is certainly the case for the various species that have been assigned to the orthocerid genus *Treptoceras* Flower, which is the dominant nautiloid taxon in the Upper Ordovician rocks exposed in the Kentucky-Indiana-Ohio region.

Other studies of existing nautiloid species indicate that taxa were defined on the basis of criteria that took into account the nature of preserved epizoan organisms encrusting the nautiloid shell, stratigraphic occurrence, and type of preservation. Some species have been defined entirely on the basis of external shell features (shell shape, rate of shell expansion, size, external ornament), while others are known only from internal molds or isolated siphuncles. This often results in the naming of at least two taxa, representing the external and internal features of a single species. Additionally, as has been noted by Aronoff (1979) and Dzik (1984), there have been very few studies that have evaluated intraspecific variation in fossil nautiloids. This is due to the lack of the large numbers of specimens needed to characterize this variation and the description of most nautiloid species on the basis of one or a few specimens that are typically incomplete. All of this indicates that Paleozoic nautiloid species are probably oversplit and that consistent criteria for defining nautiloid species are largely lacking.

Morphological features traditionally used to differentiate longiconic nautiloid taxa (orthocerids, endocerids, actinocerids) consist primarily of the external characteristics of the conch. These are features like shell size, external shell ornament, shell cross sectional shape, rate of shell expansion (apical angle), cameral length, and suture type (fig. 4). The studies of Hyatt (1900), Teichert (1933), Flower (1939, 1946), and Flower and Kummel (1950) recognized the importance of the morphology of the siphuncle as a tool to distinguish various longiconic taxa. This morphology takes into account such attributes as siphuncle segment position within the shell, segment shape, segment diameter relative to total shell diameter, septal neck type, connecting ring structure, and the presence and structure of endosiphuncular deposits. A descriptive terminology developed to further define these features. These descriptive terms, unfortunately, were often ambiguous and came to mean different things to different people.

An attempt to correct this problem was made by Aronoff (1979), who proposed to standardize measurements of various features of longiconic nautiloid shells, especially siphuncle segment position and segment shape, and use these quantitatively to define longiconic taxa. His terminology is used here. Siphuncle position is measured using the segment position ratio (SPR). This indicates the position of the mid-axis of the siphuncle segment (fig. 5B), measured from the venter, relative to the total diameter of the shell at the same point. SPR values range from dorsal (SPR greater than (>) 0.75), to middorsal (SPR=0.75), supracentral (SPR values between 0.50 and 0.75), to central (SPR=0.50), subcentral (SPR values between 0.25 and 0.50), midventral (SPR=0.25), and ventral (SPR less than (<) 0.25) (fig. 5A). The shape of the siphuncle segment is expressed as the segment compression ratio (SCR), the ratio of segment height to segment length (fig. 5B). SCR values range from depressed elongate (SCR less than 1.0, height is < length), to equiaxial (SCR=1.0; height is equal to length), to compressed (SCR > 1.0; height is greater than length) (fig. 5C). These measurements are taken along the dorsal-ventral plane of the nautiloid shell and are used in conjunction with the traditional morphological features listed above (figs. 4 and 5) to define the longiconic nautiloid taxa described here. In this paper, nautiloids are subdivided into the following groups on the basis of size: small (less than 30 cm in length), medium (30–100 cm in length), and large (in excess of 100 cm). With respect to the rate of shell expansion, the following descriptive terms will be used: gradually expanding (apical angle less than 6 degrees); moderately expanding (apical angle 7–9 degrees); and rapidly expanding (apical angle greater than 9 degrees).

Differentiating members of the cyrtoconic/breviconic Oncocerida is likewise beset with a number of difficulties. This is particularly true of the Oncoceratidae, the most abundant oncocerids in these Ordovician strata. Genera and species have been differentiated primarily on the basis of shell outline, which takes into account the amount of shell curvature, shell cross section shape, and features of the body chamber and the aperture. As pointed out by Flower (1946), these features can be dramatically altered by deformation affecting the shell during the course of fossilization. This is not a problem with the well-preserved, generally uncrushed silicified material from the Tyrone and Lexington Limestones (pls. 19 and 20), but is a significant factor complicating generic and specific identifications of matrix-filled internal molds in the more argillaceous Cincinnatian Provincial Series strata (pls. 20, 21, and 22). More of a problem in the case of the silicified material is the incomplete nature of most specimens, the collections consisting largely of small-diameter (less than 10 millimeters (mm) in diameter) adapical portions of the phragmocone. Most published Ordovician oncocerid taxa have been described from more adoral portions of the phragmocone plus the associated body chamber. Adapical portions of many of these described species are unknown. Consequently, trying to determine the affinities of these adapical oncocerid shell fragments is difficult, if not impossible. Other features used to distinguish oncocerid taxa are siphuncle segment diameter and shape, septal necks, connecting ring structure, cameral length, and suture pattern.

Tarphycerid taxa are distinguished by their general shell shape (evolute coiled, serpenticone, or gyrocone), whorl cross section, siphuncle position, thickness of the connecting rings, and septal neck type. Other features of taxonomic importance are the length of the body chamber, aperture configuration, cameral length, suture pattern, and external shell ornament. Unfortunately, most of the Ordovician tarphycerids described here are fragmentary, and species identifications are very tentative, if any assignment can be made at all.

SYSTEMATIC DESCRIPTIONS

Descriptions of the nautiloid species described here are organized in the following manner. First, there is a diagnosis summarizing the distinctive features of the taxon. This is followed by the taxonomic description, the first paragraph of which describes the morphological features of the shell that can be observed in hand specimen (fig. 4). These features are arranged in a hierarchy from

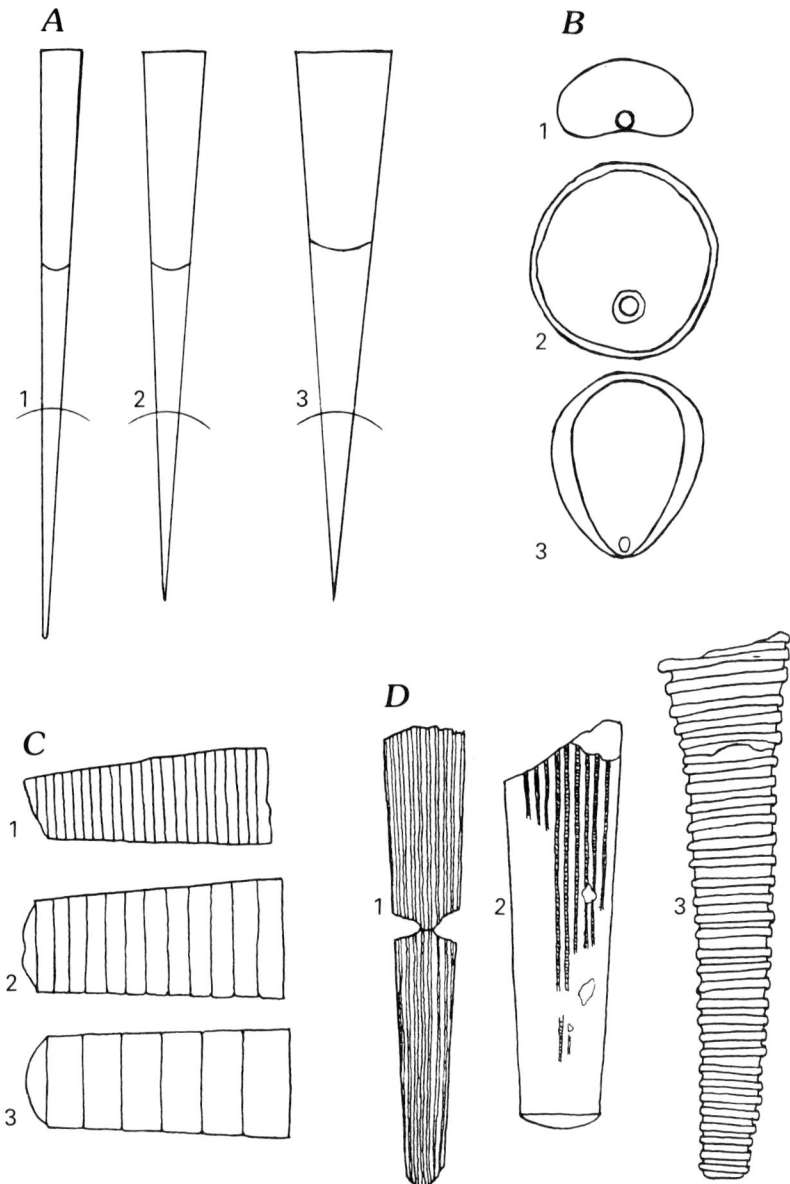

FIGURE 4.—External morphological features used to differentiate nautiloid taxa described and illustrated here. *A*, Rate of shell expansion (apical angle): 1, gradually expanding orthoconic longicone (apical angle less than 6 degrees); 2, moderately expanding (apical angle 6.5–8.5 degrees); 3, rapidly expanding (apical angle more than 9 degrees). *B*, Shell cross-sectional shape: 1, depressed (*Trocholites*); 2, circular (*Richmondoceras*); 3, compressed (*Maelonoceras*). *C*, Cameral length: 1, short cameral lengths, the cameral length/diameter (CL/D) less than one-fifth (*Cartersoceras*); 2, moderate cameral lengths, CL/D one-fifth to one-fourth (*Treptoceras*); 3, long cameral lengths, CL/D greater than one-third (*Pleurorthoceras*). *D*, Shell ornament: 1, raised longitudinal lirae (*Isorthoceras*); 2, incised longitudinal striae (*Polygrammoceras*); 3, transverse annulations (*Monomuchites*).

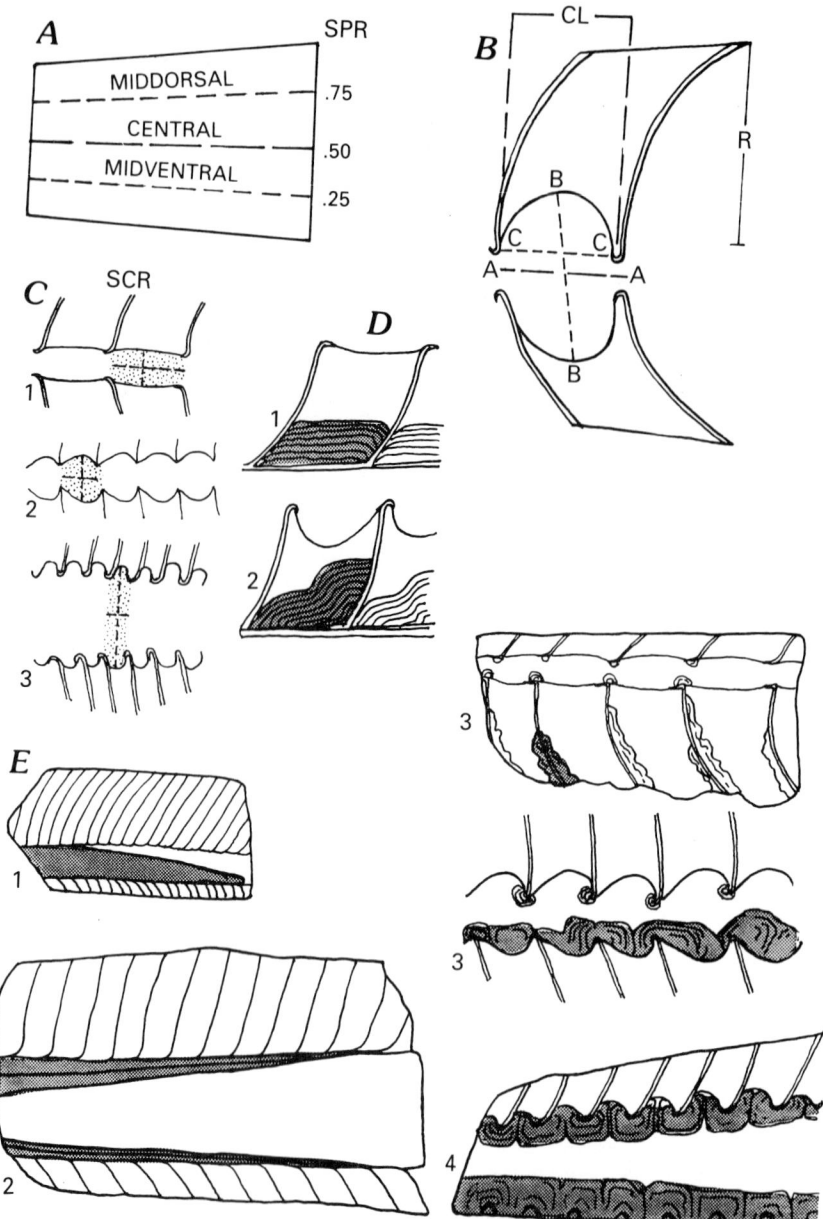

FIGURE 5.—Internal morphological features used to differentiate nautiloid taxa described and illustrated here. *A*, Siphuncle segment position ratio (SPR). *B*, Siphuncle segment measurements: CL, cameral length; AA, segment axis; BB, segment height; CC, segment length; R, radius. *C*, Siphuncle segment compression ratio (SCR): 1, SCR less than 1.0, height less than length, depressed (*Ordogeisonoceras*); 2, SCR=1.0, height equal to length, equiaxial (*Treptoceras*); 3, SCR greater than 1.0, height greater than length, compressed (*Actinoceras*). *D*, Cameral deposits: 1, planar mural deposits (*Pleurorthoceras*); 2, mural-episeptal deposits (*Treptoceras*); 3, hyposeptal and episeptal deposits (*Ordogeisonoceras*). *E*, Types of endosiphuncular deposits: 1, endosiphuncular rod (*Murrayoceras*); 2, endocones (*Cameroceras*); 3, parietal deposits (*Treptoceras*); 4, annulosiphonate deposits (*Actinoceras*) (*A*, *B* modified from Aronoff, 1979).

the more general to the more specific: (1) general shell morphological type (orthoconic longicone, cyrtoconic longicone, cyrtoconic brevicone, etc.), (2) maximum observed shell diameter and estimated maximum length, (3) apical angle (angle of shell expansion), (4) cross-sectional shape of shell, (5) suture pattern, (6) cameral length, (7) nature of the body chamber including periphract type, and (8) external shell ornament. The second paragraph describes the internal features of the shell, primarily the attributes of the siphuncle (fig. 5), which are (1) position of the siphuncle within the shell (SPR), (2) siphuncle segment diameter relative to shell diameter, (3) siphuncle shape (SCR), (4) nature of septal necks, (5) connecting ring structure, (6) endosiphuncular deposits if present, and (7) cameral deposits if present.

These descriptive paragraphs are followed by sections on the known stratigraphic and geographic occurrences of the species, a list of the specimens studied, and remarks. The latter section compares the species with other members of the same genus, describes observed variation within the specimens studied, and makes observations on the relative abundance of the species within the stratigraphic unit(s) in which it is found.

SYSTEMATIC PALEONTOLOGY

PHYLUM MOLLUSCA LINNAEUS
CLASS CEPHALOPODA CUVIER
SUBCLASS ORTHOCERATOIDEA KUHN

Generally small, predominantly orthoconic, less commonly cyrtoconic nautiloids that employed a variety of solutions to solve the problem of buoyancy regulation, including retention of cameral fluid within the camerae, crowding of septa to reduce cameral chamber volume, and various types of calcareous ballast, secreted within cameral chambers and (or) within the siphuncle tube. Connecting rings thick, complex in primitive forms, becoming thin and homogeneous in later stocks. Septal necks achoanitic, loxochoanitic, orthochoanitic, suborthochoanitic, or cyrtochoanitic. Periphract dorsomyarian; body chambers tubular, comparatively long. Central stock from which all other cephalopod groups evolved.

Stratigraphic range.—Upper Cambrian to Triassic.

ORDER ELLESMEROCERIDA FLOWER

Predominantly small, generally smooth, compressed cyrtoconic or orthoconic nautiloids with comparatively short phragmocones equipped with closely spaced septa and ventral tubular siphuncles with short elliptochoanitic septal necks and primitively thick, structurally complex connecting rings. Methods of buoyancy control typically employed were an abbreviated phragmocone, the crowding of septa, a moderate- to large-diameter, probably tissue-filled siphuncle, and the possible retention of cameral fluid within cameral chambers with the siphuncle tube sealed off by diaphragms in some taxa.

Stratigraphic range.—Upper Cambrian to Upper Ordovician.

Remarks.—This order contains the most primitive nautiloids; typically small, compressed, endogastric cyrtocones and orthocones with closely spaced septa and large-diameter, tubular ventral siphuncles. Siphuncles are characterized by thick, complex, layered connecting rings and short orthochoanitic septal necks. However, as indicated by Furnish and Glenister (1964), a multitude of morphological forms are included in the order such that a central defining morphological theme is hard to pinpoint. In all probability, the order is an artificial grouping of early evolutionary experiments in nautiloid morphology. Chen and Teichert (1983) have split off several of the Cambrian groups into separate orders, among them the Plectronocerida and the Protactinocerida. Other distinctive groups within the Ellesmerocerida that may constitute separate order-rank taxa are the longiconic Baltoceratidae and Protocycloceratidae, the strongly compressed Basslerocerida, and the Bathmoceratidae and Cyrtocerinidae, both characterized by greatly thickened, complex connecting rings.

Ellesmerocerids are most abundant in Upper Cambrian and Lower Ordovician strata. The order declined significantly at the end of the Early Ordovician (Arenigian), represented in the Middle Ordovician by the longiconic baltoceratids, the bizarre bathmoceratids, and the breviconic cyrtocerinids. Upper Ordovician relict taxa are the long-ranging genus *Cyrtocerina* Billings and the problematic slender cyrtocone *Shideleroceras* Flower, both from the Cincinnatian Provincial Series in the Kentucky-Indiana-Ohio region (Flower, 1946).

FAMILY BALTOCERATIDAE KOBAYASHI

Slender, smooth orthoconic longicones, subcircular to depressed in cross section, with moderately large, ventral siphuncles that come to assume a more central position in some advanced forms. Siphuncles lacking diaphragms, having connecting rings thinner, less complex in structure than those of other members of the order. Septal necks orthochoanitic or suborthochoanitic. Calcareous deposits, both endosiphuncular and cameral, well-developed. Endosiphuncular deposits consisting of tubular calcareous rods that are typically depressed circular in section, lie against the ventral wall of the siphuncle, are blunt or acute adorally, thickening adapically, and eventually completely fill the adapical portions of the siphuncle tube. Cameral deposits typically mural-episeptal, restricted to the adapical half of the phragmocone.

Stratigraphic range.—Lower Ordovician (Ibexian Provincial Series) to Middle Ordovician (Rocklandian Stage).

Remarks.—Good descriptions and illustrations of the structures that are characteristic of this family are found in the works by Flower (1964a, 1964b), Hook and Flower (1977), and Dzik (1984). The evolution of nautiloids with orthoconic longicone shells resulted in a variety of devices to control buoyancy and rotational instability inherent to shells with this morphology. As noted by Hook and Flower (1977), the late Ibexian was a time of intense morphological experimentation among small orthoconic longicones in which many of these devices were utilized. The most successful solution found in these longicones was the development of calcareous deposits concentrated adapically and ventrally, counterbalancing the buoyant effect of the gas-filled camerae and allowing these longicones to maintain horizontal equilibrium. This calcareous ballast approach would be utilized by longiconic nautiloids throughout the Paleozoic and into the succeeding Triassic (Crick, 1988).

The Baltoceratidae constitute the earliest development of this calcareous ballast approach, in which deposits were secreted both within the camerae and the siphuncle tube. This family forms the transitional group linking the ancestral Ellesmerocerida with the succeeding Orthocerida. Members of the Orthocerida differ from the Baltoceratidae primarily in the more central position of the siphuncle, compared with the more ventral position of these structures in the latter group, and in the presence of thin, homogeneous connecting rings, compared with the layered rings in these baltoceratids. Both orders have small- to moderate-size (less than 100 cm in length) orthoconic longicone shells with comparatively long body chambers, nearly tubular siphuncle segments with orthochoanitic septal necks, endosiphuncular deposits, and cameral deposits.

As noted by Hook and Flower (1977), endosiphuncular deposits may occur as the only device to regulate buoyancy and provide rotational stability, or they may occur in combination with cameral deposits. In the Baltoceratidae, the Protocycloceratidae, and the Troedssonellidae and Narthecoceratidae in the Orthocerida, endosiphuncular deposits are volumetrically more significant and played a greater role in regulating buoyancy than in the Orthoceratidae, the Geisonoceratidae, and other members of the Orthocerida. Endosiphuncular deposits in the Orthoceratidae and other orthocerids usually are less extensively developed, consisting of septal annuli and (or) parietal linings of connecting ring walls.

GENUS *CARTERSOCERAS* FLOWER

Type species.—*Cartersoceras shideleri* Flower, 1964a, p. 121; by original designation; Carters Limestone, Rocklandian Stage, Middle Ordovician, Tennessee.

Description.—Small, moderate to rapidly expanding baltoceratids, depressed ovate in cross section, having venter flattening during ontogeny. Camerae short; sutures sloping from the dorsum to the venter adapically, leading to the development of a broad lobe ventrally. Body chamber unknown. Shell surface smooth.

Siphuncle initially subcentral, moving to a more ventral position during ontogeny. Segments expanded, equiaxial to slightly compressed, having suborthochoanitic septal necks. Connecting rings thickening adorally with lamellar microstructure. Siphuncle with an endosiphuncular rod that tapers adorally to an acute point. Mural-episeptal cameral deposits.

Stratigraphic and geographic occurrence.—Middle Ordovician (Rocklandian Stage), Eastern North America (Laurentia).

Remarks.—Flower (1964a) erected this genus for a species formerly assigned to *Murrayoceras* Foerste that possessed siphuncle segments that are convex in outline compared with the subtubular, depressed segments characteristic of true *Murrayoceras*. In his paper, Flower indicated that there is some gradation from *Murrayoceras* to *Cartersoceras* in terms of segment shape and suggested that this characteristic is indicative of the close phylogenetic relation between these taxa. Restudy of the internal structures of the type species of *Murrayoceras*, *M. murrayi* (Billings) and other members of the genus is necessary to ascertain whether the differences between these taxa are significant enough to define two separate genera. For the purposes of this study, specimens with convex siphuncle segments are placed in *Cartersoceras* Flower; those with more tubular, depressed segments are retained in *Murrayoceras* Foerste.

CARTERSOCERAS SHIDELERI FLOWER, 1964A

Plate 1, figures 1–6, 11

Cartersoceras shideleri Flower, 1964a, p. 122, pl. 26, figs. 3, 4, 8–11.

Diagnosis.—Small, moderately expanding species of *Cartersoceras* with cameral lengths one-fourth to one-eighth the diameter of the shell and an endosiphuncular rod that is circular in cross section.

Description.—Small (less than 25 cm in length and up to 3.0 cm in diameter), moderately expanding (apical angle 8–9 degrees) orthoconic longicones that are depressed ovate in section, the venter becoming distinctly flattened with ontogeny. Sutures sloping adapically from the dorsum to the venter to form a broad ventral lobe. Camerae short with cameral lengths decreasing with ontogeny from one-fourth the diameter of the shell at a diameter of 10 mm to nearly one-eighth the diameter of the shell at diameters of 25 mm or greater. Body chamber incompletely known. Shell exterior smooth.

Siphuncle subcentral adapically (SPR=0.40 at a diameter of 10 mm), becoming midventral with ontogeny (SPR=0.25 at a diameter of 25 mm). Siphuncle segments slightly expanded, compressed, somewhat heart-shaped, the SCR values increasing with ontogeny from 1.2 at a shell diameter of 10 mm to 1.66 at a shell diameter of 20 mm. Connecting rings thickening adorally, having a lamellar microstructure. Endosiphuncular rod filling the siphuncle tube adapically and tapering to an acute point adorally. Rod circular in cross section. Adoralmost cameral deposits mural-episeptal, well-developed ventrolaterally.

Stratigraphic range and geographic occurrence.—Middle Ordovician, the Rocklandian part of the Tyrone Limestone in the Blue Grass region of central Kentucky and the equivalent Carters Limestone in the Nashville dome region of central Tennessee.

Studied material.—Holotype (USNM 302549) and paratype (USNM 302550) from the Carters Limestone, Beech Grove, Tenn.; USNM 158627 (pl. 1, fig. 3), 158631 (pl. 1, figs. 1, 2, 4–5), 158633 (pl. 1, fig. 6), 158635, 468656 (pl. 1, fig. 11), 468657–468661, plus 14 additional silicified specimens (USNM 468662), all from the 6034-CO collection, upper half of the Tyrone Limestone, Little Hickman section A, Jessamine County, Ky.

Remarks.—The bulk of the silicified specimens of *Cartersoceras* from the USGS 6034-CO collection appear to be assignable to the type species of the genus *C. shideleri* Flower. The holotype and paratype are sectioned specimens from the Carters Limestone in Tennessee. The Kentucky material is similar to the type material in the angle of expansion, depth of camerae at similar diameters, siphuncle diameter at similar diameters, and outline of the siphuncle segments. Siphuncle segment position is indeterminable in the type material, as both specimens are thin sections ground into the enclosing limestone matrix. Septal necks and connecting rings are not preserved in the etched silicified Kentucky material, although cameral deposits and the endosiphuncular rod are well preserved in these specimens (pl. 1, figs. 3, 6). Septal necks and connecting rings are, however, present in a sectioned specimen (USNM 468656) from the 6034-CO collection (pl. 1, fig. 11; fig. 6) and further validate assignment of these Kentucky specimens to the *C. shideleri*.

Cartersoceras shideleri differs from the associated *C. popei* n. sp. in being a less rapidly expanding form, in possessing somewhat longer camerae at comparable shell diameters, in possessing less compressed siphuncle segments at comparable diameters, and in the circular cross

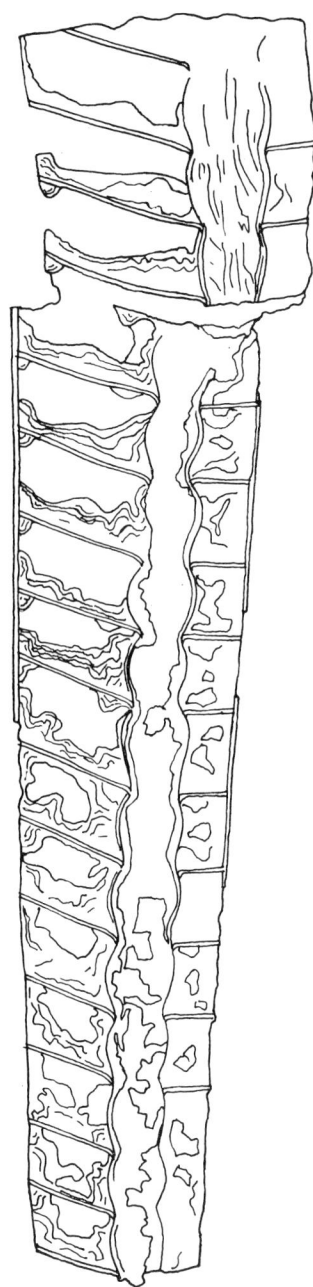

FIGURE 6.—Camera lucida drawing of a dorsal-ventral section through a specimen of *Cartersoceras shideleri* from the 6034-CO collection in the Tyrone Limestone (USNM 468656). Note thin cameral deposits and endosiphuncular rod. Compare with plate 1, figure 11.

section of the endosiphuncular rod (pl. 1, figs. 3–6) compared with the plano-convex section of this feature in *C. popei* (pl. 1, figs. 9, 10). *Cartersoceras shideleri* differs from the baltoceratid *Murrayoceras* cf. *M. murrayi* (Billings) from the same strata in being a more rapidly expanding shell with distinctly expanded siphuncle

TABLE 6.—Comparisons of the morphological features of the three species of baltoceratid nautiloids common to the USGS 6034-CO collection from the upper half of the Tyrone Limestone, Jessamine County, Ky.

[Data from silicified specimens listed in text. CL/D, ratio of cameral length to shell diameter; SPR, segment position ration; SD/D, ratio of siphuncle diameter to shell diameter; mm, millimeter]

Feature	*Cartersoceras shideleri*	*Carteroceras popei*	*Murrayoceras* cf *m. murrayi*
Apical angle, degrees	9	10	6
Diameter 25 mm			
CL/D	0.12	0.12	
SPR		.20	
SD/D		.20	
Diameter 20 mm			
CL/D	.15	.15	0.16
SPR	.30	.25	.16
SD/D	.25	.25	.16
Diameter 15 mm			
CL/D	.175	.12	.175
SPR	.32	.29	.205
SD/D	.235	.26	.175
Diameter 15 mm			
CL/D	.20	.133	.20
SPR	.33	.33	.23
SD/D	.266	.30	.23

segments in contrast to the nearly tubular siphuncle segments in *M.* cf. *M. murrayi* (pl. 1, figs. 14, 16). Table 6 compares the morphological features of these three baltoceratid taxa from the Tyrone Limestone.

Another species assigned to *Cartersoceras* by Flower (1964a) are specimens identified by Wilson (1961) as *Sactoceras? ottawaense* (Billings) from the "Leray-Rockland beds" (Middle Ordovician, Rocklandian) exposed in the Ottawa region of Ontario and Quebec. This species has a similar rate of shell expansion (apical angle 9 degrees) as well as similar cameral lengths and siphuncule diameters compared with the Kentucky material. Study of more extensive, better preserved material from the Ottawa River section may indicate that these Canadian specimens and the Kentucky material represent the same species. If this proves to be the case, the Canadian species would have priority.

Within the 6034-CO collection from the Tyrone Limestone, *C. shideleri* is the most abundant nautiloid species encountered, represented by 25 etched specimens consisting of portions of the phragmocone 4 cm or more in length, numerous smaller, fragmentary specimens, and 15 specimens still in matrix. Recognizable specimens typically consist of adapical portions of the phragmocone 40–87 mm in length. Complete body chambers are not known from specimens from this silicified collection. *C. shideleri* occurs associated with a diverse nautiloid fauna consisting of less abundant representatives of the baltoceratids *C. popei* n. sp. and *Murrayoceras* cf. *M. murrayi* (Billings), the orthocerids *Pojetoceras floweri* n. sp., *Proteoceras tyronensis* (Foerste), and *Monomuchites annularis* n. sp., uncommon specimens of *Vaginoceras*, *Actinoceras*, and *Ormoceras*, plus small oncocerids and fragmentary specimens of the tarphycerid *Plectoceras*.

As indicated above, species of *Cartersoceras* appear to be restricted to Middle Ordovician (Rocklandian) shallow marine carbonate facies across eastern and southern portions of Laurentia. No specimens referable to the genus were identified from collections from the overlying Lexington Limestone in the central Kentucky area.

CARTERSOCERAS POPEI NEW SPECIES

Plate 1, figures 7–10

Diagnosis.—Rapidly expanding species of *Cartersoceras* with cameral lengths from one-fifth to one-eighth the diameter of the shell, highly compressed siphuncle segments (SCR values up to 2.25), and an endosiphuncular rod that is plano-convex in cross section, bisected by a dorsal groove.

Description.—Small (up to 20 cm in length and shell diameters up to 3.0 cm), rapidly expanding (apical angle 10 degrees), orthoconic longicones, depressed ovate in cross section, the venter becoming flatter with ontogeny. Sutures sloping from the dorsum to the venter adapically, forming a broad ventral lobe. Camerae short, with cameral length decreasing with ontogeny, ranging from one-fifth the diameter of the shell at a diameter of 10 mm to one-eighth the diameter of the shell at a diameter of 20 mm or more. Body chamber unknown. Shell exterior smooth.

Siphuncle subcentral in position (SPR=0.45) at a diameter of 10 mm, becoming ventral with ontogeny (SPR=0.20 at diameters of 25 mm or more). Siphuncle segments expanded, heart-shaped, compressed, the SCR values increasing with ontogeny from 1.75 at a diameter of 10 mm to 2.25 at shell diameters of 15 mm or more. Endosiphuncular rod completely filling the siphuncle tube adapically and tapering adorally to an acute end. Rod flattened dorsally throughout its length and adapical sections marked by a longitudinal medial groove, giving the rod a notched, heart-shaped cross section adapically and a plano-convex section adorally. Adoralmost cameral deposits mural-episeptal, well-developed ventrolaterally.

Stratigraphic range and geographic occurrence.—Upper half of the Tyrone Limestone (Rocklandian Age portion) exposed at the Little Hickman section A,

Jessamine County, Ky. (collection 6034-CO), which is the type locality. (See locality register (appendix) for USGS collection localities.)

Studied materials.—Holotype (USNM 158625) (pl. 1, figs. 7, 9) and paratypes USNM 158624 (pl. 1, fig. 10), 158632, 468663, and 468664 (pl. 1, fig. 8), and five other specimens (USNM 468665), all from the USGS 6034-CO collection.

Remarks.—Comparisons with the type species, *C. shideleri*, are given above. The consistent nature of the variation between these specimens in the 6034-CO collection indicates that these differences are real and not the result of silicification or other taphonomic processes. Morphological comparisons of these species are presented in table 6. Within the USGS 6034-CO collection, *C. popei* is less abundant than *C. shideleri*, making up slightly less than one-third of the identified specimens of *Cartersoceras* in this collection. It shares with the latter species the same faunal associates.

No other species of *Cartersoceras* known to the writer possesses the dorsally flattened endosiphuncular rod characteristic of *C. popei*, although the cross sectional shape of this feature is known with certainty only for the type species. *Cartersoceras popei* is distinguished at once from the associated baltoceratid *Murrayoceras* cf. *M. murrayi* in its more rapidly expanding shell and in the expanded, compressed nature of the siphuncle segments, in contrast to the nearly tubular segments in *M.* cf. *M. murrayi*.

Etymology.—This species is named in honor of John K. Pope, a scholar of the Ordovician paleontology of the Cincinnati, Ohio, region.

GENUS *MURRAYOCERAS* FOERSTE

Type species.—*Orthoceras murrayi* Billings, 1857, p. 332; designated by Foerste, 1926, p. 312; Black River Formation, Blackriveran Stage, Middle Ordovician, Ontario.

Description.—Small to medium-sized, gradual to moderately expanding orthoconic longicones, depressed ovate in cross section, more flattened on venter, having dorsum evenly rounded. Sutures sloping adapically from the dorsum to the venter, forming a broad, shallow ventral lobe. Camerae short. Body chamber poorly known, apparently lengthy, at least one-fourth the total length of the shell. Shell exterior smooth or marked with fine, raised transverse growth lines.

Siphuncle subcentral initially, becoming ventral with ontogeny. Siphuncle moderately large in diameter, segments subtubular. Septal necks short, aneuchoanitic to orthochoanitic. Connecting rings thick, laminated, typically concave or sinuate. Siphuncle with a tubular endosiphuncular rod that tapers adorally to an acute end. Adoralmost cameral deposits episeptal, well-developed ventrolaterally.

Stratigraphic and geographic occurrence.—Middle Ordovician (Whiterockian? to Rocklandian Stages) in Eastern North America (Laurentia).

Remarks.—Flower (1964a) distinguished *Murrayoceras* from *Cartersoceras* on the basis of the shape of the siphuncle segments, which are tubular or subdued sinuate in *Murrayoceras*, but are distinctly expanded, convex in *Cartersoceras*. Flower (1964a) noted that some species of *Murrayoceras* exhibited ontogenetic changes in the nature of the connecting rings, from distinctly concave adapically to sinuate adorally. Study of a suite of 50 silicified specimens from the 6034-CO locality and comparisons with other figured species of both genera indicate that species assigned to *Cartersoceras* consist of smaller, more rapidly expanding shells compared with established species of *Murrayoceras*. Additionally, during ontogeny, specimens of *Cartersoceras* become more strongly depressed than *Murrayoceras* and have a flatter venter.

Typical *Murrayoceras* appears earlier than *Cartersoceras* and is most abundant in older Middle Ordovician (Blackriveran) carbonates in New York and adjacent portions of Ontario and Quebec. *Cartersoceras* is apparently restricted to younger Rocklandian carbonates, especially in Kentucky, Tennessee, the upper Mississippi valley, and the Ottawa Lowland in Canada.

MURRAYOCERAS CF. *M. MURRAYI* (BILLINGS), 1857

Plate 1, figures 12–17

Orthoceras murrayi Billings, 1857, p. 332.
Jovellania murrayi (Billings). Foord, 1891, p. 338, figs. 50a–c.
Murrayoceras murrayi (Billings). Foerste, 1926, p. 312.
Murrayoceras murrayi (Billings). Foerste, 1928a, p. 314, pl. 75, fig. 5.
Murrayoceras murrayi (Billings). Foerste, 1932, p. 121, pl. 21, fig. 2.
Murrayoceras murrayi (Billings). Wilson, 1961, p. 40, pl. 9, figs. 6, 7, pl. 11, figs. 3, 4.

Diagnosis.—Small, slender species of *Murrayoceras*, depressed ovate in cross section, having cameral lengths one-fourth to one-eighth the diameter of the shell.

Description.—Small (at least 30 cm in length and up to 2.8 cm in diameter), slender (apical angle 6–6.5 degrees), orthoconic longicones, depressed ovate in cross section, having venter becoming flatter during ontogeny. Sutures straight, transverse dorsally and laterally, turning adapically at the ventral margins, forming a broad, shallow ventral lobe. Camerae short, decreasing in length with ontogeny, the cameral lengths ranging from one-fourth the diameter of the shell at a diameter of 10 mm to one-eighth the diameter of the shell at diameters in excess of 25 mm. Body chamber unknown. Exterior marked by fine transverse growth lines.

Siphuncle subcentral (SPR=0.30) at shell diameters less than 15 mm, becoming ventral (SPR=0.175) at diameters greater than 20 mm. Siphuncle segments subtubular, the segment diameter ranging from one-fourth the diameter of the shell at a diameter of 10 mm to one-sixth the diameter of the shell at diameters greater than 15 mm. Septal necks and connecting rings poorly known. Endosiphuncular rod filling adapical portions of the siphuncle and tapering adorally to an acute end. Rod extending adorally to shell diameters up to 20 mm in diameter. Adoralmost cameral deposits consisting of ventral and lateral planar episeptal deposits.

Stratigraphic and geographic occurrence.—Upper half of the Tyrone Limestone (Rocklandian Age part) exposed at the Little Hickman A section, Jessamine County, Ky.

Studied materials.—Specimens studied were a suite of silicified specimens from the USGS 6034-CO collection: USNM 158637, 468672 (pl. 1, fig. 16), 468673 (pl. 1, fig. 17), 468674 (pl. 1, figs. 12, 13), and 468675 (pl. 1, figs. 14, 15).

Remarks.–Silicified specimens of *Murrayoceras* from the Tyrone Limestone in Kentucky appear to be most similar to the type species of the genus, *M. murrayi* (Billings) from the Middle Ordovician "Black River Formation" exposed on St. Joseph Island, Lake Huron, Ontario. Both the Tyrone Limestone specimens and a syntype of *M. murrayi* (Geological Survey of Canada specimen number 1723) have slender, depressed, longiconic shells, the cameral lengths ranging from one-fifth the diameter of the shell at diameters less than 20 mm to one-eighth the diameter of the shell at diameters greater than 25 mm and subtubular siphuncle segments with diameters one-fifth to one-seventh the diameter of the shell. Detailed comparisons between the Kentucky material and the Canadian types are difficult, as the silicified specimens from the Tyrone Limestone represent phragmocones 20 mm or less in diameter, whereas the Canadian material is represented by more adoral portions of the phragmocone greater than 17 mm in diameter. Conclusive assignment of the Kentucky material to *M. murrayi* awaits comparisons of larger suites of more complete, comparable specimens from both localities.

The Kentucky specimens of *Murrayoceras* are less similar to *M. carletonense* Foerste, a slender species from the Lowville Formation (Middle Ordovician, Blackriveran) exposed in southern Ontario. *Murrayoceras carletonense* possesses more closely spaced septa in comparison with to the Tyrone Limestone specimens, which have sutures that slope in a more pronounced manner from the dorsum to the venter. This Canadian species, however, is also based on the more adoral portions of the phragmocone, broader in diameter than is known for the Kentucky specimens, making conclusive comparisons difficult.

Murrayoceras cf. *M. murrayi* is apparently an uncommon species in the USGS 6034-CO collection, represented by fewer than two dozen fragmentary silicified specimens. This species is much less common than associated species of the genus *Cartersoceras*. Species of *Cartersoceras* can be distinguished from *M.* cf. *M. murrayi* in their more rapidly expanding shell form and in the distinctly expanded nature of the siphuncle segments in these specimens.

ORDER ORTHOCERIDA KUHN

Small to large (rarely in excess of 2 m in length), orthoconic or cyrtoconic longicones with smooth to elaborately ornamented shells. Siphuncles central or subcentral in position, comparatively narrow in diameter relative to the diameter of the shell. Thin homogeneous connecting rings and orthochoanitic, suborthochoanitic, or cyrtochoanitic septal necks. Dominant method of buoyancy regulation is the secretion of calcareous ballast in adapical portions of the phragmocone, typically consisting of both cameral and endosiphuncular deposits. Cameral deposits are volumetrically more significant and served as the primary buoyancy control structures in most members of the order, consisting of complex mural, episeptal, and hyposeptal deposits. Endosiphuncular deposits typically consisting of septal annuli, parietal deposits, or siphuncle linings. Body chambers lengthy, commonly making up to one-fourth the total length of the shell. Periphract dorsomyarian. Protoconchs typically globular.

Stratigraphic range.—Lower Ordovician to Upper Triassic.

Remarks.—The higher level systematic classification of this group is difficult due to the seemingly simple, conservative morphology of many members and the fragmentary nature of most described taxa (W.C. Sweet, oral commun., 1990). The current classification (Teichert and others, 1964) is often confusing and inconsistent. It is beyond the scope of this paper to attempt to tackle the reorganization of this order, but several taxa are redefined or emended in an attempt to develop a more consistent classification. Taxonomy is based on a combination of morphological features such as general shell shape and size; body chamber length relative to phragmocone length; length of cameral chambers; position, shape, and diameter of siphuncle segments; features of septal necks; and the morphology and development of cameral and endosiphuncular deposits. External ornament of the shell is thought to be less significant than previous classifications have indicated, significant at the genus level but no higher.

Sweet (1964) and Teichert (1988) divided the Orthocerida into two superfamilies: the orthochoanitic Orthoceratacea and the cyrtochoanitic Pseudorthoceratacea.

Members of the Orthoceratacea are known from the Lower Ordovician (upper Ibexian) and are the dominant orthocerids in the Ordovician and succeeding Silurian. The development of cyrtochoanitic forms from existing orthochoanitic stocks occurred early in the history of the order, the oldest known pseudorthoceratacceans dating from the early Middle Ordovician (late Whiterockian). Members of the Pseudorthoceratacea are locally common in the latter half of the Ordovician, and then become the dominant orthocerids in the Devonian and succeeding Carboniferous. Wade (1988) has regarded the differences between these two superfamilies to be sufficient to warrant placing the Pseudorthoceratacea in a separate order, the Pseudorthocerida. For the purposes of this paper, both the Orthoceratacea and the Pseudorthoceratacea are retained in the order Orthocerida.

FAMILY ORTHOCERATIDAE M'COY

Smooth to ornamented, small to medium-sized, gradually expanding, slender longicones, circular in cross section; straight transverse sutures and long camerae with widely spaced, strongly curved septa. Cameral length typically from one-third to greater than one-half the diameter of the shell. Body chambers tubular, one-fourth the total length of the shell. Aperture straight, transverse, lacking a hyponomic sinus.

Siphuncle central or subcentral with narrow, tubular or slightly expanded siphuncle segments that are longer than wide. Septal necks short, orthochoanitic or suborthochoanitic. Connecting rings thin, homogeneous. Endosiphuncular deposits restricted to adapical portions of the phragmocone, consisting of small annuli and (or) parietal deposits. Cameral deposits developed adapically, consisting of mural-episeptal and hyposeptal deposits that largely fill cameral chambers ventrally prior to their development dorsally. Protoconchs globular.

Stratigraphic range.—Lower Ordovician (Ibexian Provincial Series) to Upper Triassic.

Remarks.—Sweet (1964) subdivided this family into four subfamilies. These were the Orthoceratinae, which are distinguished by body chambers marked by deep longitudinal furrows; the Michelinoceratinae, which consisted of smooth Orthoceratidae with central or excentric siphuncles and well-developed cameral deposits; the Kionoceratinae, which is made up of taxa with well-developed longitudinal external ornament; and the Leurocycloceratinae, consisting of transversely ornamented forms. Body chambers are poorly known for many members of this family, such that the prevalence of the furrowed body chambers in these nautiloids is not known. The Kionoceratinae currently consists of genera with both expanded, cyrtochoanitic siphuncle segments (*Kionoceras* Hyatt, *Ohioceras* Shimizu and Obata, and *Polygrammoceras* Foerste) as well as more typical Orthoceratidae with tubular siphuncles (*Parakionoceras* Foerste), and seem to be a totally artificial grouping of taxa. The Leurocycloceratinae consists of taxa with tubular siphuncle segments and distinctive transverse annuli. *Leurocycloceras* Foerste, however, has long camerae and long septal necks in contrast to the short camerae and short orthochoanitic septal necks characteristic of *Anaspyroceras* Shimizu and Obata. The internal features of *Metaspyroceras* are poorly known and require further documentation.

Michelinoceras Foerste and other genera with slender shells, narrow tubular siphuncle segments, very long camerae, and long septal necks (that is, *Arkonoceras* Flower, *Leurocycloceras* Foerste, and *Plagiostomoceras* Teichert and Glenister) are viewed as being sufficiently distinct from other orthoceratids to constitute a separate family, the Michelinoceratidae. In contrast to the primarily Ordovician distribution of the Orthoceratidae, members of the Michelinoceratidae are most numerous in the succeeding Silurian and Devonian. The wide septal foramina and broad, subtubular siphuncle segments typical of *Kionoceras* suggest placement of this genus in the family Geisonoceratidae. The familial affinities of *Anaspyroceras*, *Metaspyroceras*, *Ohioceras*, *Parakionoceras*, and *Polygrammoceras* remain unknown pending better documentation of the key internal features of these genera.

POJETOCERAS NEW GENUS

Type species.—*Pojetoceras floweri*, n. sp., Tyrone Limestone, Rocklandian Stage, Middle Ordovician, Kentucky.

Description.—Small, slender, gradually expanding orthoconic longicones, circular in cross section with straight, transverse sutures. Camerae long, with cameral lengths averaging one-third the diameter of the shell, becoming slightly shorter with ontogeny. Body chamber unknown. Shell exterior smooth.

Siphuncle supracentral to subcentral in position, segments tubular, typically twice as long as wide. Siphuncle segments narrow in diameter, with short, orthochoanitic septal necks and thin, homogeneous connecting rings. Endosiphuncular deposits developed adapically along the ventral siphuncle wall, consisting of thin parietal deposits that form adoral to septal necks. Cameral deposits well-developed, consisting of complex mural-episeptal and hyposeptal deposits that form ventrally prior to their development dorsally.

Stratigraphic and geographic occurrence.—Lower Ordovician (Ibexian Provincial) to Middle Ordovician (Shermanian Stage), North America (Laurentia).

Remarks.—This genus is erected for a group of small, smooth orthoceratids with narrow, supracentral to subcentral, tubular siphuncle segments that were originally described from the Early Ordovician of the Western United States. Flower (1962) and Hook and Flower (1977) referred these forms with some reservations to the genus *Michelinoceras* Foerste. *Michelinoceras* is based on *Orthoceras michelini* Barrande (=*O. grande* Meneghini) from the Upper Silurian (Ludlovian) of Czechoslovakia and Sardinia. From descriptions and good illustrations of this species (Barrande, 1866–1870; Serpagli and Gnoli, 1977), *M. michelini* differs significantly from these Ordovician orthoceratids in the possession of much longer septal necks, longer camerae at comparable diameters, and narrower, more tubular siphuncle segments. Additionally, endosiphuncular deposits, which are well developed in *Pojetoceras*, are currently not known for *Michelinoceras* based on *M. michelini*.

Pojetoceras is probably most similar to true *Orthoceras* Bruguiere, based on *O. regularis* Schlotheim from the early Middle Ordovician "*Orthoceras* Limestone" of the Baltic region. Both taxa are slender, gradually expanding longicones with narrow central or subcentral, tubular siphuncle segments equipped with short orthochoanitic septal necks. Cameral deposits, consisting of thin mural-episeptal deposits, are known for *O. regularis* (Flower, 1964b, pl. 4, figs. 6, 7), but are restricted to portions of the phragmocone less than 10 mm in diameter. Endosiphuncular deposits are not known for the type species of *Orthoceras*, but related species illustrated by Dzik (1984, fig. 34, pl. 22, fig. 8) indicate the presence of small annuli and thin parietal deposits in adapical portions of the phragmocone. The body chamber of *Orthoceras* is long and tubular and marked by three distinctive linear furrows. The body chamber is currently unknown for species of *Pojetoceras*. *Pojetoceras* differs from *Orthoceras* in the possession of shorter camerae and somewhat broader siphuncle segments at comparable diameters. There is, additionally, more extensive development of both cameral and endosiphuncular deposits in *Pojetoceras* compared with their weak development or absence at similar diameters in *Orthoceras*. Although the adoral portions of specimens of *Pojetoceras* are currently unknown, maximum observed diameters (up to 1.5 cm) in large collections from Kentucky indicate that *Pojetoceras* is a smaller form than typical *Orthoceras* (known lengths in excess of 50 cm and having shell diameters of up to 2.5 cm). Figure 7 is a graphic plot comparing cameral length relative to shell diameter for *Pojetoceras floweri* and *Orthoceras regularis*.

Pojetoceras is also similar in gross shell morphology to the Late Ordovician orthoceratid *Pleurorthoceras* Flower. Both of these taxa have slender, gradually expanding shells with comparatively long camerae and central to subcentral, narrow, tubular siphuncle segments. *Pojetoceras* differs from *Pleurorthoceras* in pos-

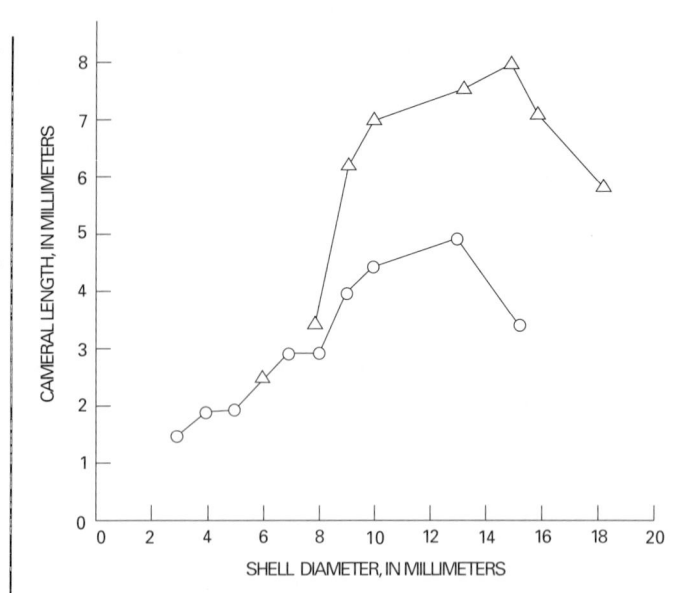

FIGURE 7.—Bivariate plot of cameral length and shell diameter for specimens of *Pojetoceras floweri* (circles) and *Orthoceras regularis* (triangles). Data for *P. floweri* from paratypes USNM 468677–468679 and suite of specimens USNM 468684. Data for *O. regularis* from Flower (1962). Note the longer camerae in *O. regularis* compared with cameral lengths at comparable diameters in specimens of *P. floweri*.

sessing broader siphuncle segments relative to shell diameter; in having short orthochoanitic septal necks, compared with the suborthochoanitic septal necks in the latter genus; in the development of complex episeptal and hyposeptal cameral deposits in *Pojetoceras*, compared with the simple mural-episeptal deposits in *Pleurorthoceras*; and in the presence of endosiphuncular deposits in *Pojetoceras* and their absence in species of *Pleurorthoceras*. Species of *Pleurorthoceras* also appear to grow to a larger size compared with species of *Pojetoceras*, reaching lengths in excess of 60 cm and shell diameters up to 3.0 cm. Maximum lengths known for species of *Pojetoceras* are less than 30 cm and shell diameters up to 1.5 cm.

As currently known, species assignable to *Pojetoceras* range from the Lower Ordovician (late Ibexian Provincial Series) to the Middle Ordovician Shermanian Stage. The genus appears to be restricted in its distribution to cratonic carbonate facies in Laurentia. Early Ordovician species assigned to *Pojetoceras* are *Michelinoceras floridaense* Hook and Flower, *M. melleni* Hook and Flower, and a number of other unnamed, fragmentary specimens of "*Michelinoceras*" described and illustrated by Hook and Flower (1977) from the late Ibexian Florida Mountains Formation in southwest New Mexico. Other Lower Ordovician representatives of the genus are a number of fragmentary specimens referred to as *Michelinoceras* spp. and collected from the Wah Wah Limestone in the Ibex area of Utah (Hook and Flower, 1977).

Younger Middle Ordovician examples of *Pojetoceras* are "*Orthoceras*" *shumardi* Billings from the Mingan Formation (Chazyan Stage) exposed in the Mingan Islands, Quebec (Foerste, 1938); the type species *P. floweri* n. sp. from the Tyrone Limestone (Rocklandian) in Kentucky; a species referred to as "*Michelinoceras* sp. 3" from the "Braeside Beds" (Blackriveran) exposed in the Ottawa region of Ontario (Steele and Sinclair, 1971); and *Orthoceras beltrami* Clarke from the "Prosser Member of the Trenton" at Wykoff, Minn. (Foerste, 1932). Undescribed species that are possibly assignable to *Pojetoceras* have been collected by the writer from the Stones River Group (Middle Ordovician, Blackriveran) in central Tennessee and the Mountain Lake Member of the Bromide Formation (Blackriveran) in the Criner Hills, in Oklahoma.

Etymology.—This genus is named for John Pojeta, Jr., in recognition of his work organizing and directing the Kentucky Ordovician paleontology program and for his encouragement and support of this writer's study of the nautiloid faunas collected as part of this program.

POJETOCERAS FLOWERI NEW SPECIES

Plate 2, figures 1–9

Diagnosis.—Small, slender, orthoconic longicone with cameral lengths from one-half to one-fifth the diameter of the shell and subcentral depressed siphuncle segments whose diameters are one-sixth to one-ninth the diameter of the shell and that contain, adapically, thin ventral parietal deposits.

Description.—Small (observed lengths of up to 25 cm and having diameters of up to 1.5 cm), gradually expanding (apical angle 3–3.5 degrees) orthoconic longicones, circular in cross section with straight transverse sutures. Camerae lengths ranging from one-half the diameter of the shell at diameters less than 5 mm to nearly one-fourth the diameter of the shell at diameters greater than 13 mm. Body chamber unknown. Shell exterior smooth.

Siphuncle supracentral (SPR=0.60) at shell diameters less than 7 mm, becoming central (SPR=0.50) at diameters of 7–12 mm, and then subcentral (SPR less than 0.50) in adoral portions of the shell greater than 12 mm in diameter. Siphuncle segments tubular to subtubular, narrow in diameter, from one-sixth to one-ninth the diameter of the shell, the segment diameter decreasing relative to shell diameter with ontogeny. Septal necks short, orthochoanitic. Connecting rings thin, homogeneous. Endosiphuncular deposits consisting of thin parietal deposits lining ventral portions of adapical siphuncle segments that fuse to form a continuous ventral lining with ontogeny. Dorsal endosiphuncular deposits unknown. Cameral deposits complex, botryoidal episeptal and hyposeptal deposits that fill ventral and lateral portions of cameral chambers with ontogeny. Thin, planar episeptal deposits dorsally that form later than ventral and lateral deposits. Variably developed ventral medial mural delta.

Stratigraphic range and geographic occurrence.—Tyrone Limestone (Rocklandian Age part), exposed in Jessamine County, Ky.

Studied material.—Holotype (USNM 468676) (pl. 2, figs. 1, 9), eight paratypes: USNM 468677 (pl. 2, figs. 2, 3), 468678 (pl. 2, figs. 4, 6), 468679 (pl. 2, fig. 5), 468680, 468681 (pl. 2, fig. 7), 468682, 468683 (pl. 2, fig. 8), and 468684, and a suite of fragmentary silicified specimens (USNM 468685), all from USGS collection 6034-CO from the upper half of the Tyrone Limestone exposed at the Little Hickman section A, Jessamine County, Ky., which is the type locality. The holotype (USNM 468676) is a dorsal-ventral thin section prepared by R.H. Flower (pl. 2, figs. 1, 9; figure 8) that illustrates the key internal features of *P. floweri*.

Several additional small silicified fragments (468685A) from USGS collection D1138-CO from a section 29 ft (10 m) above the base of the Tyrone Limestone at West Marble Creek, Jessamine County, Ky., are also tentatively assigned to *P. floweri*.

Remarks.—*Pojetoceras floweri* is the second most abundant nautiloid species identified from the USGS 6034-CO collection from the Tyrone Limestone. It is second in abundance only to the small, depressed baltoceratid *Cartersoceras shideleri*. This abundance is measured in terms of identifiable fragments present in collections studied by the writer. When added together, specimens of *Cartersoceras* and *Pojetoceras* constitute approximately 70 percent of the identifiable nautiloid material in the 6034-CO collection. Specimens of *P. floweri* studied consist of silicified portions of phragmocones 5–75 mm in length. Composite specimens reconstructed using this fragmentary material indicate that *P. floweri* was a very slender orthoconic longicone that reached lengths of up to 25 cm and shell diameters up to 1.5 cm (see pl. 2, figs. 3–5).

Pojetoceras floweri is most similar to a species figured by Steele and Sinclair (1971, pl. 11, figs. 11, 12) and identified as *Michelinoceras* sp. 3. This species was represented by only three fragmentary silicified phragmocones that indicate a small, slender orthoconic longicone with comparatively long camerae. Cameral lengths were equal to one-half the diameter of the shell at shell diameters of 5–6 mm. The siphuncle was described as small in diameter and centrally located, the septal necks and connecting rings are not preserved. *Michelinoceras* sp. 3 was collected from the Braeside Limestone in the Ottawa valley in eastern Ontario, the strata are roughly equivalent in age to the Tyrone Limestone in Kentucky. These Canadian specimens may, after further study, prove to be assignable to *P. floweri*.

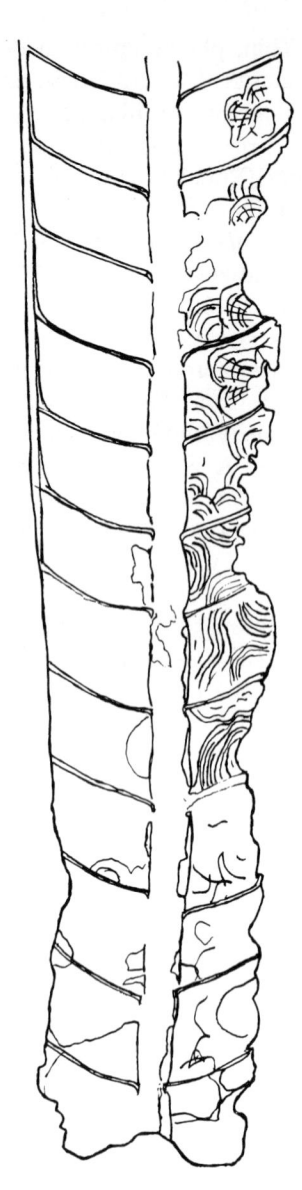

FIGURE 8.—Camera lucida drawing of sectioned holotype of *Pojetoceras floweri* (USNM 468676) from the 6034-CO collection in the Tyrone Limestone. Compare with plate 2, figures 1, 9. Venter to the right. Note complex development of ventral cameral deposits and thin adapical parietal deposits.

Pojetoceras floweri is also similar to *Orthoceras shumardi* Billings from the Middle Ordovician (Chazyan) Mingan Formation exposed in the Mingan Islands in the Gulf of St. Lawrence, Quebec (Foerste, 1938, pl. 17, fig. 3). This Chazyan species appears to be assignable to *Pojetoceras* on the basis of its very slender, gradually expanding longiconic shell, deep camerae, and narrow-diameter, supracentral to central tubular siphuncle segments. *Pojetoceras shumardi* appears to differ from *P. floweri* in having deeper camerae and a more dorsally located siphuncle at comparable shell diameters.

Also tentatively assigned to *Pojetoceras* are a number of species described by Hook and Flower (1977) from older Early Ordovician (Ibexian) carbonates in New Mexico. These species, "*Michelinoceras*" *floridaense* and "*M.*" *melleni*, are both known from single specimens, both consisting of phragmocone fragments 50 mm in length and less than 8 mm in maximum diameter. Both specimens exhibit the diagnostic features of *Pojetoceras*. Detailed comparisons with the type species, *P. floweri*, cannot be made until more adoral portions of these New Mexico species are known. Numerous other specimens described by Hook and Flower (1977) as undescribed species of *Michelinoceras* are also too fragmentary to compare with *P. floweri*.

Etymology.—This species is named in honor of the late R.H. Flower, who did the initial study of the 6034-CO fauna and who did much to encourage this writer in his studies of Ordovician nautiloids.

GENUS *PLEURORTHOCERAS* FLOWER

Type species.—*Orthoceras clarksvillense* Foerste, 1924b, p. 220; designated by Flower, 1962, p. 35; Waynesville Formation, Richmondian Stage, Upper Ordovician, Ohio.

Description.—Medium-sized, slender, gradually expanding orthoconic longicones; circular in section with straight, transverse sutures. Cameral lengths from one-third to nearly one-half the diameter of the shell, with strongly curved, widely spaced septa. Body chamber long and tubular. Shell exterior smooth.

Siphuncle slightly ventral of center, narrow in diameter, the subtubular segments only slightly expanded within camerae, twice as long as wide. Septal necks short, suborthochoanitic. Endosiphuncular deposits unknown. Cameral deposits well developed, initially mural, becoming mural-episeptal with ontogeny and deposits thickening ventrally to form a midventral boss.

Stratigraphic range and geographic occurrence.—Upper Ordovician (Maysvillian to Richmondian Stages), Laurentia.

Remarks.—Flower (1962) erected this genus for Late Ordovician orthoceratids that differed from earlier forms, such as true *Orthoceras*, by the presence of suborthochoanitic rather than orthochoanitic septal necks and through the development of strictly mural cameral deposits compared with the episeptal deposits characteristic of earlier orthoceratids. Study of Flower's figured specimens (MU 268T, 269T) as well as his figures (1962, pl. 6, fig. 4), indicates that cameral deposits are initially mural but become episeptal with ontogeny, forming a distinctive transverse adapical groove ventrally that is visible in internal molds of the camerae (pl. 2, fig. 10).

Pleurorthoceras differs from all associated taxa in the Cincinnatian Provincial Series exposed in the Kentucky-Indiana-Ohio area, in the combination of a slender, gradually expanding shell, long camerae, and a narrow, subtubular subcentral siphuncle. The type species *P. clarksvillense* is an uncommon form, occurring as fragmentary specimens in argillaceous wackestone facies in the upper part of the Bull Fork Formation (Upper Ordovician, Richmondian) in southwestern Ohio and adjacent portions of southeastern Indiana. The only other species currently placed in *Pleurorthoceras* is *P. selkirkense* (Whiteaves) from the Selkirk Member of the Red River Formation (Upper Ordovician, Maysvillian) and the overlying Stony Mountain Formation (Upper Ordovician, Richmondian) in the Winnipeg region of Manitoba.

PLEURORTHOCERAS CLARKSVILLENSE (FOERSTE), 1924B

Plate 2, figures 10–15

Orthoceras clarkesvillense Foerste, 1924b, p. 220, pl. 42, figs. 1A–B.
Michelinoceras clarkesvillense (Foerste). Foerste, 1932, p. 72.
Pleurorthoceras clarksvillense (Foerste). Flower, 1962, p. 36, pl. 6, figs. 1–4.

Diagnosis.—Species of *Pleurorthoceras* distinguished by cameral lengths that range from one-half the diameter of the shell at diameters of 15 mm or less to one-third the diameter of the shell at diameters of 30 mm; narrow, subtubular, central to subcentral siphuncle segments one-eighth the diameter of the shell.

Description.—Medium-sized (lengths in excess of 65 cm and having shell diameters of up to 3.0 cm), slender, gradually expanding (apical angle of 4–5 degrees), orthoconic longicones; circular in cross section with straight, transverse sutures. Cameral lengths decreasing with ontogeny from one-half the diameter of the shell at shell diameters of 15 mm or less to one-third the diameter of the shell at a diameter of 30 mm. Body chamber tubular, one-fourth the total length of the shell, aperture unknown. External shell surface smooth.

Siphuncle central in position (SPR=0.50) at shell diameters of 15 mm or less, then becoming subcentral with ontogeny (SPR=0.45 at diameters of 20 mm and 0.30 at a diameter of 30 mm). Segments only slightly expanded, subtubular, the SCR values ranging from 0.285 (depressed) in adapical portions of the shell 15 mm or less in diameter to 0.33 (depressed) in portions of the shell 20–30 mm in diameter. Siphuncle segments narrow, from one-eighth to one-tenth the diameter of the shell. Septal necks short, suborthochoanitic. Connecting rings thin, homogeneous. Known portions of the siphuncle devoid of endosiphuncular deposits. Ventral mural deposits well-developed in adapical portions of large specimens at shell diameters less than 22 mm. Deposits becoming episeptal adapically with ontogeny. Adoralmost dorsal cameral deposits thin, mural.

Stratigraphic range and geographic occurrence.—Upper half of the Bull Fork Formation ("Waynesville" and "Liberty" biofacies) (Upper Ordovician, Richmondian Stage) in Warren and Clinton Counties in southwestern Ohio and in Union, Franklin, and Dearborn Counties in southeastern Indiana.

Studied material.—Holotype, USNM 48255 (pl. 2, fig. 12), internal mold of a phragmocone 130 mm in length, "Waynesville Formation," Blanchester, Clinton County, Ohio; hypotypes MU 268T (pl. 2, figs. 13–15) and MU 269T (pl. 2, fig. 10) figured by Flower (1962), "Liberty Formation," Addisons Creek, Preble County, Ohio; MU 444T (pl. 2, fig. 11), "Liberty Formation," Oldenburg, Franklin County, Ind.; and measured suites of specimens: MU 29853 (nine specimens), "Liberty Formation," Addisons Creek, Preble County, Ohio; MU 29854, "Liberty Formation," Liberty, Union County, Ind.; MU 29855 (five specimens), "Liberty Formation", Oldenburg, Franklin County, Ind., and MU 29856 (five specimens), "Liberty biofacies" within the upper part of the Bull Fork Formation, Caesars Creek dam site, Warren County, Ohio.

Remarks.—*Pleurorthoceras clarksvillense* is an uncommon species, there being only 23 specimens known to the writer, most from the Shideler Collection at Miami University, Oxford, Ohio. The species is apparently restricted to the upper 15 ft of the "Waynesville biofacies" and the succeeding 20 ft of the overlying "Liberty biofacies" within the upper half of the Bull Fork Formation exposed in Clinton, Warren, and Preble Counties in Ohio. Additional material has been collected from the same strata in Franklin, Union, and Dearborn Counties in Indiana. The species has not yet been identified from equivalent strata in northern or central Kentucky.

Specimens of *P. clarksvillense* are commonly holoperipherally encrusted by the bryozoan *Spatiopora* sp. along with the "worm tube" *Cornulites flexuosus* (Hall). The exteriors of specimens from the Addison's Creek and Oldenburg localities, including body chambers, are entirely covered with these encrusting bryozoans. Evidence from these specimens and associated specimens of the proteoceratid *Treptoceras* indicates that these bryozoans were preferentially encrusting live nautiloid hosts (Baird and others, 1989).

Pleurorthoceras clarksvillense is distinct from the other described species of *Pleurorthoceras*, *P. selkirkense* (Whiteaves) from the Selkirk Member of the Red River Formation (Upper Ordovician) of southern Manitoba, in possessing shorter camerae and narrower siphuncle segments at comparable shell diameters. At shell diameters of 15 and 20 mm, cameral lengths in *P. selkirkense* are from five-sixths to two-thirds the diameter of the shell (Flower, 1962, pl. 5, figs. 1–17). Camerae at the same

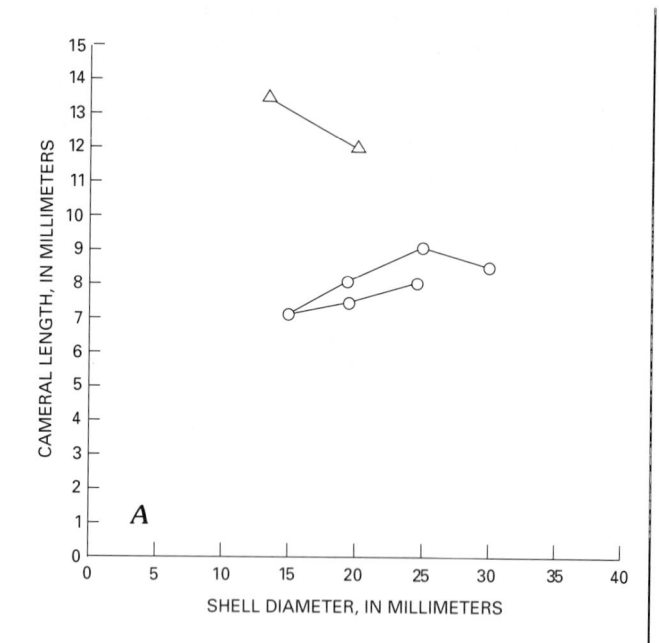

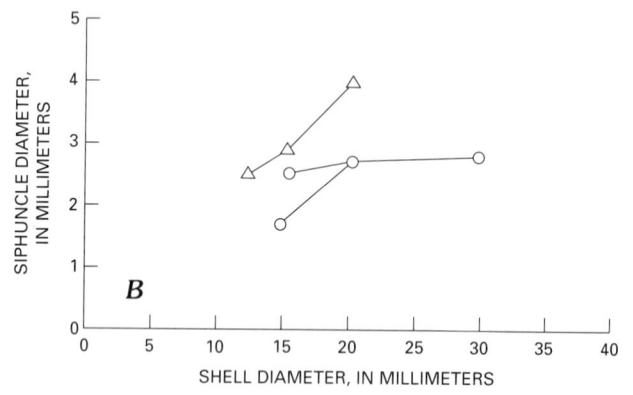

FIGURE 9.—Bivariate plots comparing specimens of *Pleurorthoceras clarksvillense* (circles) with a specimen of *P. selkirkense* (Whiteaves) from the Red River Formation in Manitoba (triangles). *A*, Cameral length versus shell diameter. *B*, Siphuncle diameter versus shell diameter. Data for *P. selkirkense* from Flower (1962).

diameters in *P. clarksvillense* average from one-half to three-eights the diameter of the shell. Siphuncle segments average one-fifth the diameter of the shell in specimens of *P. selkirkense* figured by Flower, compared with segment diameters that average one-eighth the diameter of the shell at comparable shell diameters in *P. clarksvillense*. Differences between these two species are presented graphically in figure 9.

Pleurorthoceras clarksvillense is somewhat similar to *Ordogeisonoceras amplicameratum* (Hall) from older Shermanian and Edenian strata in the Cincinnati, Ohio–Covington, Ky., area in the gradual expansion of the shell and the development of lengthy camerae. It differs from this species in being a smaller shell with subcentral siphuncle segments and short suborthochoanitic septal necks compared with the middorsal segments with moderately long orthochoanitic septal necks characteristic of *O. amplicameratum*. Siphuncle segments in adapical portions of *O. amplicameratum* develop small annulosiphonate deposits that are absent in *P. clarksvillense*. Cameral deposits in the Richmondian species consist of mural-episeptal deposits compared with the massive ventral hyposeptal and episeptal deposits and thin dorsal episeptal deposits developed in *O. amplicameratum*.

Pleurorthoceras clarksvillense differs from the Middle Ordovician (Rocklandian) orthoceratid *Pojetoceras floweri* in being a larger shell with simpler mural-episeptal cameral deposits and no endosiphuncular deposits. The smaller Tyrone Limestone species has complex episeptal and hyposeptal cameral deposits and, adapically, ventral parietal deposits lining the siphuncle connecting the ring wall.

FAMILY GEISONOCERATIDAE ZHURAVLEVA

Medium to large (up to 2 m in length), straight to slightly curved, generally gradually expanding orthoconic longicones, circular to slightly depressed in cross section with typically straight, transverse sutures. Cameral length moderate to long, one-fourth to one-half the diameter of the shell. Body chamber comparatively short, typically less than one-fourth the total length of the shell, distinctly constricted just adapical of the straight, transverse aperture. Shells smooth to elaborately ornamented with longitudinal, oblique, and (or) transverse elements.

Siphuncle segments moderately wide, diameters one-sixth to one-fourth the diameter of the shell, typically middorsal to central in position. Segments only slightly expanded, depressed, subtubular, with septal foramina wide relative to the diameter of the siphuncle segment. Thin connecting rings with moderately long, orthochoanitic to suborthochoanitic septal necks. Endosiphuncular deposits limited in volume compared with cameral deposits, consisting of annuli and parietal deposits that are typically continuous with cameral deposits. Well-developed cameral deposits adapically, consisting of planar dorsal episeptal deposits and massive, botryoidal hyposeptal and episeptal deposits ventrally.

Stratigraphic range.—Middle Ordovician (Chazyan Stage) to Middle Devonian (Eifelian Stage).

Remarks.—This seems to be a natural grouping of related genera that form the dominant group of orthocerid nautiloids in the Silurian. The family originally was defined on the basis of the occurrence of an orthochoanitic or suborthochoanitic siphuncle coupled with endosiphuncular deposits consisting of small annuli and (or) parietal deposits that grow adorally against connecting ring walls (Flower, 1939; Sweet, 1964). Subsequent study of other orthocerid taxa, in particular, the Orthoceratidae (Hook and Flower, 1977; Dzik, 1984), has indicated that these characteristics are not unique to the Geisonoceratidae.

The Geisonoceratidae, however, differ from *Orthoceras* and other orthoceratids in being larger, less slender forms that possess more expanded, broader, less tubular, dorsally located siphuncle segments with broad septal foramina, generally shorter cameral chambers, and more extensive development of cameral and endosiphuncular deposits at comparable stages of growth. This combination of features is used here to define members of this family.

A distinctive feature of this family is the supracentral to middorsal position of the siphuncle. Previous workers have inferred that siphuncles in members of this family were subcentral to midventral in position (Sweet, 1964). Specimens of *Geisonoceras rivale* (Barrande), *Harrisoceras laurelense* Flower, *Jonesoceras jonesi* (Barrande), *Mesnaquaceras curviseptum* Flower, *Temperoceras temperans* (Barrande), *Virgoceras palemon* (Barrande), and the species described below all demonstrate the more massive development of cameral deposits on the side of the shell opposite the siphuncle (pl. 3, figs. 4, 5, 10). These deposits are assumed to be developed more massively ventrally in order to lower the center of gravity, further stabilizing these longiconic shells. If this assumption is correct, then the siphuncles in most members of this family are supracentral in position. Similar supracentral siphuncles are found in *Proteoceras* Flower and in early growth stages of some species of *Treptoceras* Flower, both members of the Proteoceratidae.

The oldest members of this group are *Stereoplasmoceras longicameratum* from the Middle Ordovician (Llandeilan) "Cephalopod Shale" in Norway (Sweet, 1958) and *Mesnaquaceras curviseptum* from the lower Chazyan Day Point Limestone in New York (Flower, 1955b). Other Ordovician Geisonoceratidae are a new genus described below, currently known from uppermost Middle Ordovician (Shermanian) limestones and shales in New York, Kentucky, and the Canadian Arctic. The Silurian acme of this family consists of the genera *Geisonoceras* Hyatt, *Columenoceras* Barskov, *Harrisoceras* Flower, *Jonesoceras* Barskov, *Kionoceras* Hyatt, *Protokionoceras* Grabau and Shimer, *Temperoceras* Barskov, *Vericeras* Kolebaba, and *Virgoceras* Flower. These taxa are all known from Llandoverian to Pridolian carbonate facies in Eastern and Central North America (Foerste, 1928c) and from portions of Europe (Barrande, 1874) and Sardinia (Serpagli and Gnoli, 1977). Well-preserved material is also known from Devonian carbonates exposed in the Atlas Mountains of Morocco. In North America, Devonian representatives are poorly known, the *Geisonoceras teicherti* Flower from the lower Middle Devonian Onondaga Limestone of New York being the only well-documented species (Flower, 1939).

ORDOGEISONOCERAS NEW GENUS

Type species.—*Orthoceras amplicameratum* Hall, 1847, p. 205; Trenton Limestone, Shermanian Stage, Middle Ordovician, New York.

Description.—Large, gradually expanding orthoconic longicones, circular in cross section with straight, transverse sutures. Cameral lengths nearly one-half the diameter of the shell adapically, decreasing in length relative to shell diameter with ontogeny. Body chamber long and tubular, aperture unknown. Shell with a subdued ornament of fine longitudinal lirae.

Siphuncle situated close to the dorsal margin of the shell with elongate, depressed segments at diameters from one-fifth to one-seventh the diameter of the shell. Septal necks moderately long, orthochoanitic with broad septal foramina, nearly the diameter of the siphuncle segments. Connecting rings thin, segments only slightly expanded within camerae. Endosiphuncular deposits consisting of small annuli developed both dorsally and ventrally, restricted to adapical portions of the phragmocone. Well-developed cameral deposits adapically, the adoralmost deposits consisting of massive botryoidal hyposeptal and episeptal deposits ventrally and planar episeptal deposits dorsally. Dorsal deposits forming later than ventral deposits.

Stratigraphic and geographic occurrence.—Middle Ordovician (Shermanian Stage) to Upper Ordovician (Edenian Stage), North America (Laurentia).

Remarks.—This genus was erected for several large, gradually expanding orthoconic longicones that are known from latest Middle Ordovician (Shermanian) platform carbonates in northern Kentucky, New York, Ontario, and Baffin Island and range into equivalent or slightly younger mudstone facies of the Utica Shale in New York and Ontario and the Kope Formation in the Cincinnati, Ohio, region. These longiconic nautiloids are distinguished by their comparatively long camerae and dorsal siphuncles with slightly expanded, elongate siphuncle segments.

Ordogeisonoceras is similar internally to later Silurian taxa such as *Geisonoceras* Hyatt, differing from this genus only in the more dorsal position of the siphuncle in *Ordogeisonoceras* (pl. 3, figs. 4, 5, 8, 10). Externally, *Ordogeisonoceras* has a subtle ornament of fine longitudinal lirae with alternating heavier and finer elements. *Geisonoceras* is a slightly endogastrically curved longiconic shell marked by broad, low, transverse bands. *Ordogeisonoceras* is also similar to *Stereoplasmoceras longicameratum* Sweet from the older Middle Ordovician (Llandeilan) "Cephalopod Shale" of Norway. It differs from this form in the apparent greater curvature of the ventral septal necks in *S. longicameratum*. *Ordogeisonoceras* apparently became extinct in the Late Ordovician (Edenian), leaving a gap of about 20 m.y. before the recurrence of members of the Geisonoceratidae in carbonate facies in Laurentia in the late Llandoverian.

Etymology.—The genus takes its name from its occurrence in Ordovician strata and its similarity to the younger Silurian genus *Geisonoceras* Hyatt.

ORDOGEISONOCERAS AMPLICAMERATUM (HALL), 1847

Plate 3, figures 1–8

Orthoceras amplicameratum Hall, 1847, p. 205, pl. 51, figs. 1a–g.
Orthoceras amplicameratum Hall. James, 1886, p. 237.
Orthoceras ludlowense Miller and Faber, 1894b, p. 139–140, pl. 7, figs. 1–2.
Geisonoceras amplicameratum (Hall). Ruedemann, 1926, p. 90–91, pl. 11, fig. 1; pl. 12, figs. 2, 3.
Michelinoceras ludlowense (Miller and Faber). Flower, 1946, p. 107, 108.
Michelinoceras ludlowense (Miller and Faber). Flower, 1957b, p. 833, fig. 3B.

Diagnosis.—Very large, gradually expanding orthoconic longicone distinguished by cameral lengths that decrease with ontogeny from one-half to one-sixth the diameter of the shell, middorsal subtubular siphuncle segments whose diameters average one-sixth the diameter of the shell, marked by a subdued external ornament of alternating fine and heavy longitudinal lirae.

Description.—Large (lengths in excess of 1.25 m and having shell diameters up to 10.5 cm), gradually expanding (apical angle 4–4.5 degrees), orthoconic longicones, circular in cross section with straight transverse sutures. Cameral length slightly less than one-half the diameter of the shell at diameters of 30 mm or less, decreasing with ontogeny to minimal cameral lengths one-sixth the diameter of the shell at a diameter of 90 mm. Body chamber at least one-fifth the total length of the shell; aperture unknown. Subdued external ornament of alternating heavy and fine longitudinal lirae, six fine lirae occurring for each heavy lira.

Siphuncle segments supracentral (SPR=0.68-0.72) at shell diameters less than 35 mm, becoming middorsal to dorsal in position (SPR greater than or equal to ($\geq$) 0.75) at shell diameters in excess of 35 mm, and middorsal (SPR=0.75) at the maximum known shell diameter (90 mm). Segments subtubular, slightly expanded within camerae, longer than wide (SCR=0.40-0.50), diameters averaging one-sixth the diameter of the shell. Septal necks moderately long, orthochoanitic with broad septal foramina from two-thirds to three-fourths the maximum diameter of the segment. Connecting rings very thin, often not preserved. Small (less than 1 mm in diameter) annulosiphonate deposits occur dorsally and ventrally in adapical portions of phragmocones less than 35 mm in diameter. Cameral deposits well-developed in adapical portions of shells 40 mm or less in diameter, the adoralmost deposits consisting of botryoidal hyposeptal and episeptal deposits ventrally and planar dorsal episeptal deposits. Cameral deposits forming ventrally prior to their development dorsally.

Stratigraphic range and geographic occurrence.—Point Pleasant Tongue of the Clays Ferry Formation in northern Kentucky and the Kope Formation in the northern Kentucky–Cincinnati, Ohio, region (Shermanian and Edenian Stages). Elsewhere the species occurs in the Glendale and Poland Members of the Denley Limestone (Middle Ordovician, Shermanian Stage) in New York; Utica Shale (Upper Ordovician, Edenian-Maysvillian) in New York; and Whitby Formation (Upper Ordovician, Edenian) in southern Ontario.

Studied material.—Syntypes of *Orthoceras amplicameratum* Hall, AMNH 29686, AMNH 29687, AMNH 29688 (pl. 3, figs. 6, 7), Trenton Limestone, Middleville, N.Y.; syntypes of *O. ludlowense* Miller and Faber, UC 6457 (pl. 3, figs. 1–3) and UC 357 (two specimens), base of the "Hudson River Group" at Ludlow, Ky.; UC 18224, label just reads "Cynthiana"; and UC 18229 (pl. 3, figs. 4, 5, 8), "50 foot level" above the Ohio River, Cincinnati, Ohio.

Specimen AMNH 29688 is here designated the lectotype of *Orthoceras amplicameratum* Hall (1847); specimens AMNH 29686 and 29687 are designated paralectotypes of *O. amplicameratum* Hall.

Remarks.—While it is not possible to confirm the nature of the siphuncle segments in the types of *O. amplicameratum*, comparisons of the type specimens of *Orthoceras amplicameratum* Hall with those of *O. ludlowense* Miller and Faber indicate that these taxa are one and the same, based on similarities in shell size, apical angle, cameral length, and siphuncle position. These similarities are presented graphically in figure 10.

Ordogeisonoceras amplicameratum is a rare or uncommon species in uppermost Middle Ordovician (Shermanian) and lower Upper Ordovician (Edenian) strata exposed in the northern Kentucky–Cincinnati, Ohio, region. The species is most common in the planar-bedded fossiliferous limestone and shale of the Point Pleasant Tongue of the Clays Ferry Formation exposed along the Ohio River valley in northern Kentucky. In these strata, *O. amplicameratum* is found as well preserved, but incomplete internal molds, the septa, endosiphuncular, and cameral deposits replaced with calcite. These fragmentary individuals occur in association with more common specimens of the small orthocerid *Isorthoceras albersi* (Miller and Faber) and the small tarphycerid *Trocholites faberi* Foerste.

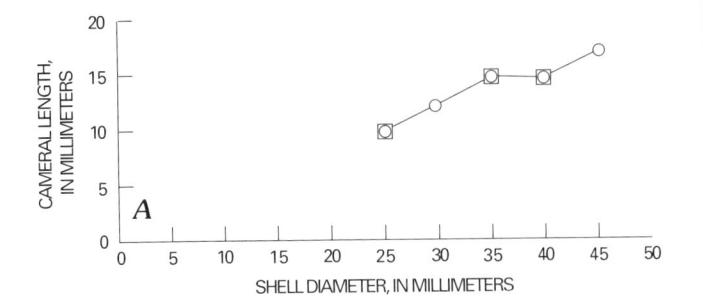

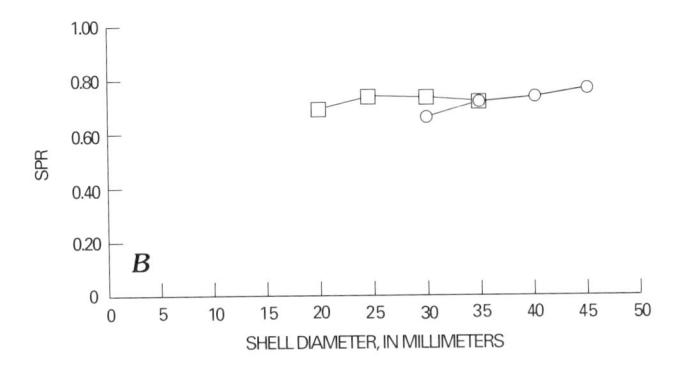

FIGURE 10.—Bivariate plots comparing specimens of "*Orthoceras*" *amplicameratum* Hall and "*Orthoceras*" *ludlowense* Miller and Faber. *A*, Cameral length versus shell diameter. *B*, Segment position ratio (SPR) versus shell diameter. Data for "*O.*" *amplicameratum* (squares) from lectotype and paralectotype AMNH 29688 and 29687. Data for "*O.*" *ludlowense* (circles) from syntypes UC 6457 and 18229 (see pl. 3).

Ordogeisonoceras amplicameratum appears to be a more common species in interbedded dark limestone and shale facies within the Denley Limestone (Middle Ordovician, Shermanian) and the black shale facies of the Utica Shale (Upper Ordovician) in northern New York and adjacent portions of southern Ontario. In these shelf-edge facies, *O. amplicameratum* occurs in association with abundant small longiconic nautiloids assignable to *Isorthoceras* Flower, the small coiled nautiloid *Trocholites*, plus the monoplacophoran *Sinuites*, the trilobites *Flexicalymene* and *Isotelus*, and a number of small strophomenid and orthid brachiopods. This is the same fauna associated with the species in the northern Kentucky area.

The only other species recognized by the writer as belonging to *Ordogeisonoceras* is a specimen described as "*Orthoceras* sp." by Foerste (1928d) and listed by him as coming from strata "corresponding to the Red River Formation" (Upper Ordovician, Edenian and Maysvillian) exposed in the Putnam Highlands in western Baffin Island in the Canadian Arctic. This specimen (University of Michigan, Museum of Paleontology no. 10261) differs from *O. amplicameratum* in having a more gradually expanding shell (apical angle 3 degrees) and longer camerae (cameral length greater than one-half the diameter of the shell at a shell diameter of 50 mm) at comparable shell diameters.

FAMILY PROTEOCERATIDAE FLOWER

Smooth or ornamented, small to medium-sized (less than 65 cm in length), moderately to rapidly expanding orthoconic or cyrtoconic longicones. Circular to slightly depressed in cross section, straight transverse sutures and short to moderately long camerae with septa exhibiting only weak curvature. Body chambers typically one-fourth the total length of the shell. Periphract dorsomyarian. Body chamber constricted just adapical to aperture.

Siphuncle segments typically supracentral in position initially, moving to a more ventral position with ontogeny. Segments expanded equiaxial or compressed in early and intermediary stages, decompressing with ontogeny, producing subtubular, depressed segments adorally. Septal necks of moderate length, recurved, sometimes recumbent, cyrtochoanitic most of the length of the phragmocone, often becoming suborthochoanitic in adoralmost septa. Connecting rings thin, homogeneous in structure. Endosiphuncular deposits typically well developed adapically, consisting of annuli that grow primarily forward to form parietal deposits that fuse to form ventral endosiphuncular linings adapically. Cameral deposits consisting of lobate mural-episeptal and hyposeptal deposits that develop ventrally prior to their dorsal development. Medial mural delta formed in most taxa.

Stratigraphic range.—Middle Ordovician (Chazyan Stage) to Upper Silurian (Ludlovian Series).

Remarks.—This family was formed for the reception of Ordovician orthocerids that developed compressed siphuncle segments and recurved cyrtochoanitic septal necks. The morphology of this group parallels that of the Pseudorthoceratidae, the predominant orthocerid family in younger Devonian and Carboniferous strata. Flower (1962) indicated that the Proteoceratidae differ from the Pseudorthoceratidae in (1) the form of the endosiphuncular deposits, which thicken where the segment curves adorally from the septal foramen versus more uniform development in the Pseudorthoceratidae, and (2) the ontogenetic decompression of the siphuncle segments in the Proteoceratidae compared with the lack or weak expression of such decompression in the Pseudorthoceratidae. Unfortunately, the complete or nearly complete specimens necessary to document ontogenetic differences in siphuncle segment shape are largely lacking for both of these families. Well-preserved, nearly complete material is known for *Cyrtactinoceras* Hyatt, *Proteoceras* Flower, and *Treptoceras* Flower within the Proteoceratidae, and all three illustrate well the ontogenetic decompression of the siphuncle segments (pl. 5, figs. 1–4). The few members of the Pseudorthoceratidae in which a significant portion of the phragmocone is known (*Adnatoceras* Flower, *Mooreoceras* Miller, Dunbar, and Condra, *Paramooreoceras* Zhuravleva, and *Pseudorthoceras* Girty) do not exhibit this pronounced decompression of siphuncle segments and have endosiphuncular deposits that lack the adoral thickening typical of *Proteoceras* and *Treptoceras*. Thus, it is suggested that Flower's criteria for differentiating these families might be valid. More complete specimens representative of both families are needed to further validate their distinctiveness. Two genera, *Isorthoceras* Flower and *Gorbyoceras* Shimizu and Obata, originally placed in the Proteoceratidae by Flower, appear to lack siphuncle segment decompression (Aronoff, 1979; Flower, 1946) and are placed here in the Pseudorthoceratidae. The annulated orthocerid *Monomuchites* Wilson is retained in the Proteoceratidae, although questionably, since there is a lack of knowledge concerning the shape of its siphuncle segments, its endosiphuncular deposits, and how both change with ontogeny.

GENUS *PROTEOCERAS* FLOWER

Type species.—*Oonoceras perkinsi* Ruedemann, 1906, p. 499; designated by Flower, 1955b, p. 821; upper part of the Chazy Limestone, Chazyan Stage, Middle Ordovician, New York.

Description.—Small, moderate to rapidly expanding orthoconic longicones that are slightly curved endogastrically in adapical portions of the shell. Circular in cross section with straight transverse sutures. Camerae short to moderate in length, typically decreasing with ontogeny relative to shell diameter. Body chamber one-fourth the total length of the shell, slightly constricted just adapical of aperture. Exterior smooth, marked dorsally by fine longitudinal color bands.

Assuming an endogastrically curved shell, the siphuncle is initially middorsal, moving to a more central position with ontogeny. Expanded globular siphuncle segments adapically, decompressing with ontogeny, the adoral segments nearly tubular, longer than wide. Septal necks cyrtochoanitic, recurved with free brims. Connecting rings thin, homogeneous. Endosiphuncular deposits consisting of annuli that thicken adorally adjacent to septal necks and that grow forward with ontogeny to form thin parietal deposits that fuse into an endosiphuncular lining. Well-developed cameral deposits adapically, consisting of planar mural-episeptal deposits dorsally and thick, botryoidal mural-episeptal deposits ventrally.

Stratigraphic range and geographic occurrence.—Middle Ordovician (Chazyan to Rocklandian Stages) in Eastern North America (Laurentia).

Remarks.—This genus was erected by Flower (1955b) for small cyrtochoanitic longicones from early Middle Ordovician strata in Eastern North America. These nautiloids are remarkable for the pronounced changes in the shape and position of the siphuncle segments with ontogeny, coupled with the presence of endosiphuncular deposits that consist of annuli that are thickest just adoral of septal necks.

There is some uncertainty as to the orientation of shells assigned to this genus. Flower (1955b) believed these shells were endogastrically curved, the siphuncle located dorsally initially, moving to a more central position with ontogeny. As evidence of this orientation, Flower cited (1) the more massive development of cameral deposits adjacent to the concave side of the shell, opposite the siphuncle, and (2) the similar thickening of endosiphuncular deposits and their earlier development on the side of the siphuncle facing the concave portion of the shell. These ballast deposits are assumed to be thicker ventrally in order to lower the shell's center of gravity, further stabilizing these longiconic shells.

Dorsally situated siphuncles are not typical of most members of the Orthocerida. Most taxa assigned to this order have either central or subcentral siphuncles. One other member of the Proteoceratidae, *Treptoceras* Flower, exhibits supracentral siphuncle segments in early growth stages, the siphuncle moving to a subcentral position with ontogeny. Some species of the annulated orthocerid *Monomuchites*, also placed in the Proteoceratidae by Flower (1962), have siphuncle segments that are supracentral in position (pl. 9, figs. 3, 4, 17, 19). Members of the Geisonoceratidae, described above, also exhibit more massive development of cameral deposits on the side of the shell opposite the siphuncle, suggesting dorsal siphuncle segments.

Proteoceras is currently known from Chazyan carbonates in the Champlain basin in Upstate New York and adjacent portions of Vermont and Quebec. Flower (1955b) suggested that Middle Ordovician (Blackriveran) species described by Foerste (1932, 1933) as members of the genus *Sactoceras* Hyatt might also be assignable to *Proteoceras*. These species were described from the "Black River Formation" exposed on St. Joseph Island in Lake Huron, Ontario. Similarly, *Sactoceras tyronensis* (Foerste) from the Tyrone Limestone in central Kentucky is also placed here in *Proteoceras*, extending the stratigraphic range of the genus into the Rocklandian.

PROTEOCERAS TYRONENSIS (FOERSTE), 1912

Plate 4, figures 1–7

Orthoceras tyronensis Foerste, 1912, p. 139, pl. 10, figs. 5a–b.
Sactoceras tyronensis (Foerste). Foerste, 1933, p. 6, pl. 10, fig. 5.
Sactoceras josephianum Foerste, 1933, p. 2–4, pl. 9, figs. 2a–b, 3a–b.
Sactoceras josephianum Foerste. Wilson, 1961, p. 64–65, pl. 25, fig. 9.

Diagnosis.—A slender, slightly endogastrically curved species of *Proteoceras* with cameral lengths from one-third to one-fifth the diameter of the shell and siphuncle segments that change position from supracentral to subcentral with ontogeny and that, adapically, are globular, equiaxial to slightly compressed in shape at diameters from one-fourth to one-fifth the diameter of the shell.

Description.—Small (phragmocone lengths of at least 20 cm and having diameters up to 2.4 cm), slender (apical angle 6–7 degrees), slightly endogastrically curved orthoconic longicones; circular in cross section with straight transverse sutures. Cameral lengths decreasing with ontogeny from one-third the diameter of the shell at diameters less than 10 mm to one-fifth the diameter of the shell adorally at shell diameters in excess of 20 mm. Body chamber lengthy, tubular; aperture unknown. Shell surface smooth.

Siphuncle segments central adapically (SPR=0.50 at diameters of 10 mm or less), then becoming supracentral (SPR=0.53–0.57) for much of the known length of the phragmocone, and then subcentral (SPR=0.47–0.48) in adoral portions greater than 15 mm in diameter. Segments globular, equiaxial at diameters of 10 mm or less, becoming compressed (SCR=1.14–1.16) at shell diameters between 10 and 15 mm. Segment shape in more adoral portions of shell unknown. Segment diameters between one-fourth and one-fifth the diameter of the shell. Septal necks cyrtochoanitic with recurved, free brims. Endosiphuncular deposits developed adapically, consisting of ventral parietal deposits that fuse to form a continuous endosiphuncular lining. Cameral deposits developed adapically, the adoralmost deposits consisting of lobate, botryoidal episeptal deposits ventrally and a well-developed medial mural delta (pl. 4, fig. 5).

Stratigraphic range and geographic occurrence.—Rocklandian part of the Tyrone Limestone, Inner Blue Grass region of central Kentucky; "Black River Formation," Middle Ordovician (Blackriveran? Stage), St. Joseph Island, Lake Huron, Ontario; and "Leray-Rockland beds" (Middle Ordovician, Rocklandian Stage), Ottawa region, Ontario and Quebec.

Studied material.—Holotype, USNM 87109, Tyrone Limestone, 30 ft below the top of the formation, High Bridge, Ky.; USNM 158629 (pl. 4, figs. 1–3), 468686, 468687, 468688 (pl. 4, fig. 6), 468689 (pl. 4, figs. 5, 7), and 468690 (suite of 10 specimens), all from USGS 6034-CO collection, Little Hickman section A, Jessamine County, Ky.

Remarks.—Foerste (1912) based his description of this species on several small (less than 4.0 cm in length), fragmentary specimens that were chiefly remarkable for the holoperipheral encrustation of the shells by the problematic sponge *Dermatostroma tyronensis* Foerste. The other salient features of this species, as described by Foerste, were an orthoconic shell, circular in section, its apical angles varying from 6 to 10 degrees. In the original description the siphuncle in this species was described as being "linear in form" and "about 2/3 of a millimeter in diameter" at a diameter of 7 mm. In a later paper, Foerste (1933) described the species as having subglobular siphuncle segments at shell diameters of 11 mm and a shell that was "very slightly curved lengthwise." The nautiloid specimens described here from the 6034-CO collection correspond well with Foerste's brief description of "*Orthoceras*" *tyronensis*. The general shell morphology, siphuncle segment position and shape, and the nature of endosiphuncular and cameral deposits in this species all indicate placement in the genus *Proteoceras* Flower.

Specimens of *P. tyronensis* are common within the silicified USGS 6034-CO collection from the Tyrone Limestone, 46 ft below the top of the formation at the Little Hickman section. Specimens are typically fragmentary, consisting of adapical portions of phragmocones less than 1.5 cm in diameter. *Proteoceras tyronensis* is the fifth most common longiconic nautiloid present in the 6034-CO collection, behind the two species of *Cartersoceras*, *Pojetoceras floweri*, and *Monomuchites annularis*. It shares the same faunal associates with these species.

Proteoceras tyronensis differs from the type species of *Proteoceras*, *P. perkinsi* (Ruedemann) plus the associated *P. pulchrum* Flower, both from the upper part of the Chazy Limestone (Middle Ordovician, Chazyan Stage) of New York, in having a more slender, less rapidly expanding shell and an apical angle of 6–7 degrees compared with values of 8–10 degrees for these Chazyan species. These differences are presented graphically in figure 11. *Proteoceras tyronensis* is most similar to "*Sactoceras*" *josephianum* Foerste, a slender, slightly curved species of *Proteoceras* from equivalent Blackriveran and Rocklandian strata exposed on St. Joseph Island in Lake Huron (Foerste, 1933). Both species have slightly curved, slender shells (apical angle 6–7 degrees) and siphuncles that are slightly dorsal of center with globular siphuncle segments. Curiously, the specimens of *P. josephianum* described and figured by Foerste (1933, pl. 9, figs. 2a, 3a) are also encrusted by a species of *Dermatostroma*.

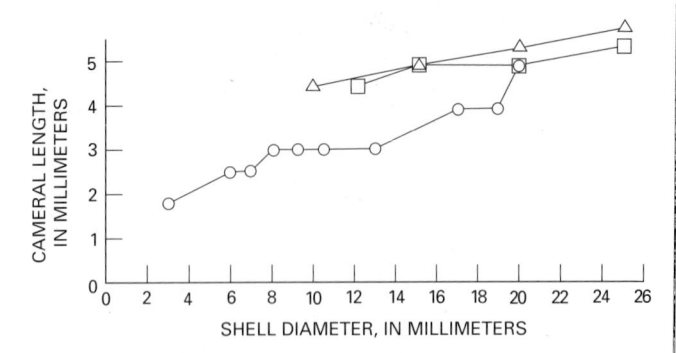

FIGURE 11.—Bivariate plot of cameral length versus shell diameter for Middle Ordovician species of *Proteoceras* from Eastern North America. Species are *P. tyronensis* (Foerste) from the USGS 6034-CO collection in the Tyrone Limestone (circles), and *P. perkinsi* (Ruedemann) (squares) and *P. pulchrum* Flower (triangles) figured by Flower (1955b) from the Chazy Limestone in New York.

Foerste (1933) described "*Sactoceras*" *josephianum* as being similar to "*Orthoceras*" *tyronensis*, remarking that it differed from this species only in the larger size of the St. Joseph Island specimens. Specimens of *P. tyronensis* from the USGS 6034-CO collection are equal in size to Foerste's figured specimens of "*S.*" *josephianum*, and it is concluded that these species are synonyms; *P. tyronensis* has priority.

Orthoceras huronense Billings and *Sactoceras pictolineatum* Foerste, both from Blackriveran strata exposed on St. Joseph Island, also appear to be assignable to *Proteoceras*. Both of these species differ from *P. tyronensis* in having more rapidly expanding shells and apical angles of 8 and 8.5 degrees, respectively. The figured specimen of *P. huronense* (GSC 1724) preserves a dorsal color pattern of seven to eight moderately thick longitudinal brown bands, whereas the specimen of *P. pictolineatum* figured by Foerste (1932, pl. 9, figs. 5a–b) has a dorsal color pattern of numerous (27+) fine, longitudinal, dark lines. Color patterns are unknown for specimens of *P. tyronensis*.

GENUS *TREPTOCERAS* FLOWER

Type species.—*Orthoceras duseri* Hall and Whitfield, 1875, p. 97; designated by Flower, 1942, p. 16; Waynesville Formation, Richmondian Stage, Upper Ordovician, Ohio.

Description.—Small to medium-sized, generally moderately expanding orthoconic longicones, circular in cross section with straight, transverse sutures. Cameral lengths ranging from one-fourth to one-seventh the diameter of the shell, typically decreasing with ontogeny relative to shell diameter. Body chamber one-fourth the total length of the shell; aperture straight, transverse, no ventral sinus. Periphract dorsomyarian. Shell basically smooth, marked by fine longitudinal lirae and transverse growth lines. Dorsum marked by longitudinal color bands.

Siphuncle supracentral or central initially, some distance ventral of center in adoral portions of large specimens. Segments expanded, heart-shaped, equiaxial initially, becoming compressed, beadlike, and then decompressing in later growth stages, the segments depressed, longer than high. Segment diameter ranging from one-fourth to one-seventh the diameter of the shell, decreasing with ontogeny. Septal necks cyrtochoanitic; dorsal brims recumbent, ventral brims recurved but free. Connecting rings thin, homogeneous. Adapical portions of phragmocone with endosiphuncular deposits consisting of annuli that thicken adorally, adjacent to septal necks and grow forward to form parietal deposits ventrally and simple annuli dorsally. Parietal deposits fuse adapically to form a continuous ventral endosiphuncular lining with ontogeny. Cameral deposits completely filling chambers in adapical half of large phragmocones; adoralmost deposits mural-episeptal, initiated ventrally and laterally before developing dorsally.

Stratigraphic and geographic range.—Upper Ordovician (Edenian to Richmondian Stages), Eastern North America (Laurentia).

Remarks.—As was indicated by Flower (1946), *Treptoceras* is the most common nautiloid taxon encountered in the Upper Ordovician section exposed in the Cincinnati, Ohio, region, which takes into account adjacent portions of Kentucky and Indiana. Many species have been described from these strata in this area (Miller, 1875; James, 1886). Most of these species, however, were described from fragmentary specimens, some representing different ontogenetic stages of the same nautiloid. The incomplete, often poorly preserved nature of most museum material, coupled with the pronounced ontogenetic changes characteristic of the genus, makes species-level systematic study of the genus difficult. To complicate matters further, Miller (1875) did not designate types, and the location of some of his figured specimens is unknown. Material that this writer believes to be the original figured specimens and that are listed as being type specimens on their museum labels are the holotype of *Orthoceras duseri* Hall and Whitfield (University of California at Berkeley, UCal 34325) and the holotype of *Orthoceras cincinnatiensis* Miller (Harvard Museum of Comparative Zoology, MCZ 3404). Specimens similar to those figured by Miller (1875, 1881) and listed as "co-types" in the collections of the University of Cincinnati (UC) are the "co-types" of *Orthoceras fosteri* Miller (UC 356) and *O. transversa* Miller (UC 1328). These specimens, in conjunction with the original species descriptions and illustrations plus well-preserved, nearly complete specimens collected by the writer, indicate the presence of at least four distinct species of *Treptoceras* in

Cincinnatian Provincial Series strata exposed in the Kentucky-Indiana-Ohio region.

In the past there has been some confusion as to the affinities of *Treptoceras*. Flower (1942, 1957a), citing the nature of the endosiphuncular deposits and the ontogenetic decompression of siphuncle segments, placed the genus in the Actinocerida. Teichert and Glenister (1953) decided that *Treptoceras* was a synonym of *Orthonybyoceras* Shimizu and Obata, the latter name having priority. Flower removed *Orthonybyoceras* (=*Treptoceras*) from the Actinocerida and placed this genus in a new family, the Proteoceratidae, within the Orthocerida. Aronoff (1979) clarified the taxonomic positions of both of these genera and demonstrated that *Orthonybyoceras* is an actinocerid distinct from *Treptoceras*, which he retained in the Proteoceratidae. Both taxa occur in the Upper Ordovician strata exposed in the Cincinnati region. *Orthonybyoceras* is an uncommon form in the Fairview Formation in northern Kentucky and in adjacent portions of Ohio and a rare element in the lower part of the Bull Fork Formation in Indiana and Ohio (pl. 18, figs. 10–17). *Treptoceras*, on the other hand, is ubiquitous to nearly every stratigraphic unit within the Upper Ordovician section exposed in the Kentucky-Indiana-Ohio area (pl. 4, figs. 8–12, pls. 5–8).

As indicated in Frey (1987a, 1989), species of *Treptoceras* are largely restricted to clastic or mixed clastic-carbonate shallow-marine facies of Late Ordovician Age in Eastern and Central North America (Laurentia). The genus has not been identified from coeval carbonate platform facies exposed in Western and Arctic North America. Species of *Treptoceras* from outside the northern Cincinnati arch region are *T. maquoketense* (Foerste) from the "Elgin Shale" in Iowa (Upper Ordovician, Maysvillian); *T. manitoulinense* (Foerste) from the Georgian Bay Formation (Upper Maysvillian and Richmondian) in southern Ontario; and species referred to as either "*Actinoceras*" *crebriseptum* (Hall) or "*Orthoceras*" *lamellosum* Hall, both from the Pulaski Formation (Maysvillian) of New York and the Georgian Bay Formation exposed at Toronto in Ontario. Additional species possibly assignable to *Treptoceras* are "*Ormoceras*" *lindsleyi* Foerste and Teichert from the Cannon Limestone (Middle Ordovician, Shermanian) of central Tennessee. This species, if indeed a species of *Treptoceras*, would mark the earliest known occurrence of the genus. More complete specimens of this Middle Ordovician species are needed to determine its affinities to *Treptoceras*.

Species of *Treptoceras* described by Flower (1942) from the "Cynthiana Limestone" (=Strodes Creek Member of the Lexington Limestone) at Cynthiana, Ky., are known only from solitary incomplete specimens sectioned horizontally, obscuring some of the diagnostic features of these species. Aronoff (1979) placed "*Treptoceras*" *persiphonatum* Flower (1942, pl. 4, fig. 2) with question in *Deiroceras*, based on its globular siphuncle segments, long septal necks, and comparatively deep camerae. He did not attempt to determine the systematic affinities of the other two species described by Flower from the same limestone lens at the same locality. The broad siphuncle segments in "*T.*" *perseptatum* Flower (1942, pl. 1, fig. 1) are suggestive of *Ormoceras* or *Orthonybyoceras*. The apparently short, recumbent septal necks in this specimen indicate placement in *Orthonybyoceras*. The narrow diameter, slightly expanded, barrel-shaped siphuncle segments and gently recurved, free cyrtochoanitic septal necks in "*T.*" *praenuntium* Flower (1942, pl. 1, fig. 2) indicate placement of this species in the Pseudorthoceratidae, possibly in *Isorthoceras* Flower.

Foerste (1893) described a number of small orthoconic longicones from the overlying Lower Silurian (Llandoverian, Aeronian) Brassfield Limestone in southwestern Ohio. He placed these species in a new genus, *Euorthoceras*, but did not formally describe the distinguishing features of this genus nor designate a type species. Restudy of the types of the two best known species referred to the genus, *E. ignotum* Foerste (USNM 88554) and *E. erraticum* Foerste (USNM 88549), reveals that these species belong to two different genera. *Euorthoceras ignotum* is a gradually expanding, slender longicone (apical angle 5 degrees) with short camerae and an internal structure similar to that of *Isorthoceras* Flower, having subcentral, very slightly expanded, barrel-shaped siphuncle segments with short, cyrtochoanitic septal necks. Mural-episeptal deposits are known from shell fragments less than 18 mm in diameter. No endosiphuncular deposits are preserved in the suite of USNM specimens. *Euorthoceras erraticum* is a more rapidly expanding form (apical angle 7 degrees) with a smooth exterior and short camerae. The siphuncle is similar to that of *Treptoceras* in that it migrates from a middorsal to a midventral position with ontogeny and possesses globular, equiaxial segments adapically that decompress to form slightly expanded, subtubular segments with ontogeny. Septal necks are short, recurved, cyrtochoanitic. Episeptal cameral deposits are present in adapical portions of the phragmocone less than 16 mm in diameter. Endosiphuncular deposits consist of parietal deposits ventrally and thin annuli dorsally, observed in the phragmocone at diameters less than 14 mm.

Euorthoceras, based on *E. ignotum*, is a member of the Pseudorthoceratidae, closely related to Ordovician taxa like *Isorthoceras* Flower. *Euorthoceras*, based on *E. erraticum*, is a member of the Proteoceratidae, closely related to, if not in fact, a species of the genus *Treptoceras* Flower. As indicated above, Foerste did not designate a type species for *Euorthoceras* (Teichert, 1940). The first species Foerste unquestionably assigned to *Euorthoceras* was *E. ignotum* (Foerste, 1893, p. 539). As such, this species is here designated the type species of *Euorthoceras*, and the genus assumes the characters of this species, namely, a slender, gradually expanding longicone with barrel-shaped siphuncle segments and short, recurved septal necks with free brims. *Euorthoceras erraticum* is removed from *Euorthoceras* and is tentatively placed in the genus *Treptoceras* Flower, pending further study. Other species assignable to *Euorthoceras*, based on *E. ignotum*, are *E.* cf. *E. virgulatum* (Hall) from the Brassfield Limestone in Ohio and southeastern Indiana and *Euorthoceras* sp. from the Ekwan River Formation (Lower Silurian, Llandoverian) of the James Bay Lowland, Quebec (Flower, 1968b). *Treptoceras? erraticum* (Foerste), if determined to be a valid species of *Treptoceras*, would be the youngest known member of the genus.

TREPTOCERAS DUSERI (HALL AND WHITFIELD), 1875

Plate 4, figures 8–12; plate 5, figures 1–6

Orthoceras duseri Hall and Whitfield, 1875, p. 97, pl. 3, figs. 2–4.

Orthoceras duseri Hall and Whitfield. James, 1886, p. 241.

Orthoceras duseri Hall and Whitfield. Cumings, 1908, p. 1036, pl. 52, figs. 2–2b.

Orthoceras duseri Hall and Whitfield. Bassler, 1915, p. 904.

Orthoceras duseri Hall and Whitfield. Parks and Fritz, 1922, p. 16–17, pl. 2, fig. 1; pl. 4, figs. 7, 8.

Treptoceras duseri (Miller). Flower, 1942, p. 16–17.

Orthonybyoceras duseri (Hall and Whitfield). Teichert and Glenister, 1953, p. 223.

Orthonybyoceras duseri (Hall and Whitfield). Flower, 1957a, p. 56–57.

Treptoceras duseri (Hall and Whitfield). Aronoff, 1979, p. 113–116, figs. 2a–c, 3d–e.

Treptoceras duseri (Hall and Whitfield). Frey, 1988, p. 81–84, pl. 1, figs. 1–6; pl. 2, figs. 1–6; pl. 3, figs. 2–4; figs. 3, 5, 6, 8.

Diagnosis.—Large, moderately expanding species of *Treptoceras* having siphuncle segments one-sixth to one-fourth the diameter of the shell that change shape with ontogeny from equiaxial to compressed and position from central to subcentral; becoming equiaxial or depressed and midventral in the adoralmost segments of large specimens.

Description.—Medium-sized (up to 46 cm in length and 5.5 cm in diameter), moderately expanding (apical angle 7.5–8.5 degrees) orthoconic longicone, circular in cross section with straight transverse sutures. Camerae short, lengths initially one-fourth the diameter of the shell, decreasing with ontogeny to one-sixth or one-seventh the diameter of the shell in adoral portions 20–35 mm in diameter. Body chamber at least one-fourth the total length of the shell; aperture straight, transverse, lacking sinus. Periphract dorsomyarian. Shell exterior essentially smooth, marked by fine transverse growth lines and fine longitudinal lirae.

Siphuncle segments central in position (SPR=0.50) at a diameter of 15 mm, migrating with ontogeny to a subcentral position (SPR=0.29–0.33) at a diameter of 27 mm, and midventral in position (SPR=0.25) in adoral portions 40 mm in diameter. Segments initially equiaxial at diameters less than 15 mm, becoming compressed (SCR=1.06–1.42) between diameters of 15 and 25 mm, equiaxial between diameters of 25 and 30 mm, then depressed for the remaining length of shell (SCR < 1.0). Segment diameter from one-fourth to one-sixth the diameter of the shell, decreasing with ontogeny. Septal necks cyrtochoanitic; dorsal brims recumbent, ventral brims recurved but free. Dorsal brims suborthochoanitic at shell diameters greater than 35 mm. Endosiphuncular deposits consisting of dorsal annuli and ventral parietal deposits, restricted to adapical half of large shells. With ontogeny, ventral parietal deposits fuse to form a continuous ventral endosiphuncular lining adapically. Adoralmost cameral deposits mural-episeptal, the deposits completely filling camerae in adapical half of large shells.

Stratigraphic and geographic occurrence.—Basal portions of the Fairview Formation, Hamilton and Brown Counties in southwest Ohio and Campbell and Kenton Counties in northern Kentucky; "Waynesville biofacies" (in the sense of Wahlman, 1992) in the Bull Fork Formation (Upper Ordovician, Richmondian) in Clinton and Warren Counties in southwest Ohio and Dearborn and Franklin Counties in southeastern Indiana. With question, from the upper part of the Clays Ferry Formation (Upper Ordovician, Edenian), Scott County, Ky.

Studied material.—Holotype, UCal 34325 (pl. 4, figs. 8–10), from "shales of the Hudson River Group near Waynesville, Ohio"; MU 417T (pl. 5, figs. 1–6) from the *Treptoceras duseri* shale unit exposed along Harpers Run, Warren County, Ohio; several suites of specimens from the *T. duseri* shale unit: MU 17170 (three specimens) from the Caesars Creek dam site, Warren County, Ohio; MU 17171 (two specimens) from Stony Hollow, north of Clarksville, Clinton County, Ohio; MU 17172 (four specimens) from the Route 42 roadcut northeast of Waynesville, Warren County, Ohio; MU 17173 (nine specimens) from Harpers Run, Warren County, Ohio; plus additional material from the Clays Ferry Formation, at Sadieville, Scott County, Ky. (USNM 468719); the basal Fairview Formation at Georgetown, Brown County, Ohio (MU 7871) and at Dent, Hamilton County, Ohio (MU 29766, pl. 4, fig. 11), and USNM 468727 from the base of the Reidlin Road section, Forest Hills, Kenton County, Ky. (pl. 4, fig. 12).

Remarks.—*Treptoceras duseri* is the type species for the genus *Treptoceras* Flower. The species is known primarily from the *T. duseri* shale unit within the "Waynesville biofacies" of the Bull Fork Formation (Upper

Ordovician, lower Richmondian) in Warren and Clinton Counties in southwestern Ohio. In this claystone unit the species is represented by abundant, largely complete, well-preserved, calcite-replaced internal molds (pl. 4, figs. 8–10; pl. 5, figs. 1–6). The distinctive preservation of nautiloids from the *T. duseri* shale unit indicates that this claystone is also the source of the holotype.

An internal mold with the general morphological features of *T. duseri* is present in USGS collection 7471-CO from the Edenian part of the Clays Ferry Formation at Sadieville, Scott County, Ky. (USNM 468719). The internal details of the siphuncle segments are not known for this specimen, and better preserved material is necessary to determine the affinities of this Clays Ferry species.

Additional well-preserved and comparatively complete material assignable to *T. duseri* is known from the interbedded claystone and thin packstone beds at the base of the Fairview Formation exposed in Brown and Hamilton Counties in the vicinity of Cincinnati, Ohio (pl. 4, fig. 11). Fragmentary material referable to this species has been collected from the same strata exposed along I–275E in Kenton County, Ky. (pl. 4, fig. 12). Less well preserved, large phragmocones (up to 50 cm in length) possibly assignable to *T. duseri* are common in the Mount Auburn Member of the Grant Lake Formation (Maysvillian) in parts of Butler and Warren Counties in Ohio.

Treptoceras duseri is most similar to the associated species, *T. fosteri* (Miller). It differs from *T. fosteri* in its more rapidly expanding shell (apical angle 7.5–8.5 degrees compared with apical angle 5.5–6.0 degrees in *T. fosteri*), the more central position of the siphuncle segments in *T. duseri* at comparable diameters (table 7 and fig. 12), and the shape of the siphuncle segments, the compressed segments in *T. duseri* lacking a well-developed zone of adnation compared with a well-developed dorsal zone of adnation in *T. fosteri* (fig. 13). The ranges of these two species overlap stratigraphically and geographically. In the *T. duseri* shale unit, sampled suites of specimens indicate that these species occur in the same beds, *T. fosteri* being slightly more abundant ($N=83$) than *T. duseri* ($N=73$) (Frey, 1989). Elsewhere, however, one or the other species is predominant, the second species typically being absent. *Treptoceras duseri* or a closely related form is the most abundant nautiloid in the Mount Auburn Member of the Grant Lake Formation, while a species similar to *T. fosteri* is predominant in the underlying Corryville Member of the same formation, both in southwestern Ohio. *Treptoceras duseri* has not been identified from strata younger than the upper part of the "Waynesville biofacies" in the Bull Fork Formation in southwestern Ohio and southeastern Indiana. *Treptoceras fosteri* or a closely related species (pl. 6, figs. 4, 8) is the most abundant nautiloid species in the "Liberty biofacies" (in the sense of Wahlman, 1992) in overlying portions of the Bull Fork Formation.

Treptoceras duseri is also similar to the associated *T. cincinnatiensis* (Miller). Both species are moderate to rapidly expanding, rather large species of *Treptoceras*, having central to subcentral siphuncle segments at comparable diameters, and well-developed adapical endosiphuncular deposits consisting of dorsal annuli and ventral parietal deposits. *Treptoceras duseri* is distinct from *T. cincinnatiensis* in the less central position of the siphuncle in the adoral portions of large specimens (table 7 and fig. 12); in the narrower diameters of the siphuncle segments at comparable shell diameters (segment heights averaging one-seventh the diameter of the shell at diameters in excess of 30 mm in *T. duseri* compared with segment heights one-fifth the diameter of the shell at the same diameters in *T. cincinnatiensis*); and in the mural-episeptal nature of last-formed cameral deposits in *T. duseri* in contrast to the episeptal and hyposeptal deposits in *T. cincinnatiensis* (compare pl. 5, figs. 1–6 with pl. 7, figs. 4–6). *Treptoceras duseri* and *T. cincinnatiensis* overlap throughout most of their respective ranges, notably in the basal Fairview Formation in northern Kentucky and in the "Waynesville biofacies" within the Bull Fork Formation in southwestern Ohio and southeastern Indiana. As indicated above, *T. duseri* is absent from the overlying "Liberty biofacies" in the Bull Fork Formation in the same area, whereas *T. cincinnatiensis* or a similar form is uncommon in these strata. Comparisons of the critical morphological features used to distinguish these Cincinnatian species of *Treptoceras* are presented graphically in figures 14–16.

Ward (1987) has remarked that species of living *Nautilus*, upon sexual maturity, reach a maximum size at which growth slows or ceases. In *Nautilus* the attainment of sexual maturity is marked by thickening of the last formed septum and is preceded by approximation of the last two or three septa. Not all specimens of *Nautilus*, however, exhibit septal approximation (Ward, 1987). Large specimens of *T. duseri* (MU 417T, pl. 5, fig. 1) and the associated species *T. cincinnatiensis* (MU 443T, pl. 8, fig. 3) show similar abrupt septa approximation adorally. At a diameter of 40 mm in *T. duseri*, cameral length decreases from 5.5 mm to lengths of 1.5 and 1.3 mm in the last formed camerae. At a diameter of 55 mm in *T. cincinnatiensis*, cameral length decreases from 9 mm to lengths of 5 mm and then 3 mm in adoralmost camerae. These examples suggest that these species of *Treptoceras* had a maximum size that could be attained, roughly shell lengths of approximately 46 cm at diameters of 5.5 cm in *T. duseri* and a length of 55 cm and a diameter of 7.5 cm for specimens of *T. cincinnatiensis*. Septal approximation was also observed in specimens of *T.* cf. *T. fosteri* from the "Liberty biofacies" in the Bull Fork Formation in

TABLE 7.—Comparisons of the morphological features of species of *Treptoceras* from the Bull Fork Formation in the Kentucky-Indiana-Ohio area showing differences in shell expansion, cameral length, siphuncle position, segment shape, and siphuncle diameter relative to shell diameter

[Data from specimens listed in text. CL/D, ratio of cameral length to shell diameter; SPR, siphuncle segment position ratio; SCR, siphuncle segment compression ratio; SD/D, ratio of siphuncle diameter to shell diameter; mm, millimeter]

Feature	*T. duseri*	*T. fosteri*	*T.* cf. *T. fosteri*	*T. cincinnatiensis*
Apical angle, degrees	7.5–8.5	6	5–6	8.5–9
Diameter 35 mm				
CL/D	0.17		0.12	0.185
SPR	.285		.23	.34
SCR	.83		1.00	1.40
SD/D	.125		.12	.20
Diameter 30 mm				
CL/D	.166	0.133	.133	.166
SPR	.30	.23	.23	.33
SCR	1.00	.875	1.10	1.20
SD/D	.166	.166	.16	.20
Diameter 25 mm				
CL/D	.19	.14	.16	.16
SPR	.36	.24	.28	.40
SCR	1.25	1.14	1.00	1.50
SD/D	.20	.16	.20	.24
Diameter 20 mm				
CL/D	.18	.15	.20	.20
SPR	.38	.30	.30	.50
SCR	1.25	1.16	1.00	1.375
SD/D	.25	.175	.225	.275
Diameter 15 mm				
CL/D	.20	.166	.25	.27
SPR	.50	.46	.31	.50
SCR	1.33	1.20	1.00	1.375
SD/D	.26	.20	.25	.30
Diameter 10 mm				
CL/D			.20	.30
SPR			.50	.55
SCR			1.25	1.20
SD/D			.25	.35

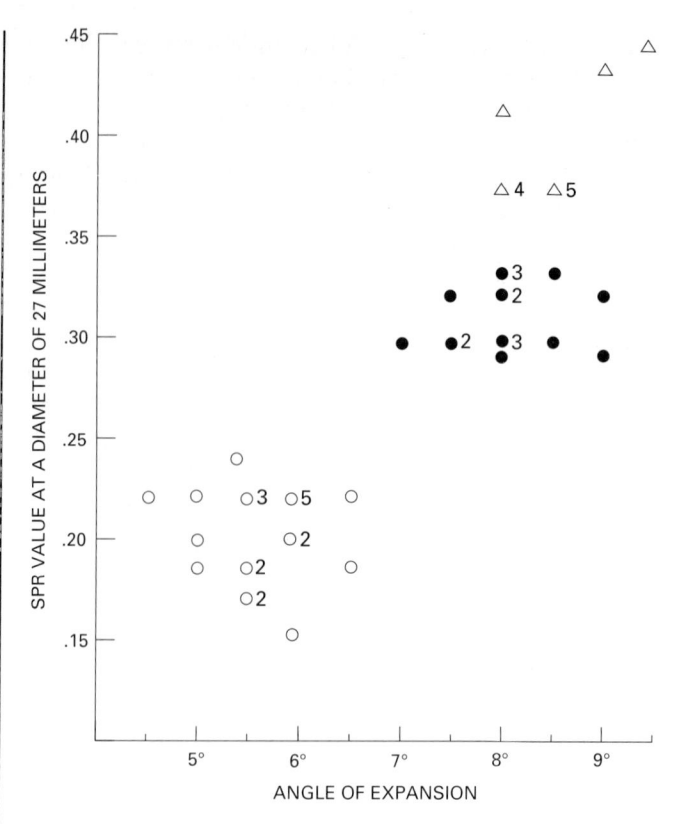

FIGURE 12.—Bivariate plot of segment position ratio (SPR) versus apical angle for species of *Treptoceras* from the Bull Fork Formation in the Cincinnati arch region of Kentucky, Indiana, and Ohio. Measurements of SPR in all specimens taken at a shell diameter of 27 mm. Species are *T. duseri*, $N=17$ (solid circles); *T. fosteri*, $N=22$ (open circles); and *T. cincinnatiensis*, $N=12$ (triangles), where N is number of specimens measured.

southwestern Ohio (MU 420T, pl. 6, fig. 4), indicating maximum lengths of 60 cm and diameters up to 5.5 cm for this species.

TREPTOCERAS FOSTERI (MILLER), 1875

Plate 5, figures 7–9; plate 6, figures 1–5, 7, 8

Orthoceras fosteri Miller, 1875, p. 127.

Orthoceras fosteri Miller. Miller, 1881, p. 319, pl. 8, figs. 7–7a.

Orthoceras duseri Hall and Whitfield. Parks and Fritz, 1922, pl. 4, fig. 7.

Treptoceras duseri (Hall and Whitfield). Frey, 1988, pl. 1, figs. 4, 5.

Diagnosis.—Medium to large, slender, gradually expanding species of *Treptoceras* with narrow siphuncle segments one-seventh to one-sixth the diameter of the shell that change shape with ontogeny from compressed to depressed and position from central to ventral.

Description.—Medium-sized (up to 45 cm in length and having shell diameters up to 4.0 cm), slender,

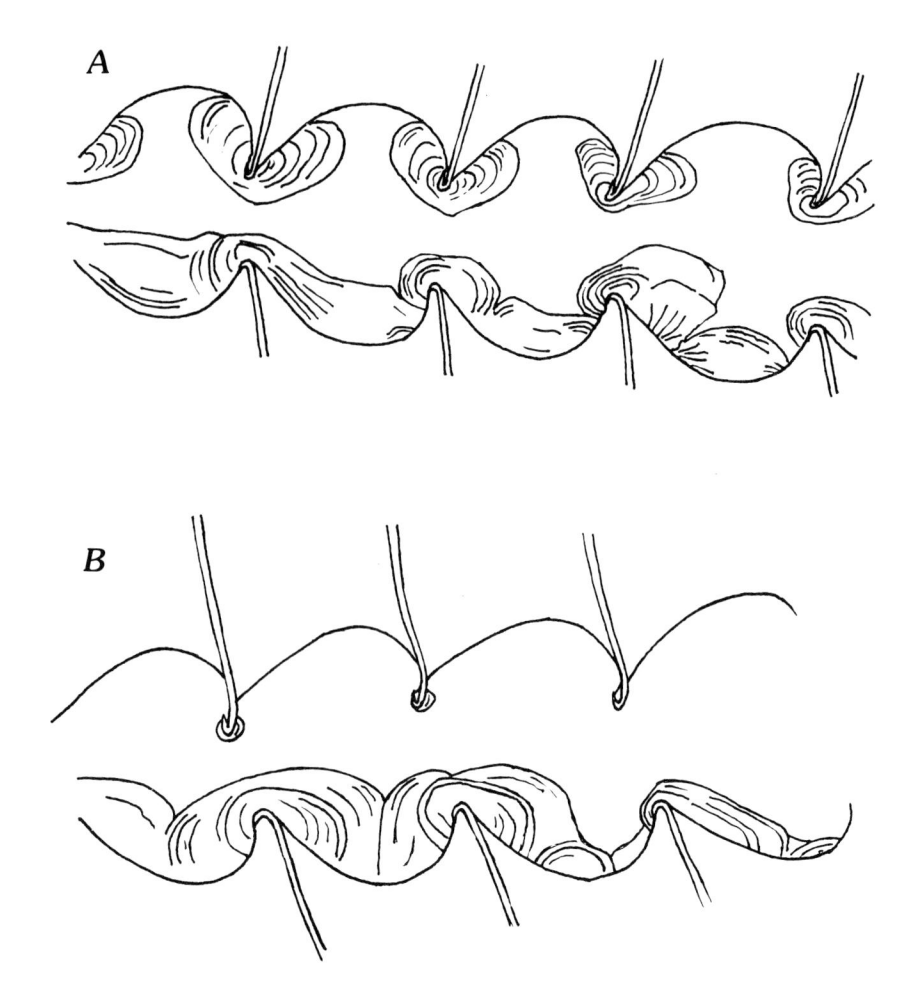

FIGURE 13.—Comparison of siphuncle segment morphology from adapical portions of specimens of *Treptoceras duseri* (A) and *T. fosteri* (B). Both sections represent portions of the phragmocone between 15 and 20 mm in diameter and 20 mm in length. Note the strong development of dorsal annuli in A and their weak development or absence in B. Also note the well-developed dorsal adnation of adoral portions of segments in *T. fosteri* (B) and the weak development of this feature in *T. duseri* (A). A, From MU 417T (pl. 5, figs. 3, 5), locality OH-8. B, From MU 17175A, locality OH-7.

gradually expanding (apical angle 5.5–6.0 degrees) orthoconic longicone, circular in cross section with straight transverse sutures. Cameral lengths short, decreasing with ontogeny from lengths one-fourth the diameter of the shell initially to one-seventh the diameter of the shell in more adoral portions of phragmocones 20–35 mm in diameter. Body chamber one-fourth the total length of the shell; aperture straight, transverse. Periphract dorsomyarian. Shell exterior essentially smooth, marked by fine transverse growth lines and longitudinal lirae. Dorsal shell surface marked by an alternating sequence of fine and heavy longitudinal color bands (pl. 6, fig. 3).

Siphuncle segments central or subcentral in position (SPR=0.45–0.50) at diameters of 10 mm, becoming ventral with ontogeny (SPR < 0.25 at diameters greater than 25 mm). Segments compressed most of the length of the phragmocone (SCR=1.15–1.25 at diameters from 10–25 mm), becoming equiaxial at diameters of 25–30 mm and depressed (SPR < 1.0) at diameters in excess of 30 mm. Segments narrow in diameter, from one-sixth to one-seventh the diameter of the shell. Septal necks cyrtochoanitic most of the length of the siphuncle, dorsal brims strongly recumbent, ventral brims recurved but free. Segments compressed, heart-shaped, having the dorsal portion of the connecting ring adnate to the adoral septum. Endosiphuncular deposits restricted to the adapical half of large specimens, dorsal annuli absent or greatly retarded in their development; ventral parietal deposits

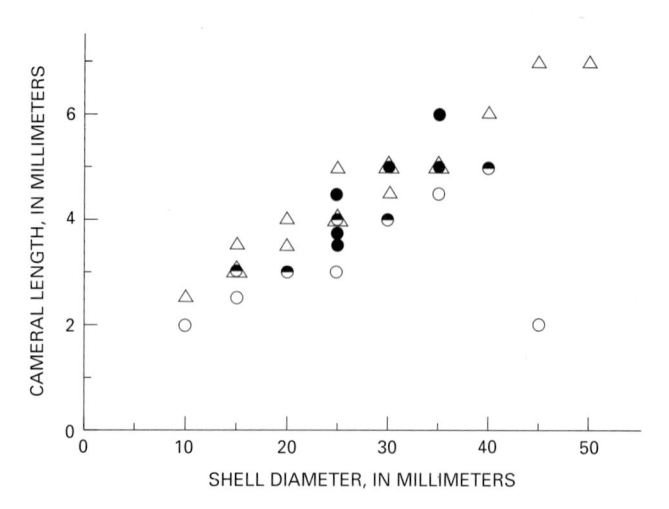

FIGURE 14.—Bivariate plot of cameral length versus shell diameter in species of *Treptoceras* from the Bull Fork Formation in the Cincinnati arch region. Species are *T. duseri* (solid circles), *T. fosteri* (open circles), and *T. cincinnatiensis* (triangles).

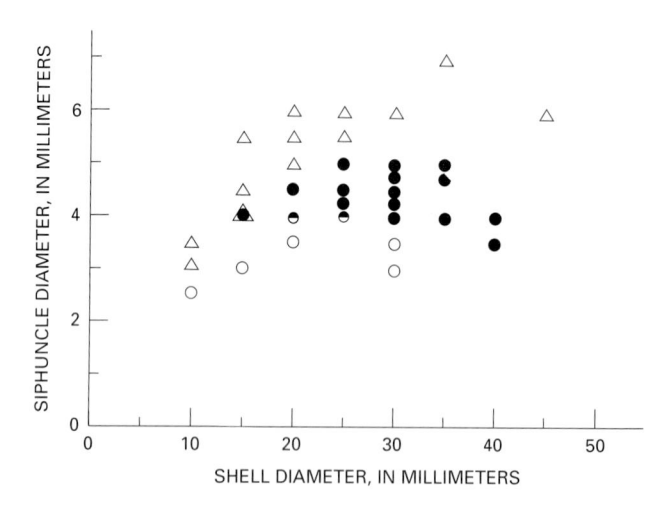

FIGURE 15.—Bivariate plot of siphuncle diameter versus shell diameter in species of *Treptoceras* from the Bull Fork Formation in the Cincinnati arch region. Species are *T. duseri* (solid circles), *T. fosteri* (open circles), and *T. cincinnatiensis* (triangles).

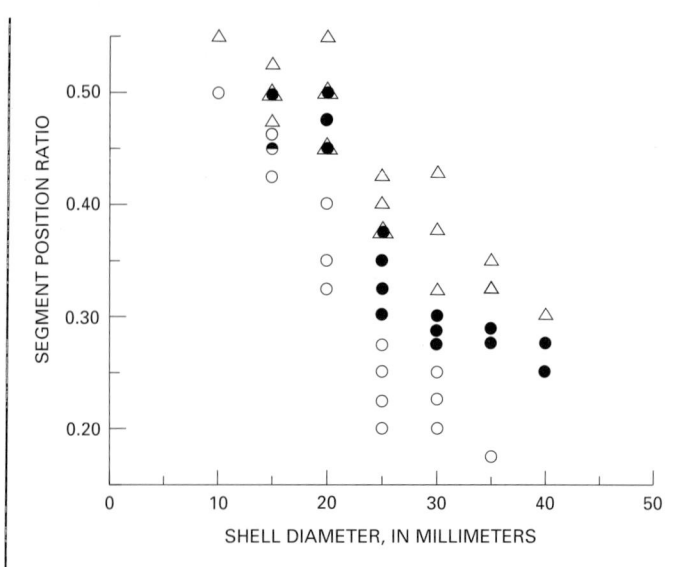

FIGURE 16.—Bivariate plot of segment position ratio versus shell diameter for species of *Treptoceras* from the Bull Fork Formation in the Cincinnati arch region. Species are *T. duseri* (solid circles), *T. fosteri* (open circles), and *T. cincinnatiensis* (triangles).

fusing to form continuous ventral endosiphuncular lining adapically with ontogeny. Adoralmost cameral deposits mural-episeptal, the deposits completely filling camerae in adapical half of large shells.

Stratigraphic and geographic occurrence.—"Waynesville biofacies" within the Bull Fork Formation (Upper Ordovician, Richmondian Stage) in Warren and Clinton Counties, Ohio, Franklin and Dearborn Counties, Ind.

Studied material.—Lectotype, UC 356 (pl. 5, fig. 8), "Richmond Formation, Waynesville beds, Clinton County, Ohio"; MU 418T (pl. 5, figs. 7, 9), *Treptoceras duseri* shale unit exposed along Harpers Run, Warren County, Ohio; hypotypes MU 419T (pl. 6, fig. 1) and MU 276T (pl. 6, fig. 3), *T. duseri* shale unit, Route 42 roadcut, Warren County, Ohio; hypotype UC 28389 (pl. 6, fig. 2), Waynesville Formation, Oldenburg, Franklin County, Ind.; and studied suites of specimens from the *T. duseri* shale unit: MU 17175 (five specimens) from the Caesars Creek dam site, Warren County, Ohio; MU 17176 (four specimens) from Stony Hollow, north of Clarksville, Clinton County, Ohio; MU 17178 (six specimens) from Harpers Run, Warren County, Ohio; and three specimens from the Dyche Collection (Miami University) De 856a–c, locality unknown.

Additional material possibly referable to this species consists of less well preserved internal molds from the "Liberty biofacies" in the Bull Fork Formation. Figured specimens are MU 420T (pl. 6, fig. 4) from Flat Fork Creek, Warren County, Ohio; and MU 422T (pl. 6, fig. 8) from a roadcut at the junction of Routes 380 and 22, Clinton County, Ohio. Additional measured specimens are MU 29858 (five specimens) from Caesars Creek dam site, Warren County, Ohio; MU 29859 (three specimens) from the roadcut at junction of Routes 380 and 22, Clinton County, Ohio; and MU 29860 (five specimens) from the area of Oxford, Butler County, Ohio.

Remarks.—"*Orthoceras*" *fosteri* was described by Miller (1875) from two fragmentary internal molds referable to this species figured in a later paper by the same author (Miller, 1881, pl. 8, figs. 7–7a). Miller distinguished the species on the basis of its large, gradually expanding shell, shallow camerae one-eighth the diameter of the shell, and an excentric siphuncle consisting of ovate segments "representing the appearance of a string of oval beads." He listed the species as occurring in "Clinton County, Ohio, in the upper part of the Cincinnati Group." The drawings of Miller's figured specimens show two fragments, one 65 mm in length has diameters of 33 and 25 mm, and a second specimen 46 mm in length has diameters of 27 and 20 mm. Both specimens have an apical angle of 6.5 degrees.

A number of incomplete specimens, listed as "cotypes" of *Orthoceras fosteri*, "Richmond Formation, Clinton County, Ohio," were found in the collections of the University of Cincinnati (UC 356). The distinctive preservation of these specimens indicates they were collected from the *T. duseri* shale unit. Among these specimens are two continuous sections of a phragmocone, one 65 mm in length has diameters of 28 and 20 mm, the second 44 mm in length has diameters of 20 mm and 13 mm. Uncompressed portions of this specimen have an apical angle of 6 degrees. This specimen may represent one or both of Miller's figured specimens, although there are minor differences in the respective shell diameters, assuming the illustrations in Miller (1881) are accurate representations of the specimens. This specimen is figured here (pl. 5, fig. 8) and is designated the lectotype of *T. fosteri* (Miller). A second, more complete specimen from the *T. duseri* shale at Harpers Run, Warren County, Ohio (MU 418T, pl. 5, figs. 7, 9), illustrates the morphology of the adapical portion of the phragmocone.

Miller (1875) remarked that *O. fosteri* most nearly resembled *O. byrnesi* Miller but differed from that species in having a larger shell with shorter camerae and siphuncle segments proportionally smaller in diameter. Miller (1875) listed *O. byrnesi* as occurring in the "hills back of Cincinnati; range unknown." No material similar to the sectioned specimen of *O. byrnesi* figured by Miller was found in the suite of specimens listed as "co-types" of *O. byrnesi* in the University of Cincinnati collections. Miller did not designate a type specimen for this species.

Studies of nearly complete specimens of Cincinnatian *Treptoceras* indicate that the cameral length and the diameter and shape of siphuncle segments vary with ontogeny. Discriminating species of *Treptoceras* requires nearly complete sectioned phragmocones. Miller's descriptions and measurements of nearly complete specimens of *T. fosteri* from the *T. duseri* shale unit (pl. 5, fig. 7) suggest that Miller's distinctions between "*Orthoceras*" *fosteri* and "*O.*" *byrnesi* might be the result of these taxa being described from different ontogenetic stages of the same species. The specimen of *O. byrnesi* figured by Miller (1875, fig. 13) represents a section of phragmocone less than 20 mm in diameter. The specimens of *O. fosteri* described by Miller in the same paper are adoral portions of larger specimens, in excess of 30 mm in diameter. The only significant difference observed in these taxa, based on Miller's figures, is the apical angle: 5 degrees in *O. byrnesi* and 6–6.5 degrees in *O. fosteri*. Specimens assigned to *T. fosteri* from the *T. duseri* shale exhibit apical angles from 5.5 to 6.0 degrees. A specimen from the Corryville Member of the Grant Lake Formation in Clermont County, Ohio (MU 11035), has an apical angle of 5 degrees and closely resembles Miller's figured specimen of *O. byrnesi* (pl. 6, fig. 6). These two species may prove to be synonymous upon further study of larger suites of complete material. This writer has elected to retain these taxa as separate species, pending a more thorough search for Miller's figured specimen of *Orthoceras byrnesi*.

Comparisons of *T. fosteri* with the associated *T. duseri* are discussed above under the remarks for the latter species. Although Miller (1881) considered these species to be synonyms, careful study of well-preserved material from the *T. duseri* shale indicates to this writer's satisfaction that they are distinct species. *Treptoceras fosteri* is a more gradually expanding species with siphuncle segments that are more ventral in their position at comparable shell diameters (table 7 and fig. 12); the segments lack dorsal annuli in adapical portions of the phragmocone. At most localities exposing the claystones of the *T. duseri* shale in southwestern Ohio, *T. fosteri* is the most abundant species of *Treptoceras* present, followed by *T. duseri* and lesser numbers of *T. cincinnatiensis*. The latter species is distinct from *T. fosteri* in its more robust, rapidly expanding shell, broader diameter siphuncle segments, and more central location of these segments at comparable diameters.

Rather poorly preserved internal molds of a similar, large (up to 60 cm in length), gradually expanding species of *Treptoceras* with siphuncle segments identical to those in *T. fosteri* occur in the overlying "Liberty biofacies" of the Bull Fork Formation in southwestern Ohio and adjacent portions of southeastern Indiana (pl. 6, figs. 2, 4, 8). These specimens differ from specimens of *T. fosteri* from the *T. duseri* shale in having more gradually expanding shells and, for the majority of these upper Bull Fork specimens, in having apical angles of 5–5.5 degrees compared with typical values of 5.5–6.0 degrees for the "Waynesville" forms. These "Liberty" specimens seem to be a more slender variant of *T. fosteri* and are very similar to Miller's species *T. byrnesi*. This variant is the most abundant nautiloid in the upper half of the Bull Fork Formation in southeastern Indiana and southwestern Ohio.

TREPTOCERAS CINCINNATIENSIS (MILLER), 1875

Plate 7, figures 1–8; plate 8, figures 1–3

Orthoceras cincinnatiensis Miller, 1875, p. 127.

Orthoceras cincinnatiensis Miller. Miller, 1881, p. 319, pl. 8, figs. 5–5a.

Treptoceras duseri (Hall and Whitfield). Frey, 1988, pl. 3, fig. 1.

Diagnosis.—Large, rapidly expanding species of *Treptoceras* with broad siphuncle segments one-fourth to one-fifth the diameter of the shell that are compressed, heart-shaped most of the length of the siphuncle, and that change position with ontogeny from supracentral to subcentral.

Description.—Medium-sized (up to 55 cm in length, having shell diameters up to 7.5 cm), rapidly expanding (apical angle 8.5–9.5 degrees) orthoconic longicone; circular in cross section with straight, transverse sutures. Cameral lengths one-fourth the diameter of the shell at diameters less than 10 mm, decreasing with ontogeny to one-seventh the diameter of the shell at shell diameters in excess of 40 mm. Body chamber one-fourth the total length of the shell; aperture straight, transverse, lacking sinus. Periphract dorsomyarian. Shell exterior essentially smooth, marked with fine transverse growth lines.

Siphuncle supracentral (SPR > 0.50) in position at diameters less than 10 mm, becoming subcentral in position with ontogeny (SPR=0.30 at a diameter of 40 mm). Segments slightly compressed (SCR=1.25) at diameters less than 10mm, becoming more compressed with ontogeny (SCR=1.33–1.50 at diameters of 10–35 mm), and slightly compressed, heart-shaped, in adoral portions of large specimens (SCR=1.20 at diameters in excess of 35 mm). Segments broad for the genus, averaging between one-fourth and one-fifth the diameter of the shell. Endosiphuncular deposits extensively developed adapically, consisting of dorsal annuli and ventral parietal deposits that fuse into a continuous lining with ontogeny. Adoralmost cameral deposits consisting of distinct episeptal and hyposeptal elements, the deposits completely filling camerae in adapical half of large shells.

Stratigraphic and geographic occurrence.—Basal Maysvillian part of the Fairview Formation in northern Kentucky; "Waynesville" and "Liberty" biofacies within the Bull Fork Formation (Upper Ordovician, Richmondian Stage) in Warren, Clinton, Butler, and Preble Counties, Ohio, and Franklin County, Ind.; Rowland Member of the Drakes Formation (Upper Ordovician, Richmondian Stage) in Nelson County, Ky.

Studied material.—Holotype, Harvard University, Dyer Collection, MCZ 3404 (pl. 7, figs. 2, 3, 5), Upper Ordovician, Cincinnati, Ohio, area; MU 423T (pl. 7, fig. 1), *T. duseri* shale, Bull Fork Formation, Stony Hollow, Clinton County, Ohio; MU 424T, MU 425T (pl. 7, fig. 8), and MU 426T (pl. 7, figs. 4, 6, 7), *T. duseri* shale, Bull Fork Formation, Harpers Run, Warren County, Ohio; MU 427T (pl. 8, fig. 1), Bull Fork Formation, Brookville Lake dam site, Franklin County, Ind.; MU 443T (pl. 8, fig. 3) and MU 445T (pl. 8, fig. 2), Bull Fork Formation, Oxford, Butler County, Ohio; USNM 468691a and 468691b, Rowland Member of the Drakes Formation, Nelsonville, Ky.; measured suites of specimens from the *T. duseri* shale unit: MU 17174 (two specimens) from Stony Run, Warren County, Ohio; MU 17179 (two specimens) from Stony Hollow at Clarksville, Clinton County, Ohio; MU 17184 (four specimens) from Caesars Creek dam site, Warren County, Ohio; MU 17185 (one specimen) from the U.S. Route 42 roadcut, Warren County, Ohio; and MU 17186 (four specimens) from Harpers Run, Warren County, Ohio. Additional measured specimens from the "Liberty biofacies" in the Bull Fork Formation are MU 29862 (one specimen) from the Dayton, Ohio, area; MU 29863 (one specimen) from the Caesars Creek dam site, Warren County, Ohio; and MU 29864 (two specimens) from Oxford, Butler County, Ohio.

Remarks.—This species was described by Miller (1875) and figured by him in a later paper (Miller, 1881, pl. 5–5a). He listed this species as occurring in strata "exposed in the hills back of Cincinnati; range unknown." Miller (1875, p. 128) referred to a specimen of this species in the C.B. Dyer collection that was 6 in. in length, and having a diameter at the small end of 3/10 in. and at the large end of 1.15 in., and having 42 camerae. This specimen (pl. 7, figs. 2, 3, 5) is preserved in the Dyer Collection at the Museum of Comparative Zoology at Harvard (MCZ 3404), is labeled as the holotype of *Orthoceras cincinnatiensis*, and is so recognized here. There is no precise stratigraphic information associated with the holotype. The distinctive preservation of this specimen indicates it is probably from the *T. duseri* shale unit in either Warren or Clinton Counties, Ohio. It should be noted, however, that similar, well-preserved calcite-replaced internal molds occur in the basal Fairview Formation in northern Kentucky and adjacent portions of the Cincinnati, Ohio, area. These strata also conceivably could be the source of the holotype.

The species is an uncommon form in the basal Fairview Formation in northern Kentucky; the incomplete specimens collected by the writer were from exposures along Interstate Route 275E in Kenton County, Ky. *Treptoceras cincinnatiensis* is most abundant in the *T. duseri* shale unit within the "Waynesville biofacies" in the lower part of the Bull Fork Formation in southwestern Ohio (pl. 7). In this shale, it is the third most abundant nautiloid species present, behind the previously described species of *Treptoceras*. Morphological comparisons with the associated *T. duseri* (Hall and Whitfield) and *T. fosteri* (Miller) are provided in the remarks sections for these species. *Treptoceras cincinnatiensis* occurs sporadically throughout the overlying portions of the Bull Fork Formation in Ohio and southeastern Indiana (pl. 8, figs. 1–3). Fragmentary silicified material assignable to the species occurs in the equivalent Rowland Member of the Drakes Formation in the vicinity of Bardstown, in north-central Kentucky.

Outside of the northern Cincinnati arch region, a similar species occurs in the Georgian Bay Formation (Upper Ordovician, Richmondian) in southern Ontario. Parks and Fritz identified this species as *Actinoceras* cf. *clouei* (Barrande) and illustrated a specimen from this strata at Toronto, Ontario (Parks and Fritz, 1922, pl. 4, fig. 5). Copper (1978, pl. 5, fig. 5) illustrated a similar form from equivalent strata exposed on Manitoulin Island that he identified as *Ormoceras piso* Billings. This species appears to be very similar in all aspects to the specimens of *Treptoceras cincinnatiensis* described here.

TREPTOCERAS TRANSVERSUM (MILLER), 1875

Plate 8, figures 4–9

Orthoceras transversa Miller, 1875, p. 129, fig. 15.
Orthoceras hindei James, 1878, p. 6, pl. 1.
Orthoceras hindei James. James, 1886, p. 240–241, pl. 4, figs. 4a–d.
Orthoceras cf. *O. transversum* Miller. Ruedemann, 1926, p. 87, pl. 12, fig. 5.

Diagnosis.—Small, slender species of *Treptoceras* distinguished by its external ornament of coarse, raised transverse lirae and its small-diameter, subcentral, slightly compressed to equiaxial siphuncle segments.

Description.—Small (up to 30 cm in length and up to 2.5 cm in diameter), slender, rather gradually expanding (apical angle 6–6.5 degrees), orthoconic longicone; circular in section with straight, transverse sutures. Cameral lengths ranging from one-fourth to one-sixth the diameter of the shell, decreasing with ontogeny. Body chamber long, approximately one-fourth the total length of the shell. External ornament consisting of coarse, closely spaced, raised transverse lirae uniformly developed the length of the shell.

Siphuncle subcentral the known length of the shell, ranging from nearly central adapically (SPR=0.44 at a diameter of 8 mm) to nearly midventral positions adorally (SPR=0.275 at a diameter of 20 mm). Siphuncle segments slightly compressed adapically (SCR=1.16 at diameters less than 15 mm) and decompressing with ontogeny, becoming equiaxial (SCR=1.0) adorally at diameters greater than 20 mm. Segments narrow, ranging from one-fifth to one-sixth the diameter of the shell. Septal necks short, recurved, cyrtochoanitic. Endosiphuncular deposits consisting of ventral parietal deposits developed at diameters less than 15 mm. Dorsal deposits unknown. Adoralmost cameral deposits of the mural type.

Stratigraphic and geographic occurrence.—The Edenian portions of the Kope and basal Fairview Formations in northern Kentucky and adjacent portions of the Cincinnati, Ohio, region and southeastern Indiana; Whetstone Gulf Shale (Upper Ordovician, Maysvillian Stage) in northwestern New York.

Studied materials.—University of Cincinnati "cotypes" are UC 1328A (pl. 8, fig. 7), UC 1328B (pl. 8, fig. 6), UC 1328C (pl. 8, figs. 4, 8), all listed as coming from "Eden Formation at Cincinnati, Ohio"; USNM 468692 (pl. 8, fig. 5) and 468693 (pl. 8, fig. 9), from the Kope Formation exposed just east of Fort Thomas, Campbell County, Ky.; UC 48259 from the "Eden Shale" at Newport, Campbell County, Ky.; plus seven unfigured specimens listed as "cotypes" (UC 1328), "Eden Shale" at Cincinnati, Ohio.

Remarks.—Miller (1875) distinguished this species on the basis of its medium-sized, gradually expanding shell and an "outer shell with transverse lines." Camerae were described as shallow, cameral lengths one-fourth to one-fifth the diameter of the shell and a siphuncle described as excentric in position. Miller indicated that the species did not extend more than 300 ft above the low-water mark of the Ohio River in the Cincinnati area, being collected from an excavation for Columbia Avenue at 150 ft above the low-water mark, and at Eden Park, 200 ft above river level. He indicated that the species was "not common by any means" in these strata. "Co-types" of *T. transversum* in the collections of the University of Cincinnati (UC 1328) consist of a number of specimens from the "Eden Formation at Cincinnati, Ohio." This collection does not contain the partially sectioned specimen figured by Miller (1875, fig. 15), although all of these specimens are obviously examples of Miller's species. The whereabouts of Miller's figured specimen remains unknown. As such, UC 1328C (pl. 8, figs. 4, 8) is here designated the lectotype for *T. transversum* (Miller). The other specimens labeled "co-types" in the University of Cincinnati collections are considered to be paratypes.

A nearly complete, shelled specimen of *T. transversum* (USNM 468692, pl. 8, fig. 5) and a specimen partially covered holoperipherally with encrusting bryozoans (USNM 468693, pl. 8, fig. 9) were collected by the writer from a thin packstone bed within the claystone and thin limestone of the Kope Formation (Upper Ordovician, Edenian) exposed at the junction of River Road and State Route 8, east of Ft. Thomas and south of the I–275E bridge, in Campbell County, Ky. Less complete specimens of *T. transversum* were collected from a claystone bed in the Kope Formation exposed 18 m above the road level in a roadcut on the north side of Orphanage Road, just west of the junction of State Route 17 and I–275E, Kenton County, Ky. Here the species was found associated with a diverse molluskan fauna that consisted of the nautiloid *Trocholites faberi* (pl. 19, figs. 6–10), several species of gastropods, the monoplacophorans *Cyrtolites* and *Sinuites*, pelecypods, and rostroconchs, as well as the brachiopods *Onniella* and *Zygospira*, and three-dimensional specimens of the graptolite *Geniculograptus typicalis*.

Fragmentary specimens of *T. transversum* were also collected from the overlying claystone and thin limestone of the basal Fairview Formation (Upper Ordovician, Edenian and Maysvillian) exposed at a roadcut along the south side of I–74, 2 mi west of Harrison, Dearborn County, Ind. In the Cincinnati arch region, *T. transversum* appears to be restricted to the Kope Formation and the overlying basal portions of the Fairview Formation exposed in adjacent portions of northern Kentucky, southwestern Ohio, and southeastern Indiana.

A species collected from the same strata at Cincinnati, Ohio, and described as *Orthoceras hindei* by James (1878) appears to be a preservational phase of *T. transversum*. These partially replaced, calcite internal molds of a small longicone have the same shell morphology as better preserved specimens of *T. transversum*. "*Orthoceras*" *hindei* James is here placed in synonymy with *T. transversum* (Miller).

Treptoceras transversum is easily distinguished from other Cincinnatian species of *Treptoceras* on the basis of its typically smaller, slender shell, marked by its unique transverse ornament, and its small-diameter, subcentral siphuncle segments. The species is distinct from similar small, slender species of the pseudoceratid *Isorthoceras* in its globular, compressed siphuncle segments compared to the depressed, barrel-shaped segments characteristic of the latter genus. The distinctive transverse external ornament in specimens of *T. transversum* also readily distinguishes it from the longitudinal lirae characteristic of *I. albersi* (Miller and Faber) and the cancellate ornament of *I. rogersensis* (Foerste) (see pl. 10), both known from the Clays Ferry Formation in northern Kentucky.

GENUS *MONOMUCHITES* WILSON

Type species.—*Monomuchites costalis* Wilson, 1961, p. 23; by original designation; "Leray beds," Rocklandian Stage, Middle Ordovician, Ontario.

Description.—Small to medium-sized, moderately to rapidly expanding orthoconic longicones, often slightly curved adapically; circular in cross section with straight, transverse sutures. Cameral lengths from one-fourth to one-seventh the diameter of the shell, decreasing with ontogeny. Body chamber long; aperture unknown. Shell exterior marked by transverse annuli, one annulation per cameral chamber. Annuli variably developed, from indistinct rounded to sharply defined elevated; separated by shallow, concave interannular spaces. Annuli and interannular spaces marked by transverse growth lines.

Siphuncle moving from supracentral to subcentral in position with ontogeny. Segments slightly expanded within camerae, depressed to compressed in shape; narrow in diameter, from one-fourth to one-sixth the diameter of the shell. Septal necks short, cyrtochoanitic, recurved with free brims. Endosiphuncular deposits poorly known, consisting of ventral parietal deposits developed adapically. Adoralmost cameral deposits consisting of trilobate, botryoidal, ventral-lateral, mura-episeptal deposits and thin, dorsal episeptal deposits.

Stratigraphic and geographic occurrence.—Middle Ordovician (Blackriveran and Rocklandian), Eastern North America (Laurentia).

Remarks.—Wilson (1961) erected this genus for Middle Ordovician annulated longicones with transverse growth lines that were previously referred to as species of *Cycloceras* M'Coy by Foerste (1932). True *Cycloceras* is a poorly known Carboniferous form based on an internal mold of a body chamber in which even the position of the siphuncle is in doubt. Sweet (1964) suggested that until better specimens of *C. annularis* (Fleming), the type species, are discovered, the name *Cycloceras* should be applied only to this species. It is probable, at any rate, that these Ordovician species are distinct from these younger Carboniferous annulated longicones.

Kobayashi (1927) erected the genus *Tofangoceras* for similar annulated longicones with narrow, expanded siphuncle segments from Middle Ordovician (Blackriveran) carbonates in Manchuria and Korea. He did not figure the exteriors of either of the two species that he referred to this genus, and his illustrations of the internal features of the genus are of a naturally weathered section that shows the expanded nature of the siphuncle segments but little else. *Tofangoceras*, as based on the type species, *T. pauciannulatum*, appears to have a more gradually expanding shell than either of the described species of *Monomuchites*, and its external ornament is stated to consist of one annulation for every two camerae, annuli being more broadly spaced than those in species of *Monomuchites*.

Monomuchites is distinguished from the associated annulated orthocerid *Gorbyoceras* Shimizu and Obata in lacking the longitudinal lirae that are characteristically developed in representatives of this genus. Similarly, *Monomuchites* can be distinguished from the annulated genus *Anaspyroceras* Shimizu and Obata on the basis of its more rapidly expanding shell form and in its lack of longitudinal lirae, although this longitudinal ornament is not well documented for the type species of *Anaspyroceras*, *A. anellus* (Conrad). Unfortunately, the nature of the siphuncle segments is poorly documented for all of these annulated longicones. Expanded globular siphuncle segments are developed in several species of *Monomuchites* and in *Gorbyoceras*. Siphuncle segments in Late Ordovician species assigned to *Anaspyroceras* by Flower (1946) are depressed, subtubular in shape.

The genus *Monomuchites* is apparently restricted to Blackriveran to Rocklandian carbonate facies throughout Eastern North America, species being known from the Watertown Limestone in New York, the "Leray-Rockland beds" exposed in the Ottawa region of Ontario and Quebec, the Tyrone Limestone in central Kentucky, and the Platteville Group in Illinois or Platteville Formation in Minnesota and Wisconsin. *Monomuchites* was a victim of the widespread extinction of nautiloid taxa that occurred at the end of the Rocklandian, and no post-Rocklandian species is known to the writer. Three species of *Monomuchites* have been identified in the USGS 6034-CO collection from the upper half of the Tyrone Formation in central Kentucky.

More complete and better preserved material is necessary to clarify the relevant morphological features of *Monomuchites* and these other Ordovician annulated longicones. Flower (1962) placed several of these genera in the Proteoceratidae, describing *Monomuchites* as having endosiphuncular deposits of the proteoceratid type and showing siphuncle segment decompression with ontogeny. He did not, however, indicate the specimens upon which these observations were based, nor has such material been illustrated. Until the details of the siphuncle and its deposits can be documented for all of these Ordovician annulated longicones, they will remain of uncertain affinities. The genus is here retained in the Proteoceratidae, on the basis of the rapidly expanding shell form and supracentral position of adapical siphuncle segments, plus proteoceratidlike endosiphuncular deposits in one of the species of *Monomuchites* described below.

MONOMUCHITES CF. *M. COSTALIS* WILSON, 1961

Plate 9, figures 16–19

Monomuchites costalis Wilson, 1961, p. 24–25, pl. 5, figs. 5–5a.

Diagnosis.—Moderately expanding species of *Monomuchites* distinguished by its narrow-rimmed, transverse annuli separated by shallow, concave interannular spaces.

Description.—Small (lengths of up to 14 cm and diameters up to 2.0 cm), moderately expanding (apical angle 7.5–8 degrees) orthoconic longicones; circular in section with straight, transverse sutures. Cameral chambers short, decreasing with ontogeny from one-fourth to one-fifth the diameter of the shell. Body chamber unknown. Shell exterior marked by low, narrow-rimmed transverse annulations, one per cameral chamber. Annuli separated by shallow concave interannular spaces typically twice as broad as the annuli. Interannular spaces bisected by sutures.

Siphuncle supracentral in position in known portions of the shell at SPR values ranging from 0.60 at a diameter of 10 mm to 0.53 at a diameter of 19 mm. Segment shape poorly known, globular, compressed (SCR=1.33) at shell diameters between 10 and 15 mm. Segments narrow, one-fourth the diameter of the shell at these diameters. Septal necks short, recurved, cyrtochoanitic. Endosiphuncular deposits unknown. Cameral deposits well-developed adapically, consisting of trilobate, botryoidal, ventrolateral, mural-episeptal deposits and thin, dorsal episeptal deposits at diameters of 10–20 mm. Adapically, there is development of a large, subcircular, central "boss" that impinges on the ventral surface of the siphuncle segment.

Stratigraphic and geographic occurrence.—Upper part of the Rocklandian portion of the Tyrone Limestone, Inner Blue Grass region of central Kentucky.

Studied material.—USNM 158667 (pl. 9, figs. 18, 19), USNM 158672 (pl. 9, figs. 16, 17), USNM 468705, and USNM 468706, all from the USGS 6034-CO collection, Jessamine County, Ky.

Remarks.—This annulated longicone is a rare element in the diverse assemblage of silicified nautiloids obtained from the USGS 6034-CO collection in the Tyrone Limestone, represented by only four specimens. These consist of silicified fragments of phragmocones less than 2.0 cm in diameter. The narrow-rimmed transverse annuli separated by shallow, concave interannular spaces and the moderate rate of shell expansion suggest assignment of these specimens to the type species of the genus, *M. costalis* Wilson, from the equivalent "Leray beds" in the Ottawa, Ontario region. The siphuncle segments preserved in Tyrone specimen USNM 468706 are somewhat more expanded than those illustrated for GSC 13844, the holotype of this species (Wilson, 1961, pl. 5, fig. 5). Further comparisons with the type material are difficult, as the second specimen (GSC 13845) figured by Wilson (1961, pl. 5, fig. 6) consists of a larger diameter, more adoral section of the shell compared with the Kentucky specimens. This writer's studies of more complete specimens of *Monomuchites* indicate that the external ornament of the shell changes with ontogeny, the annuli becoming more closely spaced and the interannular spaces decreasing in width. Pending comparisons of a larger suite of the Tyrone material with more complete specimens of the Canadian species, these Tyrone specimens are only tentatively assigned to *M. costalis*.

In terms of rate of shell expansion, cameral lengths, and siphuncle position and in the number of annuli per length of shell, these Tyrone specimens also appear to be similar to *M. decrescens* (Billings), another Canadian species from the Ottawa region. The criteria used by Wilson (1961) to distinguish this species from the associated *M. costalis* appear to result from comparisons of different ontogenetic stages of the shell. The rate of shell expansion is similar for both species, the only distinction being the number of annuli per length of shell: eight annuli in a length equal to an adoral diameter of 30 mm in *M. costalis* versus six annuli in a length equal to an adoral diameter of 25 mm in *M. decrescens* (see table 8). Comparisons using larger collections of more complete, better preserved material are necessary to determine the distinctiveness of these two Canadian species.

Monomuchites cf. *M. costalis* is distinguished from the other two species of *Monomuchites* described here from the 6034-CO collection on the basis of its narrow-rimmed annulations separated by shallow, concave interannular spaces (pl. 9, figs. 18, 19). *Monomuchites obliquum* n. sp. is a more gradually expanding (apical angle 6–6.5 degrees), slightly curved form with less well defined, rounded, and oblique annulations (pl. 9, figs. 5–8). *Monomuchites annularis* n. sp. is more similar to *M.* cf. *M. costalis* in general shell form but differs in its broader, more strongly defined annuli and interannular spaces (pl. 9, figs. 1, 11). A comparison of the distinctive external ornament used to distinguish these Tyrone species of *Monomuchites* is presented in figure 17.

MONOMUCHITES ANNULARIS NEW SPECIES

Plate 9, figures 1–3, 11–15

Cycloceras cf. *C. decrescens* (Billings). Foerste, 1932, p. 86, pl. 12, figs. 5A–B, 7 (Foerste, 1933).

Diagnosis.—Moderately expanding, slightly curved species of *Monomuchites* distinguished by its well-delineated, elevated, broad transverse annuli separated by broad, flat interannular spaces.

Description.—Medium-sized (up to 40 cm in length and shell diameters up to 5.0 cm), moderately expanding (apical angle 7–8 degrees) orthoconic longicones, slightly endogastrically curved adapically; circular in cross section with straight, transverse sutures. Cameral chambers short, decreasing with ontogeny from one-fourth the diameter of the shell at diameters of 10 mm or less to one-seventh the diameter of the shell at diameters in excess of 30 mm. Body chamber poorly known. External ornament of strongly defined, broad, elevated transverse annuli, one per cameral chamber. Interannular spaces flat, broader than annuli except in adoral portions of large specimens; marked by transverse growth lines.

Siphuncle supracentral in position in adapical portions of shells 20 mm or less in diameter, moving to a more central position at a diameter of 30 mm. Segments at these diameters expanded, globular, equiaxial to compressed; diameters typically one-fourth the diameter of the shell. Shape of segments unknown in adoral portions of the shell greater than 20 mm in diameter. Septal necks poorly known; cyrtochoanitic. Endosiphuncular deposits unknown. Cameral deposits restricted to adapical portions of the shell 20 mm or less in diameter, consisting of massive, trilobate, botryoidal ventrolateral mural-episeptal deposits and dorsal episeptal deposits.

TABLE 8.—Comparison of the morphological features of species of *Monomuchites* from the Middle Ordovician (Rocklandian) of Eastern North America

[Data from specimens figured by Wilson (1961) and suites of measured specimens from the 6034-CO collection from the Tyrone Limestone in central Kentucky. A/D, ratio of number of annuli per length to diameter; A/AIS, ratio of width of annuli to width of interannular space; CL/D, ratio of cameral length to shell diameter; SPR, segment position ration; SCR, segment compression ratio; mm, millimeter]

Feature	*M. costalis*	*M. desrescens*	*M.* cf. *M. costalis*	*M. obliquum*	*M. annularis*
Apical angle, degrees	8	8.5–9	7.5–8	6–6.5	7–8
Diameter 35 mm					
A/D	–	–	–	–	7
A/IAS	–	–	–	–	3/2.5
CL/D	–	–	–	–	.14
SPR	–	–	–	–	–
SCR	–	–	–	–	–
Diameter 30 mm					
A/D	8	–	–	–	6.5–7
A/IAS	2/2	–	–	–	2.5/3
CL/D	.13	–	–	–	.15
SPR	–	–	–	–	–
SCR	–	–	–	–	–
Diameter 25 mm					
A/D	6.5	6	–	–	5–6
A/IAS	2/2.5	2/3	–	–	2/3
CL/D	.16	.16	–	–	.16
SPR	–	–	–	–	–
SCR	–	–	–	–	–
Diameter 20 mm					
A/D	–	5	5–6	5–5.5	5–6
A/IAS	1.5/3	–	1.5/2.5	3/2	2/3
CL/D	.20	.175	.175	.175	.175
SPR	–	.55	.53–.55	.48	.55–.60
SCR	.86	.86	–	–	–
Diameter 15 mm					
A/D	–	–	5	–	5
A/IAS	–	–	1/2.5	–	1.5/2.5
CL/D	.20	–	.20	.20	.20
SPR	–	–	.60–.62	.53	.60
SCR	1.0	–	1.33	1.16	1.33
Diameter 10 mm					
A/D	–	–	–	–	4–4.5
A/IAS	–	–	–	–	1/2
CL/D	–	–	.30	–	.25–.30
SPR	–	–	–	–	.66
SCR	–	–	1.33	–	1.0

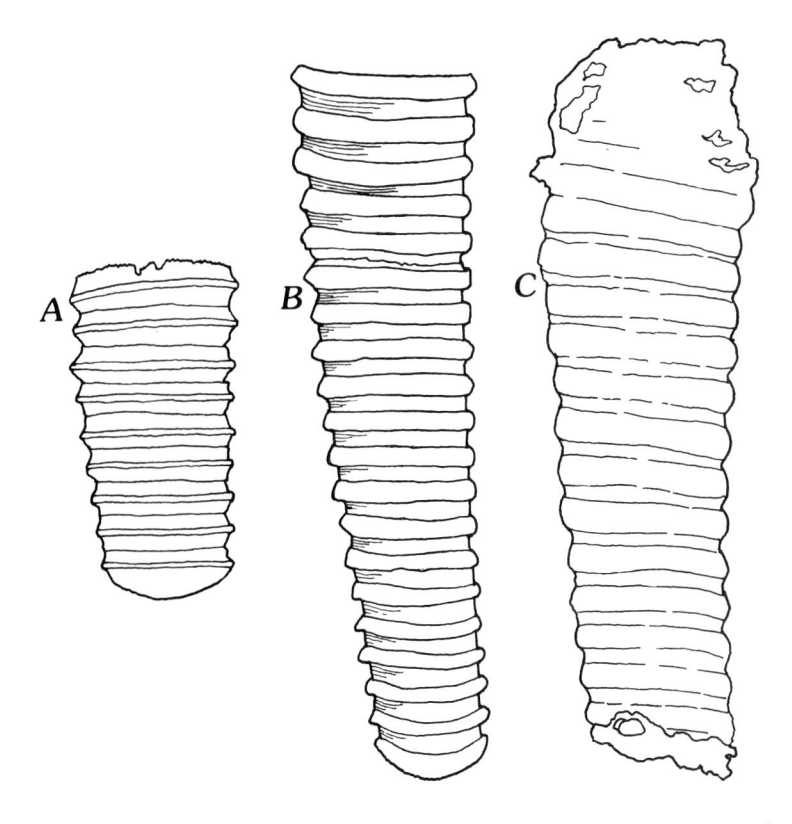

FIGURE 17.—Comparison of the external ornament characteristic of species of *Monomuchites* from the USGS 6034-CO collection in the Tyrone Limestone. *A, Monomuchites* cf. *M. costalis* (USNM 158667, pl. 9, fig. 18); *B, M. annularis* (USNM 158674, pl. 9, fig. 11); *C, M. obliquum* (USNM 158653, pl. 9, fig. 5).

Stratigraphic and geographic occurrence.—Upper part of the Rocklandian portion of the Tyrone Limestone, Inner Blue Grass region of central Kentucky; "Black River Formation" (Middle Ordovician), St. Joseph Island, Lake Huron, Ontario; Platteville Formation (Middle Ordovician, Rocklandian Stage), Minnesota and Wisconsin, or Platteville Group in Illinois.

Studied material.—Holotype USNM 158674 (pl. 9, fig. 11); paratypes USNM 158630 (pl. 9, fig. 1), 158671 (pl. 9, fig. 2), 158675, and 158676 (pl. 9, fig. 15); plus studied suite USNM 468716, all from the USGS 6034-CO collection in the Tyrone Limestone, Jessamine County, Ky.; USNM 162068 (pl. 9, fig. 14), "Platteville Formation," Dixon, Ill.; and USNM 25275 (pl. 9, fig. 13), Mineral Point, Wis.

Remarks.—*Monomuchites annularis* n. sp. is the fourth most abundant longiconic nautiloid in the 6034-CO collection from the Tyrone Limestone. It is represented in this collection by more than 25 silicified specimens 40 mm or more in length as well as by numerous smaller fragments. Specimens consist primarily of adapical portions of phragmocones less than 20 mm in diameter. These portions of the shell are less fragile than more adoral portions of the phragmocone due to the nearly complete filling of cameral chambers by cameral deposits, forming a nearly solid shell mass adapically. Unfortunately, septa and connecting rings are poorly preserved, their outlines preserved as impressions within the surrounding cameral deposits (pl. 9, fig. 15). More adoral portions of the siphuncle are, as a consequence, largely unknown.

Monomuchites annularis is the most abundant of the three species of *Monomuchites* in the 6034-CO collection. It is distinguished from the similar *M.* cf. *M. costalis* through the development of more strongly delineated, broader annuli and interannular spaces (fig. 17). *Monomuchites annularis* is less similar to *M. obliquum* n. sp., differing from this species in the development of a more rapidly expanding shell with more strongly developed transverse annuli, compared with the low, rounded, oblique annuli in *M. obliquum*.

Foerste (1932, pl. 12, figs. 5A–B, 7) described and figured *Cycloceras* cf. *C. decrescens* (Billings) from the "Black River Formation" (Middle Ordovician) exposed on St. Joseph Island, Lake Huron. Foerste described these specimens (University of Michigan 14419, 144454, and 144455) as differing from typical *Cycloceras decrescens* "only in the slightly greater prominence of their annulations." These specimens are similar to the Kentucky specimens of *M. annularis* in apical angle, the conspicuous nature of the annuli, the number and relative dimensions of the annuli per shell diameter, and in the globular, expanded nature of the siphuncle segments. These similarities indicate that these Canadian specimens are also assignable to *M. annularis*. *Monomuchites annularis* has more prominent transverse annuli and interannular spaces plus a less rapidly expanding shell compared with the type specimens of *M. decrescens* (Billings) that are figured by Foerste (1932, pl. 12, figs. 1A–C) and by Wilson (1961, pl. 5, figs. 7, 8).

Foerste (1932, pl. 12, fig. 4) described and figured "*Cycloceras* sp." from the "Platteville member of the Black River Formation" exposed at Minneapolis, Minn. Similar specimens from the Platteville Group at Dixon, Ill. (USNM 162068, pl. 9, fig. 14), and from Mineral Point, Wis. (USNM 25275, pl. 9, figs. 12, 13), that were measured as part of this study indicate that these annulated Platteville specimens are also assignable to *M. annularis*.

These occurrences of *M. annularis* indicate that this species was broadly distributed across the shallow-shelf areas of eastern Laurentia during Middle Ordovician (Rocklandian) times. This widespread distribution as well as the similar distributions of associated nautiloid taxa (*Proteoceras tyronensis*, *Ormoceras ferecentricum*) attest to the uniform, unrestricted shallow-marine shelf conditions that existed across much of this area at this time.

MONOMUCHITES OBLIQUUM, NEW SPECIES

Plate 9, figures 4–10

Diagnosis.—Gradually expanding, slightly curved species of *Monomuchites* distinguished by its low, rounded, oblique annuli and poorly delineated interannular spaces.

Description.—Small (lengths of 18 cm and diameters up to 2.0 cm), gradually expanding (apical angle 6–6.5 degrees) orthoconic longicone, endogastrically curved adapically; circular in cross section with straight, transverse sutures. Cameral chambers short, cameral lengths ranging from one-fifth to one-sixth the diameter of the shell, decreasing with ontogeny. Body chamber unknown. Exterior marked by low, indistinct, rounded, oblique annuli that slope from the dorsum to the venter in the adoral direction. Annuli separated by poorly delineated interannular spaces that are narrower than annuli. Five and a half annuli present in a length equal to an adoral diameter of 20 mm.

Siphuncle supracentral (SPR=0.53) at a diameter of 15 mm, becoming subcentral with ontogeny (SPR=0.48 at a diameter of 20 mm). Siphuncle segments narrow in diameter, one-fourth the diameter of the shell at shell diameters of 12 mm or less. Segments expanded, globular, compressed in outline (SCR=1.16) at the same diameter. Septal necks cyrtochoanitic, short, recurved, with free brims. Endosiphuncular deposits consisting of ventral parietal deposits that fuse to form a continuous ventral lining adapically. Cameral deposits consisting of botryoidal mural-episeptal deposits more extensively developed on the side of the cameral chamber opposite the siphuncle.

Stratigraphic and geographic occurrence.—Upper part of the Rocklandian portion of the Tyrone Limestone, Inner Blue Grass region of central Kentucky.

Studied material.—Holotype USNM 158653 (pl. 9, figs. 5–8), paratype USNM 158648 (pl. 9, figs. 9, 10), and paratype USNM 158643 (pl. 9, fig. 4), all from the USGS 6034-CO collection, Little Hickman section A, Jessamine County, Ky.

Remarks.—This species is a rare form within the 6034-CO collection from the Tyrone Limestone, being represented only by the type material. It is, however, distinct from associated species of *Monomuchites* in the slender, gradually expanding, endogastrically curved nature of its shell and in its low, rounded, poorly delineated oblique annulations. No other similar species of *Monomuchites* are known to this writer from equivalent Blackriveran and Rocklandian strata elsewhere in North America.

The ventral parietal deposits preserved in paratype USNM 158648 (pl. 9, fig. 10) are similar to those developed in the previously discussed proteoceratids *Proteoceras* Flower and *Treptoceras* Flower. These consist of deposits that grow forward, so that the deposit is thickest just adoral of the septal neck. These deposits fuse to form a continuous ventral lining adapically. As indicated above, endosiphuncular deposits are poorly known for most species of *Monomuchites*, making the phylogenetic affinities of this genus uncertain. This specimen of *M. obliquum* supports Flower's assignment of *Monomuchites* to the Proteoceratidae.

FAMILY PSEUDORTHOCERATIDAE FLOWER AND CASTER

Smooth to elaborately ornamented, generally small, moderately to rapidly expanding orthoconic longicones and cyrtocones; usually circular in cross section with straight, transverse sutures. Cameral chambers typically short, having cameral lengths from one-fourth to one-sixth the diameter of the shell, decreasing with ontogeny. Body chambers comparatively short, less than one-fourth the total length of the shell, constricted just adapical to aperture.

Siphuncle central to subcentral in position; segments narrow in diameter, from one-fourth to one-sixth the diameter of the shell, typically expanded equiaxial globular or depressed barrel-shaped. Expansion of siphuncle segments often localized immediately adjacent to septal necks. Septal necks cyrtochoanitic, short, recurved, but not typically recumbent, brims equal in length to septal necks. Connecting rings thin, homogeneous. Endosiphuncular deposits consisting of annuli that are initiated at the septal foramen and grow forward along connecting ring walls, eventually fusing to form a continuous lining of uniform thickness. Cameral deposits of the mural-episeptal type, lobate, initiated ventrally and laterally prior to development dorsally. Medial mural delta formed in most taxa. Initial portions of the shell typically bluntly cone-shaped.

Stratigraphic occurrence.—Middle Ordovician (Whiterockian Provincial Series) to Lower Permian (Leonardian Provincial Series).

Remarks.—This family constitutes the bulk of the superfamily Pseudorthoceratacea and is the dominant orthocerid family in the Devonian and later Paleozoic, ranging through to the end of the Early Permian. Sweet (1964), listed 42 genera as members of this family. Most of these taxa, however, are poorly known, based on fragmentary material described primarily by Flower (1939). Additional taxa were described by Zhuravleva (1978). Dzik (1984) placed many of these taxa in synonymy and transferred others to other families, recognizing but six genera. This is certainly a case of over zealous lumping of taxa but points out the need for extensive restudy and revision of the family.

Flower (1939) derived the Pseudorthoceratidae from the Geisonoceratidae through *Virgoceras* Flower, a Middle Silurian (Llandoverian-Ludlovian) genus that had the shell morphology of the Geisonoceratidae but having endosiphuncular deposits of the pseudorthoceratid type, consisting of annuli and parietal deposits that formed independent of cameral deposits. Sweet (1964) retained *Virgoceras* in the Geisonoceratidae and indicated that it exhibited endosiphuncular deposits similar to later Devonian pseudorthoceratids (Spyroceratinae). He also pointed out the presence of a number of cyrtochoanitic orthocerids from the early Middle Ordovician, antedating *Virgoceras* by 30–40 m.y. These are *Mysterioceras* Teichert and Glenister, *Stereoplasmoceras* Grabau, and a number of proteoceratids described by Flower (1955b). This suggested to Sweet a possibly polyphyletic origin for the Pseudorthoceratidae.

Dzik (1984) has also documented cyrtochoanitic siphuncle segments and pseudorthoceratid-type endosiphuncular deposits in *Clinoceras dens* Mascke from Middle Ordovician Llanvirnian (Kunda Stage) strata in the Baltic region. As indicated by Dzik, this species, with its rapidly expanding phragmocone and short adorally constricted body chamber, is similar to later pseudorthoceratids, in particular, some species of the genus *Dolorthoceras* Miller. These examples suggest that the origin of the Pseudorthoceratidae lies with these Middle Ordovician taxa rather than with the Geisonoceratidae.

GENUS *ISORTHOCERAS* FLOWER

Type species.—*Orthoceras sociale* Hall, Hall, 1877, p. 245; designated by Flower, 1962, p. 32–33; Elgin Shale, Maysvillian Stage, Upper Ordovician, Iowa.

Description.—Small, slender, gradually to moderately expanding orthoconic longicones; circular in cross section with straight, transverse sutures. Cameral chambers short, cameral lengths ranging from one-third to one-sixth the diameter of the shell, decreasing with ontogeny. Body chamber between one-fifth and one-sixth the total length of the shell, constricted just adapical of the aperture. Aperture straight, transverse. Periphract dorsomyarian. Surface ornament of subdued, transverse growth lines or raised transverse, longitudinal, or cancellate lirae.

Siphuncle central to subcentral in position having segments moving to a more ventral position with ontogeny. Segments narrow, from one-fifth to one-tenth the diameter of the shell; only slightly expanded within camerae, forming barrel-shaped segments, of which the maximum diameter is just adapical to septal necks. Septal necks cyrtochoanitic, recurved with short, free brims. Endosiphuncular deposits initiated as adorally growing annuli at septal necks, restricted to adapical portions of phragmocones. Adoralmost cameral deposits planar mural-episeptal, initiated ventrolaterally prior to their development dorsally.

Stratigraphic and geographic occurrence.—Middle Ordovician (Kirkfieldian Stage) to Upper Ordovician (Richmondian Stage), Eastern and Central North America (Laurentia).

Remarks.—Flower (1962) erected this genus for basically smooth, small, orthoconic longicones distinguished by the development of subcentral siphuncles with barrel-shaped segments, cyrtochoanitic septal necks, and endosiphuncular deposits consisting of thin parietal deposits that are greatly concentrated adapically. Besides the type species, *I. sociale* (Hall) (pl. 10, fig. 15) from the Upper Ordovician Maquoketa Formation of Iowa, Flower also placed in the genus a number of species from the "Cynthiana Limestone" in Kentucky and the middle and upper portions of the Trenton Group in New York. These latter species consist of a group of small, slender, finely ornamented longicones that are the most abundant nautiloids encountered in late Middle Ordovician (Shermanian) carbonates and dark-shale facies in Eastern North America. These species are *Isorthoceras albersi* (Miller and Faber) from the Middle Ordovician Point Pleasant Tongue of the Clays Ferry Formation in northern Kentucky (pl. 10, figs. 1–9); "*Geisonoceras*" *lineolatus* (Hall), "*G.*" *strigatum* (Hall) (pl. 10, fig. 16), "*G.*" *strangulatum* (Hall) (pl. 10, figs. 13, 14), "*G.*" *tenuistriatum* (Hall), and "*G.*" *tenuitextum* (Hall), all from the Middle Ordovician Denley and Dolgeville Limestones in New York. An additional species, *Protokionoceras oneidaense* (Walcott) from the Upper Ordovician Utica Shale in New York, also belongs in *Isorthoceras*.

Previously, these finely ornamented species have been placed in either *Geisonoceras* Hyatt or *Protokionoceras* Grabau and Shimer (Ruedemann, 1926; Titus and Cameron, 1976). These small, slender, Middle Ordovician species are definitely not assignable to either of these two genera. As indicated above, *Geisonoceras*, based on *Orthoceras rivale* Barrande from the Silurian of Eastern Europe, and *Protokionoceras*, based on *Orthoceras medullare* Hall from equivalent Silurian carbonates in the Midwestern United States, both have large, robust, gradually expanding shells with deeper cameral chambers, larger diameter, slightly expanded dorsal siphuncle segments with longer, less recurved septal necks, and annul osiphonate deposits that are continuous and contain massive botryoidal episeptal and hyposeptal cameral deposits. These small, slender Ordovician species are similar to *I. sociale* (Hall) in gross shell morphology and in details of their internal structure, especially siphuncle segment shape, septal necks, and the nature of endosiphuncular and cameral deposits. The only distinction between these taxa is in the nature of the external shell ornament. *Isorthoceras sociale* has a basically smooth shell marked by fine, transverse growth lines (pl. 10, fig. 15). These Middle Ordovician species have shells marked by fine, raised, longitudinal, transverse, or cancellate lirae. Flower (1962) indicated that these more ornate species may eventually provide the basis for a separate, allied genus. It is debatable as to whether or not details of shell surface ornament constitutes a significant criterion for differentiating genus-level nautiloid taxa. For the purposes of this paper, these more ornate taxa are retained in *Isorthoceras*.

ISORTHOCERAS ALBERSI (MILLER AND FABER), 1894B

Plate 10, figures 1–9

Orthoceras albersi Miller and Faber, 1894B, p. 140, pl. 8, figs. 1–40.

Diagnosis.—Moderately expanding species of *Isorthoceras* distinguished by an external ornament of fine, raised, longitudinal lirae.

Description.—Small (less than 25 cm in length and shell diameters up to 2.0 cm), moderately expanding (apical angle 7–8 degrees), orthoconic longicones; circular in section with straight, transverse sutures. Cameral lengths ranging from one-fourth to one-sixth the diameter of the shell, decreasing with ontogeny. Body chamber poorly known. Shell exterior marked by distinct, fine, raised, longitudinal lirae.

Siphuncle central in position (SPR=0.50) at shell diameters less than 5 mm, becoming more ventral with ontogeny having segments subcentral in position (SPR=0.33) at shell diameters of 2.0 cm. Segments elongate ovate, not quite twice as long as high, SCR values averaging between 0.75 and 0.80 (depressed). Segments narrow, one-fifth to one-eighth the diameter of the shell, decreasing in diameter relative to shell diameter with ontogeny. Septal necks cyrtochoanitic, recurved with short, free brims. Endosiphuncular deposits unknown. Adoralmost cameral deposits consisting of ventrolateral mural-episeptal deposits, the deposits completely filling cameral chambers in adapical portions of the phragmocone less than 10 mm in diameter.

Stratigraphic and geographic occurrence.—Logana Member of the Lexington Limestone (Middle Ordovician, Kirkfieldian Stage), Woodford County, Ky.; Grier Limestone Member of the Lexington Limestone (Middle Ordovician, Kirkfieldian and Shermanian Stages), Clark and Jessamine Counties, Ky.; Shermanian part of the Millersburg Member of the Lexington Limestone, Bourbon County, Ky.; and the Shermanian portion of the Point Pleasant Tongue of the Clays Ferry Formation, Kenton County, Ky.

Studied material.—Syntypes UC 360 (pl. 10, fig. 1) and UC 361 (pl. 10, figs. 2, 3), "lower part of the Hudson River Group at Ludlow, Kentucky"; MU 447T (pl. 10, fig. 4), "Point Pleasant Beds" at Cincinnati, Ohio; USNM 468695 (pl. 10, fig. 7) and 468696 (pl. 10, fig. 8), USGS 5067-CO collection, Grier Limestone Member; USNM 468697 (pl. 10, fig. 5) and 468698 (pl. 10, fig. 9), USGS 5092-CO collection, Logana Member; and USNM 468702 (pl. 10, fig. 6), USGS 5096-CO collection, Grier Limestone Member; plus suites of silicified material: USNM 468699 from USGS 5067-CO collection, Grier Limestone Member; USNM 468700 from USGS 5073-CO collection, Grier Limestone Member; USNM 468701 from USGS 5092-CO collection, Logana Member; USNM 468702 from USGS 5096-CO collection, Grier Limestone Member; USNM 468703 from USGS 6419-CO collection, Logana Member; and USNM 468704 from USGS 7353-CO collection, Millersburg Member.

Specimen UC 360 (pl. 10, fig. 1) is here designated the lectotype of *Orthoceras albersi* Miller and Faber (1894B). Specimen UC 361 (pl. 10, figs. 2, 3) is here designated a paralectotype of *O. albersi* Miller and Faber.

Remarks.—This small, longitudinally marked longiconic nautiloid is the most abundant nautiloid species in the various members of the Lexington Limestone in central and northern Kentucky. Specimens of *I. albersi* constitute 90 percent of the etched nautiloid specimens recovered from the USGS 5067-CO collection in the Grier Limestone Member (101 out of 114 specimens). The species appears to be most abundant in the deeper water, less agitated facies of the Logana Member and Grier Limestone Member within the Lexington Limestone. In these facies, *I. albersi* occurs in association with lesser numbers of the nautiloids *Gorbyoceras* cf. *G. tetreauense* (pl. 12, fig. 3), *Polygrammoceras* sp. A (pl. 12, figs. 1, 2), and

Allumettoceras cf. *A. tenerum* (pl. 20, figs. 19–22), along with other mollusks and small, twiglike bryozoans. *Isorthoceras albersi* is equally abundant in the dark-gray, mud-rich limestones of the Point Pleasant Tongue of the Clays Ferry Formation in northern Kentucky. In these limestones the species occurs in association with rare specimens of the large longiconic nautiloid *Ordogeisonoceras amplicameratum* (Hall) (pl. 3, figs. 1–5) and *Trocholites faberi* Foerste, other mollusks, and the trilobites *Flexicalymene* and *Isotelus*.

Isorthoceras albersi is similar to *Orthoceras strigatum* Hall from the Middle Ordovician (Shermanian) Denley Limestone and equivalents in north-central New York. A comparison of these Kentucky specimens with Hall's type specimens (AMNH 29691 and 29692) indicates that these specimens are similar in general shell morphology and in possessing an external ornament of fine, raised, longitudinal lirae (compare pl. 10, figs. 5–9 with pl. 10, fig. 16). The internal features of Hall's species are not known, as his material consists of matrix-filled composite molds. Ruedemann (1926, pl. 10, fig. 13) figured a sectioned specimen of a closely related species, "*Geisonoceras*" *tenuistriatum* (Hall). This specimen reveals narrow, slightly expanded siphuncle segments similar to those present in *I. albersi*. The type specimens of *I. strigatum* differ from the Kentucky specimens of *I. albersi* only in their more slender, gradually expanding shells with an apical angle of 6 degrees versus apical angles of 7 or 8 degrees for *I. albersi*. Pending study of New York material preserving the features of the siphuncle segments, the Kentucky specimens will remain in *I. albersi*, although it may eventually be determined that this species is a synonym of *I. strigatum* (Hall), Hall's name having priority. A comparison of the morphological features of *I. albersi*, *I. rogersensis*, and several of the species described by Hall (1847) is presented in figure 18.

ISORTHOCERAS ROGERSENSIS (FOERSTE), 1914A

Plate 10, figures 10–12

Orthoceras rogersensis Foerste, 1914a, p. 143–144, pl. 1, figs. 17a–b.
Treptoceras praenuntium Flower, 1942, p. 19, pl. 1, fig. 2.

Diagnosis.—Slender, gradually expanding species of *Isorthoceras* distinguished by an external ornament of fine, raised, cancellate lirae.

Description.—Small (lengths of 25 cm having shell diameters up to 2.4 cm), slender, gradually expanding (apical angle 6.5 degrees) orthoconic longicones; circular in section with straight, transverse sutures. Camerae short, cameral lengths averaging between one-fifth and one-sixth the diameter of the shell, initially decreasing with ontogeny, then increasing in length at shell diameters in excess of 20 mm. Body chamber unknown. External

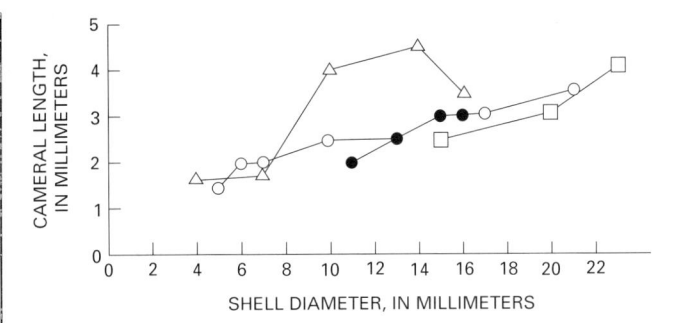

FIGURE 18.—Bivariate plot of cameral length versus shell diameter for selected species of *Isorthoceras* from Middle and Upper Ordovician strata in Eastern North America. Species are *I. albersi* from USGS collections in the Lexington Limestone (Shermanian) in Kentucky (open circles); *I. strigatum*, syntype AMNH 29692 from the Trenton Limestone (Shermanian), New York (solid circles); *I. rogersensis*, syntype USNM 87193A from the Clays Ferry Formation in Kentucky (squares); and *I. sociale*, MU 14566 from the Maquoketa Formation (Maysvillian), Iowa (triangles).

ornament of fine, raised, longitudinal and transverse lirae, equally developed, forming a fine cancellate pattern.

Siphuncle central to subcentral (SPR=0.50–0.46) at shell diameters less than 15 mm. Siphuncle segments barrel-shaped, depressed (SCR=0.73–0.85), narrow, one-fifth the diameter of the shell. Septal necks cyrtochoanitic, recurved, with short, free brims. Endosiphuncular deposits unknown. Mural-episeptal cameral deposits at shell diameters of 15 mm or less.

Stratigraphic and geographic occurrence.—Basal portion of the Clays Ferry Formation (uppermost Middle Ordovician, Shermanian Stage), Harrison and Scott Counties, Ky.

Studied material.—Syntypes USNM 87193A (pl. 10, figs. 10, 11) and USNM 87193B (pl. 10, fig. 12), "Cynthiana Limestone" exposed at railroad cut 1 mi north of Rogers Gap, 5 mi south of Sadieville, Scott County, Ky.; holotype of *Treptoceras praenuntium* Flower (UC 22696), "Cynthiana Limestone," Poindexter Quarry at Cynthiana, Harrison County, Ky.; plus uncataloged slabs containing specimens from the "Cynthiana Limestone" at Rogers Gap, Ky., in the Shideler collection at Miami University.

Specimen USNM 87193A (pl. 10, figs. 10, 11) is here designated the lectotype of *Orthoceras rogersensis* Foerste (1914A). Specimen 87193B (pl. 10, fig. 12) is here designated a paralectotype of *O. rogersensis* Foerste.

Remarks.—*Isorthoceras rogersensis* is a locally common species in the argillaceous, thin limestone beds within the basal Clays Ferry Formation in north-central Kentucky. In these strata, *I. rogersensis* occurs as calcite-replaced or matrix-filled internal molds associated with gastropods, pelecypods, crinoids, inarticulate brachiopods, and the trilobites *Flexicalymene* and *Isotelus*. *Isorthoceras rogersensis* differs from the preceding species, *I. albersi* (Miller and Faber), in being a slightly larger and more gradually expanding shell, but is primarily distinguished by its cancellate external ornament compared with the distinctive longitudinal ornament of the latter species.

Flower (1942) described *Treptoceras praenuntium* from the "Cynthiana Limestone" exposed at Cynthiana, Ky. As indicated above, this species is not assignable to *Treptoceras*, as it has narrow, barrel-shaped siphuncle segments. This species is similar to *I. rogersensis* in its gradual rate of expansion (apical angle 6.5 degrees), short cameral chambers one-sixth the diameter of the shell, barrel-shaped, depressed siphuncle segments, and occurrence at the top of the Lexington Limestone in north-central Kentucky. These similarities indicate assignment of "*Treptoceras*" *praenuntium* to *I. rogersensis*. Specimens of the former species preserving the external shell ornament are necessary to further confirm the affinities of these taxa. Unfortunately, "*T.*" *praenuntium* is known only from the holotype, and the bed that was the source of this specimen has been quarried away.

Isorthoceras rogersensis is also similar to a species described by Hall (1847) as "*Endoceras proteiforme* variety *tenuitextum*," from the Middle Ordovician Trenton Group exposed at Middleville, N.Y. Comparisons of Hall's types (AMNH 29717 and 29718) with the types of *I. rogersensis* indicate a similar general shell morphology (apical angle 6–7 degrees) and an external ornament of fine cancellate lirae. Unfortunately, the internal features of Hall's species are unknown. Pending discovery of specimens of *I. tenuitextum* preserving critical features of the siphuncle segments, these species are retained as separate species. As was the case with the previous species, if these Kentucky and New York specimens prove to be the same, Hall's species name would have priority.

A third species referable to *Isorthoceras* is known from uncommon fragmentary specimens from the *Treptoceras duseri* shale unit within the "Waynesville biofacies" in the Bull Fork Formation (Upper Ordovician, Richmondian) in southwest Ohio (Frey, 1988). This is a small (less than 20 cm in length), slender (apical angle 5.5–6.5 degrees) orthoconic longicone with very short cameral chambers (cameral lengths averaging one-seventh the diameter of the shell) and narrow, barrel-shaped subcentral siphuncle segments. Endosiphuncular deposits consist of thin parietal deposits. Cameral deposits are mural-episeptal. Although the external ornament of this Richmondian species is unknown, the short camerae typical of this species distinguish it from both *I. albersi* and *I. rogersensis*. More complete material is needed to more fully describe this Richmondian species.

GENUS *GORBYOCERAS* SHIMIZU AND OBATA

Type species.—*Orthoceras gorbyi* Miller, 1894, p. 322; designated by Shimizu and Obata, 1935, p. 4; Saluda Formation, Richmondian Stage, Upper Ordovician, Indiana.

Description.—Slender, small to medium-sized, gradually to moderately expanding orthoconic longicones that adapically may be curved endogastrically. Shells typically circular in cross section, laterally compressed in early growth stages of some species. Sutures straight, transverse to slightly oblique, sloping from the dorsum to the venter in the adapical direction. Cameral chambers short, cameral lengths from one-fourth to one-ninth the diameter of the shell, decreasing with ontogeny. Body chamber long and slender, contracting toward the aperature in some species. Aperture simple, transverse. Periphract dorsomyarian. Shell exterior marked by prominent transverse, rounded annuli, oblique in some species; as well as fine, raised transverse and longitudinal lirae.

Siphuncle subcentral with expanded segments that vary in shape, ranging from depressed barrel-shaped to equiaxial globular in outline. Segments narrow, one-fourth to one-sixth the diameter of the shell. Septal necks cyrtochoanitic, recurved, but with free brims. Endosiphuncular deposits known for a few species, consisting of thin parietal deposits restricted to adapical portions of large individuals. Mural-episeptal cameral deposits adapically.

Stratigraphic and geographic occurrence.—Middle Ordovician (Rocklandian Stage) to Upper Ordovician (Richmondian Stage), North America (Laurentia).

Remarks.—This genus is one of several poorly defined annulated longiconic taxa proposed by Shimizu and Obata (1935). Their description of *Gorbyoceras* was brief: "(shell) possessing peculiar annulations and distant longitudinal striae; the annulations sloping sharply downward and weakening towards the venter." In the same paper, they also described *Hammelloceras* (type species *Spyroceras hammelli* (Foerste)) as "Externally…somewhat similar to *Kionoceras* except that it has prominent annulations and an ormoceratid siphuncle"; as well as *Porteroceras* (type species *Spyroceras porteri* (Schuchert)), distinguished as being "related to *Hammelloceras*, differing from it in possessing more prominent ornamentation and a sactocerid siphuncle." *Gorbyoceras*, *Hammelloceras*, and *Porteroceras* are all annulated longicones with longitudinal lirae, cyrtochoanitic septal necks, and expanded siphuncle segments. All three are from Upper Ordovician strata in North America. Flower (1946) and Sweet (1964) both placed these taxa in synonymy, with *Gorbyoceras* having priority.

Flower (1946) described seven species of *Gorbyoceras* from Upper Ordovician (Richmondian) strata in Indiana and Ohio and placed in the genus a number of Ordovician species previously referred to as *Spyroceras*

Hyatt, a Devonian genus based on *Orthoceras crotalum* Hall. These are "*Spyroceras*" *ferum* (Billings) and "*S.*" *microlineatum* Foerste from the Richmondian of Anticosti Island, Quebec; "*S.*" *parksi* Foerste from the Richmondian Georgian Bay Formation in southern Ontario; plus "*S.*" *calvini* Foerste and "*S.*" *perroti* (Clarke), both from the Maquoketa Formation of Iowa. Other Upper Ordovician species assignable to *Gorbyoceras* are "*S.*" *geronticum* (Foerste and Savage) and *G. giganteum* Nelson from the Shamattawa and Chasm Creek Limestones exposed in the Hudson Bay region of Manitoba. Also placed in *Gorbyoceras* are a number of similar forms from older Middle Ordovician strata: *G. tetreauense* Wilson from the "Leray-Rockland beds" (Rocklandian) in the Ottawa region of Canada and *G. iowaense* (Foerste) from the Decorah Shale (Rocklandian) in Iowa.

The familial affinities of *Gorbyoceras* are currently uncertain. Flower (1962) and Sweet (1964) placed this genus in the Proteoceratidae on the basis of its cyrtochoanitic septal necks and occurrence in Ordovician strata. The features of the siphuncle segments and the presence and types of endosiphuncular deposits developed in species assigned to *Gorbyoceras* remain largely unknown. No species known to this writer have been documented to show the siphuncle segment decompression characteristic of the Proteoceratidae. Flower (1946) has illustrated siphuncle segments for *G. crossi* Flower, *G. curvatum* Flower, *G. duncanae* Flower, and *G. hammelli* (Foerste), all from the Cincinnatian Provincial Series in the Cincinnati arch region. *Gorbyoceras crossi* and adoral portions of a specimen he referred to as *G. hammelli* both have expanded, globular, equiaxial segments and ventral parietal deposits that appear to be similar to those developed in the proteoceratid *Treptoceras* (Flower, 1946, pl. 1, figs. 2, 7). *Gorbyoceras curvatum* (Flower, 1946, pl. 5, fig. 3), *G. duncanae* (Flower, 1946, pl. 1, fig. 1), and adapical portions of *G. hammelli* illustrated here (pl. 12, fig. 7) exhibit less broadly expanded, barrel-shaped segments similar to those of the pseudorthoceratid *Isorthoceras*. The parietal deposits evident in the figured specimen of *G. hammelli* also are similar to those documented for various members of the Pseudorthoceratidae. All of the species of *Gorbyoceras* studied here possess septal necks that are cyrtochoanitic with recurved, short, free brims, similar to those typical of pseudoceratid genera. As ontogenetic decompression has not been demonstrated for any species of *Gorbyoceras* known to this writer, herein this genus is removed from the Proteoceratidae and placed with some uncertainty in the Pseudorthoceratidae. Further study may indicate that *Gorbyoceras* as presently known contains two distinct genera, one with expanded, globular siphuncle segments that may belong in the Proteoceratidae, and another with depressed, barrel-shaped segments that belongs in the Pseudorthoceratidae. Study of more complete, better preserved material is needed to more accurately determine the affinities of the species currently assigned to this genus.

GORBYOCERAS GORBYI (MILLER), 1894

Plate 12, figures 12, 13

Orthoceras gorbyi Miller, 1894, p. 322, pl. 10, fig. 2.
Orthoceras gorbyi Miller. Cumings, 1908, p. 1037, pl. 52, fig. 3.
Spyroceras gorbyi (Miller). Foerste, 1928a, p. 283, pl. 61, fig. 4.
Gorbyoceras gorbyi (Miller). Shimizu and Obata, 1935, p. 4.
Gorbyoceras gorbyi (Miller). Flower, 1946, p. 145, pl. 2, figs. 9, 10; pl. 4, figs. 3, 6, 7.

Diagnosis.—Large, slender, gradually expanding, endogastrically curved species of *Gorbyoceras* distinguished by low, rounded, oblique annuli that slope from the dorsum to the venter in an adoral direction, opposite the slope of the sutures.

Description.—Long (lengths up to 71.5 cm and shell diameters up to 3.2 cm), slender, gradually expanding (apical angle 4 degrees) orthoconic longicones, curved endogastrically such that the venter is gently convex and the dorsum is concave. Shell initially circular in cross section, becoming slightly depressed with ontogeny, then becoming more circular again at diameters in excess of 25 mm. Sutures sloping from the dorsum to the venter in the adapical direction, becoming more transverse with ontogeny. Cameral chambers short, cameral lengths decreasing with ontogeny, from one-fourth the diameter of the shell at a diameter of 20 mm to one-seventh the diameter of the shell at diameters greater than 25 mm. Body chamber long and tubular, contracting slightly toward the aperture, making up at least one-fourth the total length of the shell. Aperture poorly known; periphract dorsomyarian. Shell exterior marked by low, rounded, oblique annuli that slope opposite the direction of the sutures, extending from the dorsum to the venter, directed adorally. Annuli spaced 4-5 mm apart at diameters in excess of 20 mm. Shell marked with alternating sets of coarser and finer longitudinal lirae and transverse growth lines that parallel the trend of the annuli.

Siphuncle subcentral in position, moving to a more ventral position with ontogeny; having SPR values of 0.40 at a diameter of 20 mm and SPR values of 0.28 at a diameter of 32 mm. Siphuncle segments poorly known, expanded, globular, equiaxial in known portions of the shell 25 mm in diameter. Septal necks cyrtochoanitic, short with recurved free brims. Endosiphuncular deposits unknown. Cameral deposits mural-episeptal, developed adapically at shell diameters of 25 mm or less.

Stratigraphic and geographic occurrence.—Saluda Formation (Upper Ordovician, Richmondian Stage), southeastern Indiana and adjacent portions of southwestern Ohio.

Studied material.—Hypotype (MU 397T) figured by Flower (1946, pl. 2, figs. 9, 10), Shideler Collection, Miami University, basal Whitewater Formation, Four-Mile Creek, Preble County, Ohio; MU 402T (pl. 12, figs. 12, 13); and MU 16050, Bliss Collection, Miami University, "Concrete Layer" in the Saluda Formation, Sayers Farm locality, 2 mi northwest of Hamburg, Franklin County, Ind. Also studied was the holotype of *Orthoceras gorbyi* Miller (USNM 64337), which Miller (1894, pl. 10, fig. 2) listed as coming from the "Hudson River Group, Franklin County, Indiana."

Remarks.—*Gorbyoceras gorbyi* is distinguished from associated species of *Gorbyoceras* by its long, slender, endogastrically curved shell marked by low, rounded, oblique annuli that are oriented opposite to the obliquely curved sutures. The internal features of the species are poorly known due to the poor preservation of these features in nautiloids occurring in massive-bedded, micritic limestone lithologies in the Saluda Formation. Flower (1946) described the siphuncle segments of *G. gorbyi* as globular, similar to those of the genus *Treptoceras*, with short cyrtochoanitic septal necks. He also indicated that the external annuli in this species were surpressed on the venter of large specimens. This feature is not evident in the material studied here, the annuli being developed uniformly dorsally and ventrally.

The associated *G. crossi* Flower is similar to *G. gorbyi* in possessing a long, slender shell, circular in section, and low, slightly oblique annuli. The annuli in *G. gorbyi* are more strongly developed and more strongly inclined than in *G. crossi*. Transverse, raised lirae are present in *G. gorbyi* and absent in *G. crossi*. The camerae in *G. gorbyi* are described by Flower (1946) as being shorter than those present in *G. crossi*. The material studied here indicates no significant difference in cameral lengths at comparable diameters, at least in adapical portions of phragmocones. The described differences between these species may be the result of differences in preservation rather than a representation of actual morphological differences.

Within the Kentucky-Indiana-Ohio area, *G. gorbyi* is restricted to massive-bedded, fine-grained limestone facies that occur above the lower *Tetradium* biostrome within the Saluda Formation. In this facies the species is associated with a diverse nautiloid assemblage of *G. crossi* Flower, a large species of *Cameroceras*, the breviconic ellesmerocerid *Cyrtocerina*, the oncocerids *Beloitoceras bucheri* Flower (pl. 22, figs. 4, 5), *Diestoceras indianense* (Miller and Faber) (pl. 22, figs. 6–9), and *D. shideleri* Foerste (pl. 21, figs. 1, 2, 5; pl. 22, figs. 1, 3), several species of slender oncocerid *Manitoulinoceras*, and the ascocerid *Schuchertoceras prolongatum* (Foerste). All of these species are preserved as nearly complete, uncrushed composite molds, often preserving the fine external ornament of the shell (pl. 22, figs. 4, 5, 8). This extensive nautiloid fauna was described in part by Flower (1946) and represents the peak development of the "Arctic" nautiloid fauna in the northern Cincinnati arch region.

GORBYOCERAS HAMMELLI (FOERSTE), 1910

Plate 12, figures 4, 6, 7, 10, 11

Orthoceras hammelli Foerste, 1910, p. 74, pl. 1, fig. 4.
Spyroceras hammelli (Foerste). Foerste, 1924b, p. 222, pl. 39, fig. 1; pl. 40, fig. 1.
Hammelloceras hammelli (Foerste). Shimizu and Obata, 1935, p. 6.
Gorbyoceras hammelli (Foerste). Flower, 1946, p. 151, pl. 1, figs. 2, 3, 9; pl. 2, fig. 5; pl. 3, figs. 1, 4; pl. 4, figs. 1, 2, 5; pl. 5, figs. 4, 9.

Diagnosis.—Small, straight, slender species of *Gorbyoceras* distinguished by oblique annuli that slope from dorsum to venter adorally, the annuli becoming less distinct and spaced further apart with ontogeny.

Description.—Small (lengths up to 30 cm and observed shell diameters up to 2.4 cm), slender (apical angle 5–6 degrees), orthoconic longicones; circular in section with straight, transverse sutures. Cameral length ranging from one-third to one-fifth the diameter of the shell, decreasing with ontogeny. Body chamber incompletely known; aperture unknown. Shell exterior marked by oblique, rounded annuli that slope from the dorsum to the venter in the adoral direction. Annuli prominent, spaced 2 mm apart in adapical portions of the shell less than 15 mm in diameter; becoming less distinct, low, and rounded, spaced 4–5 mm apart in adoral portions of the shell greater than 17 mm in diameter. Shell marked by distinct alternating sets of coarse and fine longitudinal lirae and intersecting, raised, transverse growth lines that parallel the trend of the annuli.

Siphuncle subcentral in position most of the length of the phragmocone, becoming more ventral in position with ontogeny, the SPR values ranging from 0.38 (subcentral) at a diameter of 9 mm to 0.24 (ventral) at a diameter of 17 mm. Siphuncle segments expanded, roughly barrel-shaped in portions of the shell 9 mm or less in diameter, becoming more compressed, globular (SCR=1.0, equiaxial) at diameters of 15 mm or greater. Segments unknown from more adoral portions of the shell greater than 17 mm in diameter. Septal necks cyrtochoanitic, short, recurved, with free brims. Endosiphuncular deposits consisting of uniformly developed, thin parietal deposits that fuse to form a continuous lining with ontogeny; restricted in development to adapical portions of the shell less than 15 mm in diameter. Mural-episeptal cameral deposits well-developed in adapical portions of shells less than or equal to 15 mm in diameter.

Stratigraphic and geographic occurrence.—Saluda Formation and its Hitz Limestone Member (Late Ordovician, Richmondian Stage), Franklin, Ripley, and Jefferson Counties, Ind., and adjacent portions of Bedford, Trimble, Oldham, and Jefferson Counties in northern Kentucky.

Studied material.—Hypotypes figured by Flower (1946), are UC 24165 from the Saluda Formation at Canaan, Ind., and 24166 from the Hitz Bed at Madison, Ind.; plus MU 408T from the top of the Saluda Formation at Versailles, Ripley County, Ind. (pl. 12, figs. 4, 7). Additional material consists of MU 409T (pl. 12, figs. 10, 11) and MU 446T (pl. 12, fig. 6), plus numerous small fragmentary specimens, all from the "Hitz Bed" exposed along the I-471 roadcut, just north of Madison, Jefferson County, Ind. Also studied was the holotype of *Orthoceras hammelli* Foerste (USNM 87190), which Foerste (1910, pl. 1, fig. 4) listed as coming from "Top of Saluda Bed, Saluda Creek, Jefferson County, Indiana."

Remarks.—This species, originally described by Foerste (1910), served as the type species for *Hammelloceras* Shimizu and Obata (1935). Flower (1946) has indicated that, as defined by these authors, there are no valid distinctions between the genera *Gorbyoceras*, *Hammelloceras*, and *Porteroceras*. As such, "*Orthoceras*" *hammelli* is retained in *Gorbyoceras* pending further study of these annulated longicones. *Gorbyoceras hammelli* is similar to the preceding species, *G. gorbyi* (Miller), in the development of a slender, longiconic shell and an external ornament of oblique, raised annuli and longitudinal and transverse lirae. Both species possess subcentral siphuncles and mural-episeptal cameral deposits. *Gorbyoceras hammelli* is distinct from *G. gorbyi* in being a smaller form with a more rapidly expanding, straight rather than gently curved longiconic shell; in possessing oblique annuli that are less steeply inclined than those of the type species; and in the weaker development of these annuli with ontogeny, a feature not evident in *G. gorbyi*, in which the annuli remain prominent the entire length of the shell.

In the "Hitz Bed," *G. hammelli* occurs as well preserved if often fragmentary specimens in which the external shell, septa, and siphuncle segments are replaced with sparry calcite. This coarse spar replacement often obscures the details of the internal features of the shell. *Gorbyoceras hammelli* occurs in the "Hitz Bed" in association with a diverse assemblage of nautiloids: *G. crossi* Flower, the actinocerid *Ormoceras hitzi* Foerste, the breviconic ellesmerocerid *Cyrtocerina madisonensis* (Miller), and a number of small oncocerids currently described as species of *Oncoceras* and *Diestoceras*. These nautiloids occur in association with the sponges *Dermatostroma* and *Stromatocerium*, a small solitary rugosan coral, rare articulate brachiopods, and a diverse fauna of well-preserved gastropods, pelecypods, and ostracodes.

GORBYOCERAS CF. G. TETREAUENSE WILSON, 1961

Plate 12, figure 3

Gorbyoceras tetreauense Wilson, 1961, p. 38, pl. 10, figs. 2–5.

Diagnosis.—Small, slender, rather rapidly expanding, laterally compressed, and endogastrically curved species of *Gorbyoceras* marked by oblique, low, rounded annuli that become less distinct with ontogeny.

Description.—Small (lengths up to 16 cm and shell diameters up to 2.0 cm), slender, moderately expanding (apical angle 8–8.5 degrees), endogastrically curved longicones with sutures initially straight, transverse; becoming slightly oblique with ontogeny, inclined from the dorsum to the venter in the adapical direction. Cameral lengths ranging from one-fourth to one-fifth the diameter of the shell, decreasing with ontogeny. Body chamber incompletely known. Shell marked by raised annuli that are transverse initially, becoming slightly oblique with ontogeny, inclined from the dorsum to the venter in the adoral direction, opposite the trend of the sutures. Annuli spaced 2 mm apart at a diameter of 7 mm, and 3 mm apart at a diameter of 15 mm. Exterior marked by alternating sets of coarse and fine longitudinal lirae.

Siphuncle central (SPR=0.50) at a shell diameter of 8 mm, becoming subcentral with ontogeny (SPR=0.375 at a diameter of 12 mm). Siphuncle segment shape and maximum diameter unknown. Endosiphuncular and cameral deposits unknown.

Stratigraphic and geographic occurrence.—Logana Member of the Lexington Limestone (Middle Ordovician, Kirkfieldian Stage), Woodford and Franklin Counties, Ky.

Studied material.—USNM 468708 (pl. 12, fig. 3) from USGS 5092-CO collection, lower part of the Logana Member, Tyrone quadrangle, Woodford County, Ky.; USNM 468707 from USGS 5073-CO collection, upper part of the Logana Member, Frankfort East quadrangle, Franklin County, Ky.; and USNM 468709 from USGS 7791-CO collection, lower part of the Logana Member, Salvisa quadrangle, Woodford County, Ky.

Remarks.—Fragmentary silicified specimens of *Gorbyoceras* from the Logana Member of the Lexington Limestone exposed in central Kentucky, west and south of Lexington, appear to be assignable to *G. tetreauense* described by Wilson (1961) from the "Leray beds" (Middle Ordovician, Rocklandian) of the Ottawa region of Quebec. This is indicated by the small, rather rapidly expanding, gently endogastrically curved longiconic shells represented in USGS collections. These specimens have an external ornament consisting of nearly transverse, raised annuli that are spaced increasingly further apart with ontogeny, prominent longitudinal lirae, and much weaker transverse lirae. The Kentucky specimens are laterally compressed ovate in cross section at diameters of

8 mm or less, becoming more circular in cross section with ontogeny. Wilson (1961) did not describe or illustrate a cross-sectional profile of her Canadian type material. The Kentucky specimens also exhibit annuli that become less strongly developed with ontogeny, similar to the younger Cincinnatian species *G. hammelli* (Foerste). This feature is not evident in Wilson's figured specimens. Wilson's type material (GSC 1274–75, 1961, pl. 10, figs. 2–5) indicates expanded globular segments at shell diameters of 8–12 mm. The nature of the siphuncle segments in the Kentucky material remains unknown, although these specimens indicate that the siphuncle is central in position at shell diameters of 10 mm or less. Pending better documentation of their internal features, these Kentucky specimens are only tentatively assigned to *G. tetreauense*.

Foerste (1910) described *Orthoceras (Spyroceras) bilineatum-frankfortense* from the Curdsville Limestone Member of the Lexington Limestone exposed at the Old Crow Distillery, southeast of Frankfort, Ky. (USGS locality 5092-CO). This species was represented by a single composite mold preserving a portion of the shell 35 mm in length having a maximum diameter of 15 mm. Foerste's figured specimen appears to be distinct from the specimens from the overlying Logana Member in the more closely spaced transverse annuli and more gradual rate of shell expansion evident in this specimen. More complete specimens of the Curdsville species are needed to more accurately compare these Kentucky species.

In the shales and thin limestone of the Logana Member, *G.* cf. *G. tetreauense* is an uncommon species in comparison with the abundant specimens of the associated orthocerid *Isorthoceras albersi* (Miller and Faber). In the USGS 5092-CO collection from the lower part of the Logana Member, *G.* cf. *G. tetreauense* makes up slightly less than 20 percent of the total number of nautiloids present. In the USGS 7791-CO collection from the same horizon, the species makes up less than 10 percent of the identified nautiloid material. In these collections, these nautiloids occur with the well-preserved fossil fauna of the monoplacophoran *Cyrtolites*, gastropods, pelecypods, and the dalmanellid brachiopod *Onniella*.

ORTHOCERIDA, FAMILY UNCERTAIN
GENUS *ANASPYROCERAS* SHIMIZU AND OBATA

Type species.—*Orthoceras anellus* Conrad, 1843, p. 334; designated by Shimizu and Obata, 1935, p. 4; "Beloit Member, Black River Formation," Blackriveran Stage, Middle Ordovician, Wisconsin.

Description.—Small, slender, gradually expanding orthoconic longicones, circular in cross section with simple, transverse sutures. Cameral chambers short. Body chamber unknown. Shell exterior distinctly marked by prominent, rather sharp, narrow, transverse annuli separated by equally deep interannular grooves. One annulus per camerae; annuli straight or sloping obliquely from the dorsum to the venter in the adapical direction. Shell surface marked by fine, raised transverse lirae.

Siphuncle central or subcentral in position, narrow, having segments tubular or expanding only very slightly within camerae. Septal necks short, orthochoanitic. Endosiphuncular and cameral deposits unknown.

Stratigraphic and geographic occurrence.—Middle Ordovician (Blackriveran Stage) to Upper Ordovician (Richmondian Stage), North America (Laurentia).

Remarks.—This genus is another of the annulated longiconic taxa proposed by Shimizu and Obata (1935). Their description was brief: "This genus is readily distinguished from *Spyroceras* by its more crowded, stronger annulations." Flower (1943), reviewing a number of these annulated genera, emended this description to include annulated longicones possessing siphuncles that were essentially tubular with straight, short, orthochoanitic septal necks. In his study of Cincinnatian cephalopods (Flower, 1946), he further included in this genus Upper Ordovician species that, in addition to the prominent annuli, also possessed alternating prominent and weak longitudinal lirae that formed nodelike projections on annular surfaces. Earlier Blackriveran species, including the type species *A. anellus*, appear to lack these longitudinal lirae. Much better preserved material, especially specimens preserving the critical features of the siphuncle, is necessary to resolve the uncertainty surrounding this genus, its morphology, and its affinities to other described taxa.

ANASPYROCERAS CF. *A. CYLINDRICUM* (FOERSTE), 1932

Plate 12, figure 5

Spyroceras cylindricum Foerste, 1932, p. 97, pl. 11, figs. 3, 6, 7.
"*Spyroceras*" *cylindricum* Foerste. Wilson, 1961, p. 35, pl. 9, figs. 1–3.

Diagnosis.—Slender species of *Anaspyroceras* distinguished by prominent transverse annuli that slope obliquely from the dorsum to the venter in the adapical direction having interannular spaces equal in width to annuli.

Description.—Small (length less than 30 cm and shell diameters up to 1.0 cm), slender (apical angle 3–3.5 degrees), orthoconic longicones; circular in cross section. Sutures and cameral lengths unknown. Body chamber unknown. Exterior marked by prominent raised annuli, averaging 1 mm in width, separated by interannular spaces of equal width. Annuli sloping obliquely from dorsum to venter in the adapical direction.

Siphuncle central in position. Shape of siphuncle segments unknown, segments apparently narrow in diameter. Septal necks unknown. Endosiphuncular deposits and cameral deposits unknown.

Stratigraphic and geographic occurrence.—Curdsville Limestone Member of the Lexington Limestone (Middle Ordovician, Kirkfieldian Stage), Curdsville, Ky.

Studied material.—USNM 48290, "E.O. Ulrich Collection," Curdsville, Ky. (pl. 12, fig. 5). Also studied was the holotype of *Spyroceras cylindricum* Foerste (AMNH 807B), which Foerste (1932, pl. 11, figs. 3, 6–7) listed as coming from "Coarse-grained part of the Trenton Formation, Watertown, New York."

Remarks.—The Curdsville Limestone Member specimen illustrated here is similar to several slender, annulated, longiconic nautiloid species distinguished by the oblique nature of the annuli. These species are "*Spyroceras*" *arcuoliratum* (Hall), "*S.*" *cylindricum* Foerste, both from the "Trenton Limestone" in New York, and "*S.*" *allumettense* Foerste from the "Leray-Rockland beds" in the Ottawa, Ontario region. All three species are based on fragmentary specimens whose internal features are unknown. Foerste (1932–1933) differentiated "*S.*" *cylindricum* from "*S.*" *arcuoliratum* on the basis of its less rapidly expanding shell, noting that, otherwise, these taxa are closely similar. Foerste distinguished "*S.*" *allumettense* from "*S.*" *arcuoliratum* by the lesser obliquity of the annuli in the former species.

The Kentucky specimen is distinct from "*S.*" *allumettense* (GSC 6830) in the greater obliquity of the annuli. The apical angle of USNM 48290 (3 degrees) is less than that of the "*S.*" *arcuoliratum* (apical angle 5 degrees) and indicates placement in "*S.*" *cylindricum*.

All of these species are similar to the type species of *Anaspyroceras*, which is *Orthoceras anellus* Conrad, in the gradually expanding slender form of the shell and in the closely spaced, raised, narrow annulations with narrow, rounded crests. Assignment of the Curdsville specimen to *Anaspyroceras* is tentative, pending discovery of additional specimens that illustrate the morphological features of the siphuncle segments of this taxon.

This form is an uncommon species within the Curdsville Limestone Member, known only from the partially silicified external mold described and illustrated here. It occurs in a coarsely crystalline, fossiliferous grainstone associated with echinoderm fragments and an unidentified rhynchonellid brachiopod.

GENUS *POLYGRAMMOCERAS* FOERSTE

Type species.—*Polygrammoceras twenhofeli* Foerste, 1928a, p. 263; by original designation; Ellis Bay Formation, Gamachian Stage, Upper Ordovician, Anticosti Island, Quebec.

Description.—Medium to large, moderate to rapidly expanding orthoconic longicones; circular in section with straight, transverse sutures. Cameral chambers short, cameral lengths ranging from one-fourth to one-seventh the diameter of the shell, decreasing with ontogeny. Body chamber incompletely known, at least one-fifth the total length of the shell; aperture unknown. Periphract dorsomyarian. Shell surface marked by fine to coarse, sharply delineated, longitudinal striae.

Siphuncle central in position adapically, becoming more ventral with ontogeny. Segments expanded within camerae, globular, equiaxial to slightly compressed; narrow in diameter, one-fifth the diameter of the shell. Septal necks cyrtochoanitic, short, recurved, with free brims. Endosiphuncular deposits unknown. Mural-episeptal cameral deposits developed adapically.

Stratigraphic and geographic occurrence.—Middle Ordovician (Shermanian) to Lower Silurian (Wenlockian Series), North America (Laurentia).

Remarks.—Foerste (1928a) erected this genus for Ordovician longicones that had an external ornament of longitudinal striae, ribs, or bars but that lacked fluting. It became common practice to place here any longitudinally striated longiconic orthocerid. Troedsson (1932) placed in *Polygrammoceras* a large number of species from the lower Middle Ordovician (Llanvirnian) "*Orthoceras* Limestone" of the Baltic region. All of these species possessed longitudinally striated shells, short orthochoanitic septal necks, and subtubular siphuncle segments. All three species of *Polygrammoceras* described by Foerste (1928a) from Anticosti Island, including the type species *P. twenhofeli*, possess expanded, beadlike siphuncle segments and cyrtochoantic septal necks. It is clear that *Polygrammoceras* should be restricted to longitudinally striated shells that possess expanded siphuncle segments similar to these Anticosti species. Troedsson's species belong elsewhere. One of his species, *P. endoceroides* Troedsson, is the type species for the genus *Troedssonella* Kobayashi.

The familial affinities of *Polygrammoceras* remain uncertain. The general shell morphology and gross form of the siphuncle segments are similar to those of the Upper Ordovician proteoceratid *Treptoceras* Flower. Unfortunately, neither complete sectioned specimens of *Polygrammoceras* nor endosiphuncular deposits are known. In the absence of evidence of ontogenetic siphuncle segment decompression and the presence of proteoceratid endosiphuncular deposits, this genus cannot be placed with certainty in the Proteoceratidae. Sweet (1964) placed *Polygrammoceras* in the subfamily Kionoceratinae within the Orthoceratidae. The features of the siphuncle segments in the type species of *Polygrammoceras* clearly are not those of a member of the Orthoceratidae. As is indicated above, the subfamily Kionoceratinae is a largely artificial grouping of taxa. Longitudinal ornament of the type present in *Polygrammoceras* evolved a number of times in the Ordovician and Silurian in a number of different orthocerid lineages.

POLYGRAMMOCERAS SP. A

Plate 12, figures 1, 2

Diagnosis.—Moderately expanding species of *Polygrammoceras* distinguished by its external ornament of narrow, longitudinal striae separated by smooth interspaces that are twice as broad as the striae.

Description.—Medium-sized (lengths of 30 cm and shell diameters up to 3.0 cm), moderately expanding (apical angle 7 degrees) orthoconic longicone; circular in cross section with straight, transverse sutures. Cameral lengths ranging from one-fourth to one-seventh the diameter of the shell, decreasing with ontogeny. Shell surface marked by narrow, longitudinal grooves incised at 1-mm intervals in specimens 20–30 mm in diameter. The intervening areas smooth, twice as broad as the striae.

Siphuncle central in position (SPR=0.50) at diameters less than 10 mm, becoming subcentral with ontogeny (SPR=0.428 at diameters of 15 mm). Siphuncle segments expanded, globose, changing with ontogeny from equiaxial (SCR=1.00) at shell diameters less than 10 mm, to slightly compressed (SCR=1.20) at diameters of 15 mm, to equiaxial in adoral portions of the shell at diameters in excess of 25 mm. Septal necks cyrtochoanitic, recurved, with free brims in tangential sections. Endosiphuncular deposits unknown. Cameral deposits consisting of mural-episeptal deposits that are lobate, botryoidal ventrally with a medial ventral mural delta. Deposits well-developed in adapical portions of shell less than 15 mm in diameter.

Stratigraphic and geographic occurrence.—Grier Limestone Member and Strodes Creek Member of the Lexington Limestone (Shermanian-Edenian), Clark County, Ky.

Studied material.—Incomplete phragmocone, 135 mm in length, (USNM 468710; pl. 12, figs. 1, 2) and additional fragmentary silicified material from the USGS 5067-CO collection, Grier Limestone Member, Ford, Clark County, Ky.; USNM 468729, Strodes Creek Member, roadcut Route 89, 5 mi east of Winchester, Clark County, Ky.

Remarks.—This species is based primarily on silicified material described above from the Grier Limestone Member in Clark County, Ky. An external mold of a similar longitudinally striated longiconic nautiloid impressed on the base of a stromatoporoid coenosteum (USNM 468729) was also collected by the writer from the higher Strodes Creek Member of the Lexington Limestone, also in Clark County, Ky. The distinctive, longitudinally incised external ornament and relatively shallow camerae, coupled with the expanded, globular siphuncle segments (pl. 12, figs. 1, 2) preserved in the Grier Limestone Member material indicate placement in the genus *Polygrammoceras* Foerste as based on the type species *P. twenhofeli* Foerste from Upper Ordovician strata exposed on Anticosti Island, Quebec.

This writer has elected not to erect a new species on the basis of this material at this time due to the lack of abundant, well-preserved specimens representative of the species as well as the limited knowledge of the important morphological features of the type species and other previously described species of *Polygrammoceras*. The Lexington Limestone species is distinct from the type species *P. twenhofeli* Foerste in possessing narrower striae with broader interspaces in contrast to the broader striae with narrow interspaces described for the Upper Ordovician type species (Foerste, 1928a). It is distinct from *P. ellisensis* Foerste (1928a) from the Ellis Bay Formation (Upper Ordovician, Gamachian) on Anticosti Island, in possessing flat, indistinct "ribs" compared with the more prominent, rounded longitudinal "ribs" characteristic of the Canadian form. Comparisons with the third species described by Foerste from the Ellis Bay Formation on Anticosti Island, *P. latolineatum* Foerste (1928a), are difficult to make due to a lack of specimens at comparable diameters, the Lexington Limestone species being represented by specimens less than 30 mm in diameter and *P. latolineatum* being known from more adoral sections of the phragmocone greater than 35 mm in diameter.

In the 5067-CO collection from the Grier Limestone Member, the species occurs in association with abundant specimens of the small orthocerid *Isorthoceras albersi* (Miller and Faber) (pl. 10, figs. 5, 6), plus the univalves *Cyrtolites* and *Lophospira*, and the pelecypods *Deceptrix* and *Modiolodon*. *Polygrammoceras* sp. A can be distinguished from the associated *I. albersi* in being a larger form with an ornament of incised longitudinal grooves in contrast to the raised longitudinal lirae characteristic of the latter species. Additionally, siphuncle segments in *P.* sp. A are expanded and globular in outline, in contrast to the depressed, barrel-shaped segments in *I. albersi*. *Isorthoceras albersi* is about 10 times more abundant than *Polygrammoceras* sp. A in these collections.

Species of the genus *Treptoceras* are similar in general shell morphology, siphuncle segment position and shape, and cameral length at comparable diameters. They can be distinguished from *Polygrammoceras* only if the exterior ornament of the shell is preserved, as *Treptoceras* lacks the distinctive incised longitudinal striae characteristic of *P.* sp. A.

POLYGRAMMOCERAS? CF. P. SP. A

Plate 11, figures 1–3

Diagnosis.—Very large, moderately expanding orthoconic longicone with subcentral, slightly expanded siphuncle segments one-sixth to one-eighth the diameter of the shell, decompressing with ontogeny; moderately long, suborthochoanitic septal necks.

Description.—Large (known portions of the shell indicating lengths up to 1.13 m and a maximum observed shell diameter of 9.0 cm), moderately expanding (apical angle 7 degrees) orthoconic longicone; circular in cross section with straight, transverse sutures. Cameral chambers short, cameral lengths ranging from one-sixth the diameter of the shell at a diameter of 70 mm to one-ninth the diameter of the shell at a diameter of 90 mm. Body chamber unknown. Shell exterior unknown.

Siphuncle subcentral in position for the known length of the shell, having a SPR value of 0.33 at a diameter of 70 mm to a value of 0.28 at a diameter of 90 mm. Segments slightly expanded, compressed (SCR=1.10) adapically at a diameter of 70 mm, becoming equiaxial (SCR=1.00), plano-convex at a diameter of 90 mm. Siphuncle segments narrow, diameter ranging from one-sixth the diameter of the shell adapically, to slightly less than one-eighth the diameter of the shell adorally; segments decompressing with ontogeny. Septal necks moderately long, suborthochoanitic both dorsally and ventrally. No endosiphuncular or cameral deposits are present in preserved portion of the phragmocone.

Stratigraphic and geographic occurrence.—Strodes Creek Member of the Lexington Limestone (Edenian), Winchester region, Clark County, Ky.

Studied material.—USNM 468694 (pl. 11, figs. 1–3) from the USGS 7310-CO collection.

Remarks.—This form is represented by a single, very large fragment of a phragmocone from the Strodes Creek Member of the Lexington Limestone. The specimen is a well-preserved, calcite-replaced internal mold that has been sectioned dorsoventrally to reveal the siphuncle segments. This specimen consists of 21 camerae and is compressed laterally as the result of sediment compaction. Indications are that the original shell cross section was circular. The siphuncle segments and some of the camerae are filled with a fine-grained matrix of blue-gray, calcareous mudstone that weathers to a buff yellow. A majority of the camerae are, however, filled with coarse, sparry calcite (pl. 11, fig. 1). A number of the septa are fractured, especially in the section of the shell between 75 and 80 mm in diameter. This fracturing is probably due to postdepositional compaction of the shell and the enclosing sediments.

The affinities of this specimen are difficult to determine due to the incomplete nature of this specimen and the lack of additional material preserving more adapical portions of the phragmocone. The large size of this specimen makes it difficult to compare with any other orthocerid from the Ordovician of the Cincinnati arch region. The only comparable species in terms of shell diameter is *Ordogeisonoceras amplicameratum* (Hall) from the Shermanian part of the Point Pleasant Tongue of the Clays Ferry Formation in northern Kentucky. This species is, however, distinct from USNM 468694 in having a more gradually expanding shell and deeper camerae at comparable diameters, a more excentric, dorsally located siphuncle, and orthochoanitic rather than suborthochoanitic septal necks.

The suborthochoanitic septal necks, lack of cameral or endosiphuncular deposits, and adoral decompression of siphuncle segments all indicate that this fragmentary specimen represents the adoral end of a very large orthocerid. Similar morphological features have been observed in much smaller specimens of the proteoceratid *Treptoceras* Flower (pl. 5, figs. 1, 7). This suggests that the 468694 specimen might represent the adoral end of an enormous species of *Treptoceras*. Species of *Treptoceras* are currently unknown from strata older than Late Ordovician (Edenian), although "*Ormoceras*" *lindsleyi* Foerste and Teichert (1930) from the Shermanian part of the Cannon Limestone in central Tennessee looks like *Treptoceras*.

As indicated above, comparably sized specimens of *Treptoceras* and *Polygrammoceras* cannot easily be differentiated unless the external shell is preserved. *Treptoceras* is basically a smooth-shelled form, in contrast to *Polygrammoceras*, which has a distinctive external ornament of sharply incised longitudinal striae. Specimens of *Polygrammoceras* sp. A have been collected from both the Grier Limestone Member and Strodes Creek Member of the Lexington Limestone in Clark County, Ky., the Strodes Creek Member in the same county being the source of the 468694 specimen. *Polygrammoceras latolineatum* Foerste from the Upper Ordovician Ellis Bay Formation on Anticosti Island approaches the USNM 468694 specimen in size, represented by an adapical portion of the phragmocone at least 40 cm in length and up to 6.0 cm in diameter. This, in conjunction with the occurrence of *Polygrammoceras* sp. A in the Strodes Creek Member in the Winchester region, suggests that this large fragmentary specimen may be an internal mold of the adoral end of a large species of *Polygrammoceras* rather than a species of *Treptoceras*. No specimens of *Treptoceras* have been collected from the Strodes Creek Member in this area by the writer. More complete material from these strata preserving the adapical portions of the phragmocone and the external shell surface is needed in order to more fully determine the affinities of this interesting specimen.

POLYGRAMMOCERAS? SP. B

Plate 12, figures 8, 9

Diagnosis.—Small, slender, gradually expanding longiconic orthocone marked by broadly spaced, narrow longitudinal furrows about one-fourth as wide as the intervening smooth spaces.

Description.—Small (15 cm in length and shell diameters up to 1.3 cm), slender, gradually expanding (apical angle 5.5 degrees) orthoconic longicone; circular in cross section with straight, transverse sutures. Sutures poorly defined, cameral lengths one-sixth the diameter of the shell at a shell diameter of 10 mm. Body chamber unknown. Shell exterior marked by broadly spaced (1 mm apart), narrow, longitudinal, sharply defined squared-off furrows about one-fourth as wide as the intervening smooth spaces.

Siphuncle central in position adapically (SPR=0.50 at a diameter of 10 mm). Nature of siphuncle segments unknown. Septal necks unknown. Endosiphuncular and cameral deposits unknown.

Stratigraphic and geographic occurrence.—Upper Rocklandian part of the Tyrone Limestone, Jessamine County, Ky.

Studied material.—USNM 468711 (pl. 12, figs. 8, 9) from the silicified USGS 6034-CO collection, Little Hickman section A, Jessamine County, Ky.

Remarks.—This form is known from a single fragment of a portion of a phragmocone, 35 mm in length, from the USGS 6034-CO collection from the Tyrone Limestone. The external ornament, consisting of sharply defined, squared-off longitudinal grooves separating broader, flat interspaces, is very similar to that described for species of *Polygrammoceras* by Foerste. The critical features of the siphuncle segments are, however, unknown for this specimen, making its placement in *Polygrammoceras* uncertain. In an unpublished manuscript, Flower placed this Kentucky specimen with a group of small, slender, longitudinally marked longicones he placed previously (Flower, 1952) in *Kionoceras*, Hyatt. These species occur in the Ordovician, ranging from the Chazyan to the Shermanian. Flower expressed doubt that these forms were related to typical Silurian *Kionoceras*, which he described as being much larger species with expanded siphuncle segments. Additional material preserving the key features of the siphuncle segments are necessary to further determine the affinities of this species.

GENUS *RICHMONDOCERAS* NEW GENUS

Type species.—*Richmondoceras brevicameratum*, n. sp.; designated here; Saluda Formation and upper part of the Bull Fork Formation, Richmondian Stage, Upper Ordovician, Indiana and Ohio.

Description.—Medium to large, slender, gradually expanding orthoconic longicones; circular in section with straight, transverse sutures. Cameral chambers short, cameral lengths decreasing with ontogeny. Body chamber long and tubular, at least one-fifth the total length of the shell, aperture unknown. Periphract dorsomyarian. Shell exterior unknown, apparently smooth.

Siphuncle midventral in position the length of the phragmocone. Siphuncle segments greatly expanded within camerae having segment diameter decompressing with ontogeny from one-third to one-sixth the diameter of the shell. Segments highly compressed, the length of the phragmocone. Septal foramina broad, averaging half the total diameter of the siphuncle segment. Septal necks cyrtochoanitic and recumbent dorsally and ventrally in more adapical portions of the phragmocone; septal neck, highly curved and short with free brims adorally in large specimens. Connecting rings thin, adnate adapically and ventrally to free part of the septa. Endosiphuncular deposits greatly retarded in their development, consisting of thin adapical parietal deposits. Episeptal and hyposeptal cameral deposits restricted to adapical portions of phragmocone.

Stratigraphic and geographic occurrence.—Upper Ordovician (Richmondian Stage) strata in East-Central North America (Laurentia).

Remarks.—This genus is erected for medium to large, gradually expanding orthoconic longicones possessing short camerae and midventral siphuncles with greatly expanded, compressed segments and the limited development of both endosiphuncular and cameral deposits. The genus is currently represented only by the type species *Richmondoceras brevicameratum* n. sp., which is restricted in its occurrence to upper Richmondian strata in southeastern Indiana and southwestern Ohio.

The affinities of *Richmondoceras* are uncertain. The recumbent cyrtochoanitic septal necks and ontogenetic decompression of siphuncle segments are features common to the order Actinocerida. The greatly retarded development of endosiphuncular deposits is, however, distinctly unlike members of this order, and the thin parietal deposits evident in specimens of *R. brevicameratum* are more similar to those occurring in members of the order Orthocerida. At least one orthocerid family, the Proteoceratidae, also exhibits siphuncle segment decompression with ontogeny and the development of parietal endosiphuncular deposits (pl. 13, figs. 7, 8). The uniformly developed recumbent septal necks and broad septal foramina in *Richmondoceras* are, however, not typical of members of the Proteoceratidae. Siphuncle segments are similar to those illustrated (Sweet, 1964) for members of the Greenland oceratidae, a problematic group of lower and middle Paleozoic orthocerids. *Richmondoceras* seems to lack the thin endosiphuncular linings that are described as being characteristic of this group. Until a larger suite of more complete specimens of *Richmondoceras* is collected and studied, the affinities of this genus will remain uncertain. Its known shell morphology is sufficiently distinct from other Cincinnatian longiconic nautiloids to warrant the erection of a new genus.

Etymology.—The genus gets its name from its occurrence in Richmondian strata exposed in southeastern Indiana and southwestern Ohio.

RICHMONDOCERAS BREVICAMERATUM NEW SPECIES

Plate 13, figures 1–8

Diagnosis.—Medium to large, slender, gradually expanding orthoconic longicone distinguished by its short camerae, expanded, compressed, midventral siphuncle segments, and weak development of both endosiphuncular and cameral deposits.

Description.—Medium to large (up to 73.5 cm in length and 6.5 cm in diameter), slender, gradually expanding (apical angle 6 degrees) orthoconic longicone; circular in section with straight, transverse sutures. Cameral chambers short having cameral lengths decreasing with ontogeny from one-sixth the diameter of the shell at a shell diameter of 20 mm to one-ninth the diameter of the shell at a shell diameter of 40 mm. Adoralmost septa in large specimens becoming approximate with ontogeny. Body chamber long, tubular, at least one-fifth the total length of the shell. Aperture unknown. Shell exterior apparently smooth.

Siphuncle position ranging from subcentral (SPR=0.40) at shell diameters of 20 mm or less, to midventral (SPR=0.25) at shell diameters of 50 mm. Siphuncle segments greatly expanded the length of the phragmocone, ranging from highly compressed (SCR=3.0) at a shell diameter of 25 mm to compressed (SCR=1.25) at a diameter of 50 mm. Segment diameter ranging from one-third the diameter of the shell at shell diameters of 20 mm or less to slightly less than one-sixth the diameter of the shell at a diameter of 50 mm. Septal foramina broad, the length of the siphuncle, ranging from nearly one-half to two-thirds the maximum diameter of the siphuncle. Septal necks cyrtochoanitic, short, recumbent most of the length of the phragmocone. Brims becoming free in adoral portions of the shell, 45–50 mm in diameter. Connecting rings adnate ventrally to both adapical and adoral septa at diameters of 35–40 mm; adnate ventrally to only the adapical septum at diameters greater than 45 mm. Endosiphuncular deposits consisting of small annuli and thin, ventral parietal deposits observed in portions of the phragmocone 15 mm or less in diameter. Cameral deposits consisting of hyposeptal and episeptal deposits occurring at shell diameters of 40 mm or less.

Stratigraphic and geographic occurrence.—Saluda Formation (Upper Ordovician, Richmondian Stage) and its equivalents exposed in Ripley County, Ind., and Preble County, Ohio; upper part of the Bull Fork Formation (Richmondian), Clinton County, Ohio.

Studied material.—Holotype MU 398T (pl. 13, figs. 1, 3, 4) from the Whitewater Formation along Laugherty Creek, north of Versailles, Ripley County, Ind.; paratype MU 399T (pl. 13, fig. 2), Saluda Formation along Four-Mile Creek, Preble County, Ohio; plus paratypes MU 400T, a suite of specimens from the uppermost Bull Fork Formation at the Cowan Lake spillway, Wilmington, Clinton County, Ohio (pl. 13, figs. 5–8).

Remarks.—This species is represented by a limited number of specimens from the fine-grained limestones of the Saluda Formation in southeastern Indiana. These specimens consist of isolated adoral portions of large individuals, often with attached portions of the body chamber (pl. 13, figs. 1, 2). *Richmondoceras brevicameratum* is, however, most abundant in the interbedded blue clay shale and thin fossiliferous wackestone of the uppermost Bull Fork Formation exposed in the vicinity of Wilmington, Clinton County, Ohio. Specimens typically consist of incomplete matrix-filled, occasionally calcite-replaced internal molds representing portions of the phragmocone and isolated, often crushed body chambers (pl. 13, figs. 5–8). Specimens from the Cowan Lake locality are often covered with holoperipheral encrustations of the sponge *Dermatostroma*. These specimens occur in association with similarly preserved specimens of the orthocerid *Treptoceras*, the endocerid *Cameroceras*, and rare specimens of the large oncocerid *Diestoceras* (pl. 21, fig. 4). This low-diversity nautiloid fauna occurs in mud-rich facies deposited in shallow-marine, mud-bottom environments that supported thickets of small, twiglike ramose bryozoans. Also abundant are the orthid brachiopod *Hebertella*, the pelecypods *Ischyrodonta*, *Pseudocolpomya*, and *Caritodens*, the large asaphid trilobite *Isotelus*, and large numbers of the ornate ostracode *Drepanella richardsoni*.

Richmondoceras brevicameratum can be distinguished from the externally similar, slender, gradually expanding longiconic orthocerid *Treptoceras fosteri* (Miller) (pl. 6, figs. 4, 5, 7, 8) by its more expanded, broader diameter siphuncle segments. In southwestern Ohio, specimens of *T. fosteri* are most common near the base of the "Liberty biofacies" in the Bull Fork Formation, whereas specimens of *R. brevicameratum* appear to be restricted to the uppermost portions of the same formation. A comparison of the morphological features of *R. brevicameratum* and *T. fosteri* is presented in figure 19.

Richmondoceras brevicameratum is a locally common species characteristic of upper Richmondian shallow-marine facies across the northern periphery of the Kentucky-Indiana-Ohio region. Currently, the species is known only from fragmentary specimens that, however, allow for a composite reconstruction of almost the entire shell greater than 15 mm in diameter. A larger suite of more complete specimens preserving the adapical portions of the phragmocone, the siphuncle segments, and their internal features are needed to more fully determine the systematic affinities of this species and this genus.

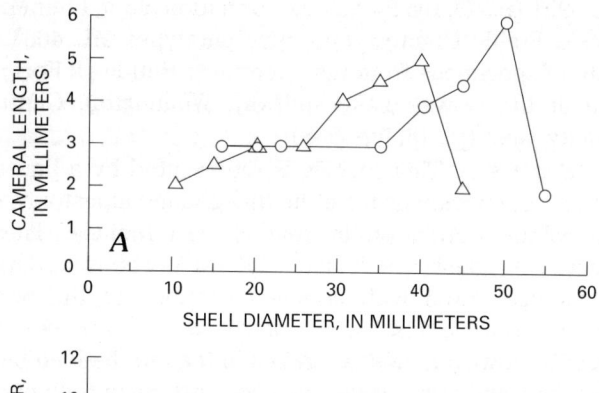

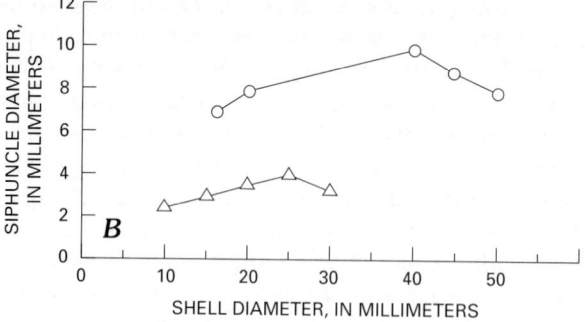

FIGURE 19.—Bivariate plots of cameral length versus shell diameter, A, and siphuncle diameter versus shell diameter, B, for specimens of *Richmondoceras brevicameratum* (circles) and *Treptoceras fosteri* (triangles) from the upper part of the Bull Fork Formation in southwestern Ohio. Specimens of *R. brevicameratum* from locality OH-10, Clinton County, Ohio. Specimens of *T. fosteri* from locality OH-7, Warren County, Ohio, and locality OH-11, Clinton County, Ohio. Note the uniformly larger siphuncle diameters typical of specimens of *R. brevicameratum*.

Etymology.—The species name denotes the uniformly short nature of the cameral chambers, which run the length of the phragmocone in this form.

SUBCLASS ENDOCERATOIDEA TEICHERT

Medium-size to very large (up to 6 m in length) orthoconic longicones and, less commonly, brevicones, both with large tubular, generally ventral siphuncles. Septal necks may be short in primitive forms, becoming elongate holochoanitic or macrochoanitic in younger taxa. Connecting rings primitively thick, becoming thin, lining the inner surfaces of septal necks in younger forms. Buoyancy regulation via calcareous ballast in the form of telescoping inverted cones (endocones) concentrated in the apical portions of the siphuncle and pierced by a narrow, open endosiphuncular tube. Cameral deposits absent.

Stratigraphic range.—Lower Ordovician to Lower Silurian.

ORDER ENDOCERIDA TEICHERT

Features of the subclass.

Stratigraphic range.—Lower Ordovician to Lower Silurian.

FAMILY ENDOCERATIDAE TEICHERT

Straight, large to extremely large longiconic shells having large ventral siphuncles that are generally from one-third to two-thirds the diameter of the shell. Septal necks holochoanitic or macrochoanitic with thin connecting rings that form linings on the inner surfaces of septal necks. Apical portions of the siphuncle inflated in some forms, occasionally filling the entire apical portion of the shell. Endosiphuncular structures generally consisting of the endosiphuncular tube and simple blades.

Stratigraphic range.—Lower Ordovician (upper Ibexian Provincial Series) to Lower Silurian (Wenlockian Series).

Remarks.—Flower (1955a) noted some of the problems surrounding the classification and identification of members of this group. The fragmentary nature of most described species results in a lack of knowledge of the morphology of the entire shell, leading to a profusion of names, each representing a particular portion of the same organism. Equally problematic is the tendency for the endosiphuncles to be completely recrystallized during diagenesis, often obliterating the critical structures within. These critical structures are the shape and form of the endocones, the position and cross-sectional shape of the endosiphuncular tube, and the number and configuration of radial plates, termed "blades." The conservative nature of members of this family also make classification and identification difficult, most taxa being long ranging and exhibiting similar, simple endosiphuncular structures. All of these problems plague the present study, especially the diagenetic alteration of the endosiphuncle and the fragmentary nature of most endoceratid material, leading to the tentative identification of most of these specimens.

Endocerids have not previously been described from the Middle Ordovician High Bridge Group and the Lexington Limestone in Kentucky and equivalent strata in Tennessee. Large endocerids have been described in a general sense from the overlying Cincinnatian Provincial Series in the area. Flower (1946) presented a review of the described taxa from the region. Most of these specimens have traditionally been referred to one of several large species of "*Endoceras*" described by Hall (1847) from late Middle Ordovician (Shermanian) strata in New York. A majority of these species are currently inadequately known. The morphology of the best known of these species, *Endoceras proteiforme* Hall, is discussed below.

GENUS *CAMEROCERAS* CONRAD

Type species.—*Cameroceras trentonense* Conrad, 1842, p. 267, by original designation; Trenton Limestone, Shermanian Stage, Middle Ordovician, New York.

Description.—Large to extremely large (up to 6 m in length) orthoconic longicones with circular to slightly depressed cross sections and straight, transverse sutures. Slight ventral lobe developed in some species. Cameral chambers typically short, from one-fourth to one-tenth the diameter of the shell, the cameral length decreasing with ontogeny relative to shell diameter. Body chamber large, up to one-third the total length of the shell; aperture simple, transverse. Periphract dorsomyarian. Shell exterior smooth. Dorsum marked by longitudinal color bands.

Siphuncle large, tubular, diameter up to 60 percent the diameter of the shell adorally; typically ventral in position, less commonly removed from the ventral shell margin. Septal necks holochoanitic; foramina broad, equal to the maximum width of siphuncle tube. Connecting rings narrow, extending from the tip of one septal neck adapically to that of the next. Endocones simple; endosiphuncular tube narrow, circular to depressed, elliptical in cross section, central to ventral in position. Number and configuration of endosiphuncular blades poorly known, reported to consist of three blades. Apical end of the siphuncle gradually tapering to an acute apex, well marked to the apical tip by septal "annuli." Siphuncle is a little more than one-third the diameter of the shell at this point.

Stratigraphic and geographic occurrence.—Middle Ordovician (Whiterockian Provincial Series) to Lower Silurian (Wenlockian Series), Laurentia, Baltica, Siberian Platform.

Remarks.—There has been much confusion concerning the relationship of this genus, based on the apical end of a large endocerid from the Trenton Limestone (Middle Ordovician) of New York, and the genus *Endoceras* Hall, which is based on more adoral portions of a large endocerid from the same strata. A good discussion of these taxa was presented by Flower (1955a). It has been suggested by a number of authors that *Cameroceras trentonense* Conrad, the type species of *Cameroceras*, was the apical end of one of the large endocerid species described by Hall in 1847. Since Conrad's genus was described in 1842, his name would have priority over *Endoceras* Hall, which would become a junior synonym of *Cameroceras*. Hall, however, did not designate a type species for his genus. Subsequently, Miller (1889) designated *Endoceras annulatum* Hall as the type species of *Endoceras*. This species differs from other large endocerids from the Trenton Limestone in possessing an annulated exterior. These annulated species, placed in the genus *Cyclendoceras* by Grabau and Shimer (1910), are properly referred to as true *Endoceras* Hall. Similar large endocerids with smooth shell exteriors are currently placed in the genus *Cameroceras* Conrad.

Flower (written commun., 1981) indicated that *Endoceras* is most typical of the Lowville and Black River Limestones in New York, where it is uncommon in younger (Kirkfieldian and Shermanian) strata. The genus is also widespread in the Platteville Formation or Group (Blackriveran and Rocklandian) of the upper Mississippi valley region and in the Upper Ordovician Red River Formation and its equivalents in Western and Arctic North America. *Cameroceras* is the common large endocerid in the Trenton Group (Kirkfieldian and Shermanian) in New York and adjacent states, and in younger Cincinnatian strata in Eastern North America (Frey, 1989). Gil (1988) has described and illustrated what appear to be the earliest representatives of *Cameroceras* from the early Middle Ordovician (Whiterockian) Juab Limestone of Utah.

The critical internal features of the siphuncle are poorly documented for *Cameroceras*. Conrad's (1842) original description of the genus was brief: "Straight, siphuncle marginal; longitudinal septum, forming a roll or involution with the margin of the siphuncle." Foerste (1924b) described *Cameroceras*, based on *C. trentonense*, as having short, rapidly tapering oblique endocones, the straight ventral part being in contact with the ventral wall of the siphuncle and the endocones being asymmetrical about the central axis of the siphuncule, thicker dorsally than ventrally. He described the endocones of *Endoceras proteiforme* Hall as being long, few in number, free from contact with the siphuncle wall, and symmetrical about the central axis of the siphuncle. Flower (1955a) described the siphuncular structure in *E. proteiforme* as consisting of endocones, neither short nor widely spaced, the endosiphuncular tube circular in section, situated close to the ventral wall of the siphuncle. The latter is the condition described for *Cameroceras* by Foerste. Flower reiterated the suggestion by Sardeson (1930) that *Cameroceras trentonense* and the various Trenton Group species of *Endoceras* described by Hall were different portions of the same nautiloid. Flower noted that endosiphuncular blades were not clearly preserved in these forms due to coarse recrystallization of the siphuncle. In the same paper, however, he illustrated (Flower, 1955a, fig. 2J) a cross section stated to represent the endosiphuncular structure typical of several Trenton Group species of "*Endoceras*." In these species the endosiphuncular tube is highly elliptical and supported by three blades. In his review of Chazyan and Mohawkian endocerids, Flower (1958) described the endosiphuncular tubes of several species of "*Endoceras*" (=*Cameroceras* of this report) as being depressed, elliptical in cross section, terminating in the ventral half of the siphuncle. Flower (1958) also indicated that *Cyclendoceras* Grabau and Shimer (=true *Endoceras*) consists of large, straight shells with variably developed, raised, transverse annuli and large siphuncle situated close to the ventral margin of the shell. Septal necks were described as being holochoanitic; endocones subcircular, terminating in a central endosiphuncular tube. Endosiphuncular blades were not known for the genus nor was the apical end of the siphuncle, although Flower speculated that the apical end in this genus was of the *Nanno* type, inflated apically to fill the apical end of the shell. Recent collections of specimens of *Endoceras* from the Upper Ordovician Beaverfoot Formation in British Columbia have confirmed this speculation.

More recent descriptions of these later Ordovician Endoceratidae are limited. Gil (1988) described and illustrated three species of *Cameroceras* from Whiterockian strata in Utah, some forms of which have slightly compressed cross sections and simple endocones that terminate in either centrally or ventrally positioned endosiphuncular tubes. Blades could not be distinguished due to coarse recrystallization. Frey (1983, 1988) described and figured *Cameroceras* cf. *C. inaequabile* (Miller) from the Upper Ordovician (Richmondian) of the Cincinnati area. This species is known from isolated recrystallized siphuncles and complete, if extensively crushed, shells up to 1 m in length, which include the body chamber and apical end of the shell. The siphuncle is large, situated close to the ventral margin of the shell, constituting 30 percent of the diameter of the shell at the apex and expanding to nearly 60 percent the diameter of the shell at adoral diameters in excess of 80 mm. Septal necks are holochoanitic. Endocones are somewhat retarded in their development, consisting of long, thin cones that terminate in a central endosiphuncular tube that becomes depressed, elliptical in cross section adapically. Clearly, better preserved, more complete material is critical to developing a more rigorous definition of *Cameroceras* as well as other members of the Endoceratidae.

CAMEROCERAS ROWENAENSE NEW SPECIES

Plate 14, figures 1, 2; figure 21*A*

Diagnosis.—Moderately expanding species of *Cameroceras* distinguished by a siphuncle that is removed from the ventral margin of the shell adorally and which has a subcentral endosiphuncular tube whose axis is situated midway between the center and the venter.

Description.—Medium-sized (portions of phragmocone up to 70 cm in length and shell diameters up to 8.0 cm), moderately expanding (apical angle 7.5–8 degrees) orthoconic longicones; circular to slightly depressed in cross section with straight, transverse sutures. Cameral chambers short, cameral lengths one-sixth the diameter of the shell at a diameter of 55 mm and one-ninth the diameter of the shell at a shell diameter of 75 mm. Body chamber unknown. Shell exterior basically smooth, marked by fine transverse growth lines.

Siphuncle large and tubular, removed from the ventral margin of the shell by a distance of 4–7 mm in shell sections that are 50–75 mm in diameter. Siphuncle makes up from one-third to slightly over half the total diameter of the shell, increasing in diameter relative to shell diameter with ontogeny. Siphuncle circular in cross section adapically, becoming depressed in section adorally, slightly wider than high. Septal necks holochoanitic. Mural portions of septa extending one-third to one-half the length of the cameral chamber. Connecting rings short, forming thin linings on the inside of septal necks. Endocones present at shell diameters up to 55 mm. Cones asymmetrical and having endosiphuncular tube subcentral; axis of tube situated nearly midway between the center and the venter. Dorsal endocone thickness roughly twice the thickness of ventral portions. Apical end of siphuncle unknown.

Stratigraphic and geographic occurrence.—Leipers Limestone (Upper Ordovician, Maysvillian Stage part), Cumberland River valley and tributary streams, Russell County, Ky. Older fragmentary material from the "Cynthiana Limestone" at Cynthiana, Ky., in the Shideler Collection at Miami University may also be assignable to this species.

Studied material.—Holotype MU 266T (pl. 14, figs. 1, 2) and four paratypes (MU 267T, A–D), plus several other less complete, fragmentary specimens, Shideler Collection location number 3.51 B2, exposures along the Cumberland River, opposite Belk Island, near Rowena, Ky.

Remarks.—The type specimens of *C. rowenaense* were discovered among specimens stored in the vault at the Department of Geology, Miami University, Oxford, Ohio. Stored with the specimens were several USNM labels written by A.F. Foerste, listing these specimens as co-types A-63M and A-63N. The name applied to these specimens by Foerste was *Endoceras rowenaense*. These specimens were apparently awaiting description by Foerste when he died in the late 1930's. Foerste's proposed name is retained here for this species.

Cameroceras rowenaense is most similar to one of several species described by Hall (1847) as *Endoceras proteiforme*. Hall's specimens were from the Trenton Limestone (Middle Ordovician, Shermanian) at Middleville, N.Y. Hall did not designate type specimens for *E. proteiforme*. Foerste (1924b) indicated that typical *E. proteiforme* was represented by material Hall illustrated on his plates 48–50, 53, and 57. Foerste selected the original of figure 4 on plate 48 as the lectotype of the species and listed the specimens figured on plate 49 as "co-types." Foerste's figures and text indicate that he placed two distinct forms under the name *E. proteiforme*. The lectotype and a "co-type," both at the AMNH, are gradually expanding longicones (apical angle 6 degrees) with dorsally located endosiphuncular tubes (Foerste, 1924b, pl. 21, figs. 1, 3; pl. 22, fig. 1; pl. 23, fig. 1*B*). A second "co-type" (also at the AMNH) is a more rapidly expanding specimen (apical angle 8.5 degrees) with a ventrally located endosiphuncular tube (Foerste, 1924b, pl. 21, fig. 2; pl. 23, fig. 2). This latter form is the same figured by Flower (1958, pl. 60, fig. 6) and identified by him as *E. proteiforme* from the "Denmark Member of the Sherman Fall Limestone" at Trenton, N.Y. The rapid expansion of shells of this type is similar to that illustrated for specimens of *Cameroceras trentonense* Conrad (AMNH 815; Hall, 1847, pl. 56, figs. 4a–c; Foerste, 1924b, pl. 24, figs. 1, 2*A–B*) from the same strata. The more rapidly expanding forms with ventral endosiphuncular tubes are placed here in *C. trentonense* Conrad, and the more gradually expanding species with dorsal endosiphuncular tubes are retained in *C. proteiforme* (Hall). *Endoceras magniventrum* Hall (1847, pl. 53, fig. 1*B*) is thought to represent the adoral portion of a large specimen of *C. trentonense* and is placed in synonymy with this species.

Cameroceras rowenaense is most similar to the species identified here as *C. trentonense* Conrad, both taxa having moderate to rapidly expanding shells (apical angles 8–9 degrees) and a siphuncle with asymmetrical endocones that have ventrally located endosiphuncular tubes. *Cameroceras rowenaense* differs from *C. trentonense* in possessing a smaller diameter siphuncle (one-third the diameter of the shell at a diameter of 50 mm) that is subcentral in position (SPR=0.30) compared with the broader (half the diameter of the shell), nearly midventral (SPR=0.26) siphuncle present in *C. trentonense* at the same diameter. The endosiphuncular tube in *C. rowenaense* is also removed from the ventral margin of the siphuncle, compared with its more marginal position in *C. trentonense* at the same diameter. A comparison of the cross sections of these two species at a diameter of 55 mm is presented in figure 20. In addition, the camerae in *C. rowenaense* are somewhat shorter than those of *C. trentonense* at the same shell diameters. A comparison of cameral lengths relative to shell diameter for these two species is presented in figure 21. A silicified siphuncle from USGS collection 4073-CO in the Grier Limestone Member of the Lexington Limestone at Tyrone, Woodford County, Ky., may be assignable to *C. trentonense* (USNM 468712, pl. 14, figs. 3, 4).

Cameroceras rowenaense is also similar to *C. inaequabile* (Miller) from younger Richmondian strata in the Kentucky-Indiana-Ohio area in its rate of shell expansion, the relative diameter of the siphuncle to the diameter of the shell, and the short length of cameral chambers. It is distinct from *C. inaequabile* in the more central position of the siphuncle and more ventral position of the endosiphuncular tube in *C. rowenaense* (compare figs. 20*A* and 20*C*). A comparison of the morphological fea-

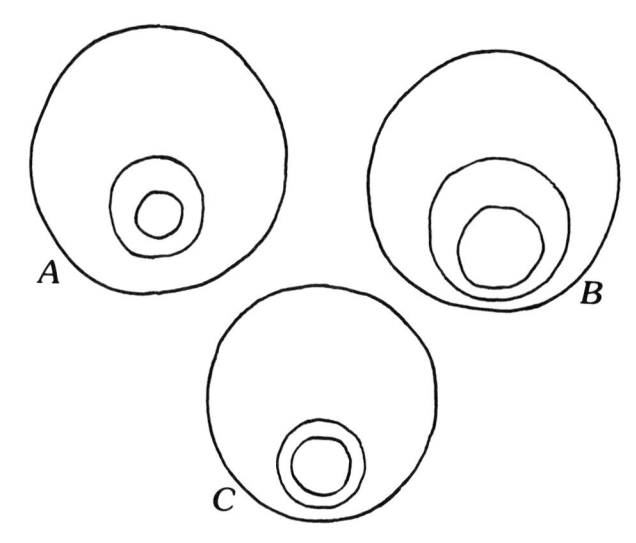

FIGURE 20.—Scale drawings comparing cross-sectional features of species of *Cameroceras* from Middle and Upper Ordovician strata in Eastern North America. Sections all from specimens 55–60 mm in diameter. *A*, *Cameroceras rowenaense*, MU 266T, Leipers Limestone (Maysvillian), Rowena, Ky.; *B*, *C. trentonense*, figured by Flower (1958, pl. 60, fig. 6), Sherman Fall Limestone, Trenton Falls, New York; *C*, *C. inaequabile*, MU 171, Bull Fork Formation, Warren County, Ohio.

tures of all three of these species of *Cameroceras* is presented in table 9.

Cameroceras rowenaense is distinguished from the large endoceratid described here from the Bull Fork Formation in Bath County, Ky. (pl. 14, figs. 5, 6) in its more rapid rate of shell expansion and in the macrochoanitic septal necks observed in the latter species. These elongate septal necks are distinct from the holochoanitic septal necks characteristic of *C. rowenaense* and other more typical species of *Cameroceras*.

As indicated above, *C. rowenaense* is currently known from the Leipers Limestone exposed in the valley of the Cumberland River in Russell County, Ky. Specimens typically consist of incomplete calcite-replaced internal molds representing portions of the phragmocone between 50 and 75 mm in diameter and often holoperipherally encrusted by the bryozoan *Atactoporella*. These specimens occur in an olive-colored fossiliferous wackestone that weather slight gray. The limestone contains large algal oncolites (see pl. 14, fig. 1) and is extensively bioturbated. Associated species are a small, geniculate species of the brachiopod *Rafinesqina*, the elongate, compressed pelecypod *Orthodesma*, trilobites, and ostracodes. A list of associated nautiloid taxa can be found in the work of Flower (1946).

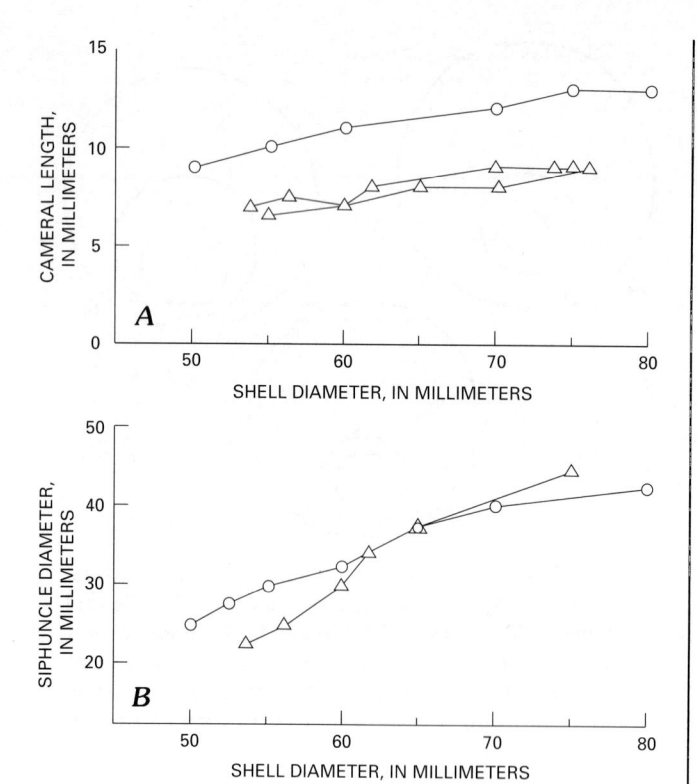

FIGURE 21.—Bivariate plots of cameral length versus shell diameter, *A*, and siphuncle diameter and shell diameter, *B*, for species of *Cameroceras* from the Ordovician of New York and Kentucky. Species are *C. rowenaense* (triangles) from the Leipers Limestone, Rowena, Ky. (MU 266T and 267T), and *C. trentonense* (circles) from the Trenton Limestone, Middleville, N.Y. (Hall, 1847).

TABLE 9.—Comparison of the morphological features of the Middle Ordovician (New York) and Upper Ordovician species of *Cameroceras* from the Cincinnati arch region of Kentucky, Indiana, and Ohio

[Data from specimens of *C. trentonense* figured by Hall (1847), type specimens of *C. rowenaense* (MU 266T, 267T), and specimens of *C. inaequabile* from the *T. duseri* shale in Warren County, Ohio (MU 17165). CL/D, ratio of cameral length to shell diameter; SPR, siphuncle segment position ration; SD/D, ratio of siphuncle diameter to shell diameter; EP, endosiphuncle tube position within siphuncle; mm, millimeter]

Feature	*C. trentonense*	*C. rowenaense*	*C. inaequabile*
Apical angle, degrees	8.5–9	7.5–8	7–8
Diameter 50–55 mm			
CL/D	.18	.12	.14
SPR	.26	.30	.22
SD/D	.56	.36	.22
EP	Ventral	Midway between center and venter.	Center

CAMEROCERAS? SP.

Plate 14, figures 5, 6

Diagnosis.—Large, gradually expanding endoceratid known from adoral portions of the phragmocone and distinguished by short cameral chambers one-tenth the diameter of the shell, a dorsally depressed siphuncle that constitutes two-thirds the total diameter of the shell, and septal necks that are macrochoanitic.

Description.—Large (phragmocone estimated to have been up to 1 m in length at a maximum observed shell diameter of 8.2 cm), gradually expanding (apical angle 6 degrees) orthoconic longicone; circular to slightly depressed in cross section with straight, transverse sutures. Cameral chambers short having cameral lengths ranging from one-eighth to one-tenth the diameter of the shell at shell diameters between 70 and 80 mm. Body chamber unknown. Shell exterior essentially smooth, marked by numerous closely spaced, fine, transverse growth lines.

Siphuncle large and tubular, removed from the ventral shell margin by a distance of 7–9 mm. Siphuncle depressed ovate in section, wider than high, becoming more flattened dorsally with ontogeny. Siphuncle constitutes two-thirds the total diameter of the shell at shell diameters of 70–80 mm. Septal necks long, macrochoanitic, their distal ends extending 2–3 mm beyond the adoral face of the adjacent adapical septum, and overlapping for this distance. Connecting rings absent. Endocones unknown. Nature of endosiphuncular structures unknown. Apical end of siphuncle unknown.

Stratigraphic and geographic occurrence.—Upper part of the Bull Fork Formation (Upper Ordovician, Richmondian Stage), Bath County, Ky.

Studied material.—USNM 468713 (pl. 14, figs. 5, 6) from the Bull Fork Formation, 50 ft below the contact with the Drakes Formation, exposed at a roadcut at the junction of I–64 and State Route 36, Owingsville, Ky.

Remarks.—This species is represented by the adoral portion of a large endoceratid, 125 mm in length and from 70 to 80 mm in diameter. The specimen has the general attributes of a species of *Cameroceras*, the most abundant endoceratid in Upper Ordovician strata exposed in the Kentucky-Indiana-Ohio area. This species, however, is distinct from all other specimens of *Cameroceras* known to the writer, in the nature of its septal necks, which are macrochoanitic rather than holochoanitic, as is typical of other members of the genus. Other endoceratids with macrochoanitic septal necks are *Hudsonoceras* Flower, *Kutorgoceras* Balashov, *Protovaginoceras* Ruedemann, and *Williamsoceras* Flower, all genera that are restricted to lower Middle Ordovician strata in North America, Baltic Europe, and portions of Soviet Asia. The affinities of this Kentucky specimen remain tentative, pending the discovery of more complete specimens, especially specimens preserving the critical apical portions of the siphuncle, providing knowledge of the endocones and other endosiphuncular structures.

GENUS *VAGINOCERAS* HYATT

Type species.—*Endoceras multitubulatum* Hall, 1847, p. 59; designated by Hyatt, 1884, p. 266; Black River Limestone, Blackriveran Stage, Middle Ordovician, New York.

Description.—Large orthoconic longicones of circular to slightly depressed cross section with straight, transverse sutures. Shell surface essentially smooth. Nature of body chamber unknown.

Siphuncle large, tubular, about one-third the diameter of the shell, situated slightly removed from the ventral shell margin. Siphuncle slightly wider than high in cross section, no ventral flattening. Septal necks holochoanitic, internally lined with thick connecting rings. Endocones compressed to cuneate in cross section, having two vertical endosiphuncular blades. Endosiphuncular tube central in position, compressed or cuneate in cross section. Siphuncle gradually tapering to an acute apex, not strongly inflated adapically.

Stratigraphic and geographic occurrence.—Middle Ordovician (Blackriveran to Kirkfieldian Stages), North America (Laurentia) and Baltic Europe.

Remarks.—Flower (1955a) provides a good description of this genus. He pointed out that, on the basis of gross shell morphology alone, there are no significant differences between this genus and *Cameroceras* Conrad. The only differences noted were in the shape and structure of the endocones and endosiphuncular tube, which are depressed, elliptical in *Cameroceras* and compressed and cuneate in *Vaginoceras*. Another distinction is in the nature of the endosiphuncular blades. *Vaginoceras* has two vertical blades, and *Cameroceras* was stated by Flower (1955a) to have one vertical blade and two oblique blades. Although the apical end of the siphuncle of *Vaginoceras* was known from only one poorly preserved specimen, Flower described it as being very similar to that of *Cameroceras*. He indicated that the annuli on isolated siphuncles would be less evident in their development compared with those of *Cameroceras* due to the thicker connecting rings in *Vaginoceras*. Unless endosiphuncular structures are preserved, it is evident that distinguishing these two genera would be difficult, if not impossible.

Vaginoceras is stated by Flower to be characteristic of Blackriveran and Rocklandian nautiloid faunas in Eastern North America but absent from the underlying Chazyan and overlying Shermanian strata. In these Blackriveran strata, *Vaginoceras* is associated with less abundant species of *Endoceras* Hall. *Cameroceras* becomes the dominant large endoceratid in younger Shermanian and Richmondian rocks in the same region. For the purposes of this paper, poorly preserved, large endoceratids from the High Bridge Group in central Kentucky are placed in *Vaginoceras*. Similarly preserved material from younger Middle and Late Ordovician strata in the area will be placed in *Cameroceras*.

VAGINOCERAS SP. A

Plate 15, figures 1–4; 8–12

Diagnosis.—Slender, gradually expanding species of *Vaginoceras*, distinguished by an external shell marked by narrow, raised, transverse bands and possessing only a slightly inflated siphuncle apex marked by low but distinct septal annuli.

Description.—Medium to large (estimated shell lengths up to 90 cm and maximum observed shell diameters of 6.5 cm), gradually expanding (apical angle 5.5 degrees) orthoconic longicones; circular in section with straight, transverse sutures. Cameral chambers short, cameral lengths averaging one-seventh the diameter of the shell at shell diameters of 45–55 mm. Body chamber tubular, at least one-sixth the total length of the shell. Shell exterior marked by indistinct, narrow, raised transverse bands 1–2 mm in width.

Siphuncle large, tubular, one-third the diameter of the shell at shell diameters of 45–55 mm. Siphuncle removed from ventral shell margin by a distance of 5 mm (SPR=0.265, nearly midventral). Septal necks holochoanitic, connecting rings unknown. Adapical portions of siphuncle slender, tubular (10 mm in diameter), circular in cross section, expanding adapically to a maximum diameter of 13 mm at 10–20 mm adoral of the siphuncle apex. Apex acute, apical tip narrowly rounded. Apical end marked by low, rounded "septal annuli," 4–5 mm wide, sloping from the dorsum to the venter in the adoral direction, annuli becoming obsolete ventrally. Adapically, endosiphuncular structures are the dorsal and ventral vertical blades and a compressed, central endosiphuncular tube 1 mm in diameter that extends to the very tip of the apex.

Stratigraphic and geographic occurrence.—Upper part of the Rocklandian portion of the Tyrone Limestone (Middle Ordovician), Jessamine County, Ky.

Studied material.—USNM 158668 (pl. 15, figs. 1–4), USNM 158652 (pl. 15, figs. 8–10), USNM 468714 (pl. 15, figs. 11, 12), plus a suite of four additional specimens (USNM 468715), all from the USGS 6034-CO collection, Little Hickman section A, Jessamine County, Ky. Additional material consists of a silicified siphuncle (MU 29868) from the Tyrone Limestone at High Bridge, Ky.

Remarks.—This endoceratid is represented in the silicified USGS 6034-CO collection by a single camerated portion of the phragmocone and attached portions of the body chamber, 23 cm in total length, diameters of 45–65 mm (USNM 158668 (pl. 15, figs. 1–4), plus a number of apical portions of the siphuncle (pl. 15, figs. 8–12). It is thought that these fragmentary specimens represent a single endocerid species, although intervening portions of the shell are not known. The adoral portion of the phragmocone might just as easily be assigned to *Cameroceras* because the endosiphuncular structures are not preserved in this specimen. The apical ends of the siphuncle, while having the external shape of *Cameroceras*, preserved some key diagnostic features of *Vaginoceras*: dorsal and ventral vertical blades with a narrow diameter and a compressed central endosiphuncular tube. This Tyrone Limestone material demonstrates clearly that the apical ends of the siphuncle in *Vaginoceras* are, as suggested by Flower, of the *Cameroceras* type, slightly expanding and then tapering to an acute tip and having camerae extending adapically to the very apex of the siphuncle.

Comparisons with other described species of *Vaginoceras* are difficult due to the incomplete nature of the Kentucky material and the equally incomplete nature of other published species. Flower (1955b) recognized only two species in *Vaginoceras*: the type species *V. multitubulatum* (Hall) and *V. longissimum* (Hall), both from the Black River Limestone in New York. Both species are known from isolated siphuncles that contained endocones. Troedsson (1926, pls. 3–5) described and figured a number of fragmentary siphuncles he assigned to Hall's species and to "*Vaginoceras* species indeterminate I and II," all from the basal Cape Calhoun Formation (Middle Ordovician, Blackriveran) in northern Greenland. Species he referred to as "*Endoceras* species indeterminate I" (Troedsson, 1926, pl. 5, figs. 3, 4; pl. 6, fig. 2) and *E. proteiforme* Hall (Troedsson, pl. 6, fig. 2) are also from the basal portion of the Cape Calhoun Formation and also appear to be species of *Vaginoceras*. The latter species is represented by a portion of the phragmocone 14 cm in length and 6.5 cm in diameter and having cameral lengths one-ninth the diameter of the shell.

The preserved Tyrone Limestone material is more closely comparable with *Vaginoceras* sp. described by Stait (1988, pl. 2, figs. 10, 11) from the Lourdes Formation (Middle Ordovician, upper Blackriveran) exposed in western Newfoundland. The figured portion of the phragmocone measures 70 mm in length and has a gradual rate of shell expansion, cameral lengths averaging one-sixth the diameter of the shell, and siphuncle diameters roughly one-third the diameter of the shell, both at shell diameters of 50 mm.

Sweet (1958) described and figured *Vaginoceras* sp. from the Middle Ordovician "Cephalopod Shale" in Norway. His figured specimen (Sweet, 1958, fig. 3; pl. 1, fig. 4) represented a portion of the phragmocone that ranged from 32 to 35 mm in diameter. Cameral lengths ranged from one-fifth to one-seventh the diameter of the shell. The apical angle is 6.5 degrees. No comparable material at these shell diameters is present in the 6034-CO collection from Kentucky.

VAGINOCERAS SP. B

Plate 15, figures 5–7

Diagnosis.—Slender, gradually expanding section of siphuncle, circular in cross section, having centrally located compressed cuneate endosiphuncular tube and dorsal and ventral vertical blades.

Description.—Slender, gradually expanding (apical angle 4 degrees), tubular section of siphuncle 9.0 cm in length, adoral diameter 20 mm and adapical diameter 17 mm. Siphuncle tube circular in cross section. Adorally with a laterally compressed tube 8 mm wide and 16 mm high that tapers rapidly adapically to a point separating dorsal and ventral vertical blades. Siphuncle surface smooth, no evidence of "septal annuli."

Stratigraphic and geographic occurrence.—Basal Curdsville Limestone Member of the Lexington Limestone (Middle Ordovician, Kirkfieldian Stage), Frankfort area, Franklin County, Ky.

Studied material.—USNM 468718 (pl. 15, figs. 5–7) from the USGS 5072-CO collection, Old Crow Distillery, Frankfort East B section.

Remarks.—Little can be inferred from this specimen beyond that the preserved endosiphuncular structures indicate that it is a species of *Vaginoceras*. The preserved portion of the siphuncle cannot be compared with the more adoral and adapical sections known for the preceding Tyrone Limestone species of *Vaginoceras*. The more gradual rate of expansion suggests this siphuncle may come from a more slender species of *Vaginoceras*, distinct from the Tyrone Limestone form. This specimen is significant in that it extends the range of *Vaginoceras* into the Kirkfieldian.

GENUS *TRIENDOCERAS* FLOWER

Type species.—*Triendoceras montrealense* Flower, 1958, p. 453, by original designation; "Chazy Beds," Chazyan Stage, Middle Ordovician, Quebec.

Description.—Large, gradually expanding orthoconic longicones that are slightly depressed in cross section. Venter broadly rounded, no conspicuous midventral flattening. Sutures straight, transverse, cameral lengths uniformly short in known portions of the phragmocone. Body chamber unknown. Shell surface apparently smooth.

Siphuncle situated at the ventral shell margin, tubular, enlarging with ontogeny from diameters one-third the diameter of the shell adapically to one-half the diameter of the shell adorally. Siphuncle depressed ovate in cross section, wider than high. Septal necks holochoanitic with thin, narrow connecting rings. Endocones short. Endosiphuncular tube having the cross sectional shape of an isosceles triangle, and having the broad, slightly rounded convex base close to the dorsal wall and the apex directed ventrally. Endosiphuncular tube situated dorsally within the siphuncle. Apical end of the siphuncle unknown.

Stratigraphic and geographic occurrence.—Middle Ordovician (Chazyan) to Upper Ordovician (Edenian? Stage), Eastern North America (Laurentia).

Remarks.—This genus is based on two rather poorly preserved fragmentary phragmocones described from an early Middle Ordovician limestone unit in Quebec (Flower, 1958). Flower's diagnosis of the genus centered on the triangular shape of the endosiphuncular tube in these specimens, although he failed to illustrate them. Specimens were described as being otherwise "typical Endoceratidae," similar to *Cameroceras* and *Vaginoceras* in all other characteristics. Because the endocones were replaced with sparry calcite in the type species, there is no evidence of the pattern of the endosiphuncular blades in this genus.

Flower (1958) also referred to this genus a form known primarily from isolated siphuncles that are not uncommon in the Upper Ordovician (Edenian) Kope Formation in the Cincinnati, Ohio-Covington, Ky., area. These specimens typically consist of "speiss," the preserved mud-infilling of the endocones, the rest of the siphuncle and the phragmocone being largely unknown. This form is tentatively placed in *Triendoceras* here and is described below.

TRIENDOCERAS? DAVISI, NEW SPECIES
Plate 16, figures 1–5

Diagnosis.—Large-diameter, rapidly expanding infillings of the siphuncle that have a distinctive subtriangular cross section adapically, having the acute angle directed ventrally and the broadly rounded base located dorsally.

Description.—Known primarily from large (up to 40 cm in length and diameters up to 4.0 cm), isolated molds of the siphuncle and endosiphuncular tube (speiss). These consist adapically of rapidly expanding (apical angle 8–9 degrees), narrow-diameter infillings of the endosiphuncular tube that expand from a narrow tube 3 mm in diameter to a diameter of 35 mm at a length of 80 mm. Subtriangular cross-sectional shape of endosiphuncular tube at these diameters, having the venter forming an acute angle and the dorsum being more broadly rounded. Siphuncle infilling becoming more gradually expanding (apical angle 4 degrees) adorally, the remainder of the length of the siphuncle. Cross-sectional shape adorally becoming circular. Adoral portions of the siphuncle bear impressions of the connecting rings; raised rings 2–3 mm wide that slope from the dorsum to the venter in the adoral direction, indicating a ventrally located siphuncle.

A portion of the phragmocone, 40 mm in length and 50 mm in width, indicates cameral lengths at this diameter of one-fifth the diameter of the shell and the presence of holochoanitic septal necks. Remainder of the phragmocone and body chamber unknown.

Stratigraphic and geographic occurrence.—Edenian part of the Kope Formation (Upper Ordovician), Cincinnati, Ohio, area and adjacent portions of northern Kentucky.

Studied material.—Holotype MU 410T (pl. 16, fig. 1) and paratype MU 412T (pl. 16, fig. 4) both from the Kope Formation at the Orphanage Road locality, Kenton County, Ky.; paratype MU 411T (pl. 16, fig. 2) from the same formation at Cincinnati, Hamilton County, Ohio; paratype MU 413T (pl. 16, fig. 5) Shideler locality 1.23A8; and paratype MU 414T (pl. 16, fig. 3) from the "Fulton Beds" at Moscow, Clermont County, Ohio. Numerous additional fragments from the Kope Formation in the greater Cincinnati area of Ohio and Kentucky were also studied.

Remarks.—This species is known primarily from matrix infillings of the siphuncle and endosiphuncle tube, often referred to as "speiss." Only a single fragment of the phragmocone preserving portions of several cameral chambers (pl. 16, fig. 4) and the septal necks was available for study. More adoral portions of these infillings commonly preserve impressions of the connecting rings, consisting of narrow rings 2–3 mm wide that encircle the siphuncle and slope from the dorsum to the venter in the adoral direction (pl. 16, fig. 1). The features of the phragmocone and endosiphuncle are otherwise unknown. Figure 22 illustrates the relation between the preserved speiss and the endosiphuncular structures of the siphuncle. One interesting phenomenon associated with these infillings of the siphuncle is the black coating that often covers the speiss, possibly indicative of a high organic content (conchiolin?) in the enclosing endocones (pl. 16, fig. 1).

This writer has elected to erect a new species for this form on the bases of the distinctive subtriangular cross section of adapical portions of the speiss, the rapidly expanding nature of the internal molds, and the restricted occurrence of siphuncles of this type to the Kope Formation in the greater Cincinnati region of southwestern Ohio and northern Kentucky. Siphuncles assigned to this species cannot be confused with those of any other endoceratid from Upper Ordovician strata in this region. *Triendoceras? davisi* is the most commonly encountered nautiloid within the Kope Formation in the Cincinnati-to that described by Flower (1958) for the Middle Ordovician (Chazyan) endoceratid *Triendoceras*. The acute angle in cross sections of these Kope endosiphuncular tubes is directed ventrally, consistent with Flower's description of these features in his specimens of *Triendoceras*. The affinities of this Edenian species remain uncertain, pending discovery of more complete specimens preserving critical portions of the phragmocone as well as further study of the type species of *Triendoceras* Flower.

Etymology.—This species is named in recognition of the contributions of Richard A. Davis, a fellow student of the nautiloid cephalopods and of the paleontology of the Ordovician of the Cincinnati, Ohio, region.

SUBCLASS ACTINOCERATOIDEA TEICHERT

Medium to large orthoconic or slightly curved cyrtoconic longicones, usually depressed in cross section with large subcentral or midventral, less commonly marginal, siphuncles. Siphuncle segments broadly expanded within camerae, diameters commonly between one-third and two-thirds the diameter of the shell. Septal necks cyrtochoanitic, often recumbent. Buoyancy regulation provided by calcareous ballast in the form of massive endosiphuncular deposits that form in and around a branching endosiphuncular vascular system, consisting of a central canal, radial canals, and perispatia. Cameral deposits typically well developed. Apical end of shell short, blunt, terminating in a large, cap-shaped initial segment that covers the apex of the shell.

Stratigraphic range.—Lower Ordovician to middle Carboniferous.

ORDER ACTINOCERIDA TEICHERT

Characters of the subclass.

Stratigraphic range.—Lower Ordovician to middle Carboniferous.

FAMILY ACTINOCERATIDAE SAEMAN

Large, straight, generally fusiform shells with blunt apices, slightly to greatly depressed in cross section, having straight transverse or slightly sinuous sutures. Siphuncle segments large, subcentral to ventral in position, segments strongly expanded within camerae. Segments often decompressing with ontogeny. Septal necks long with comparatively short brims. Endosiphuncular canals well developed, of the curved type; remainder of the siphuncle filled with endosiphuncular deposits consisting of coalesced annulosiphonate deposits. Cameral deposits consisting of both episeptal and hyposeptal deposits.

Stratigraphic range.—Middle Ordovician (Blackriveran Stage) to Lower Silurian (Llandoverian Series).

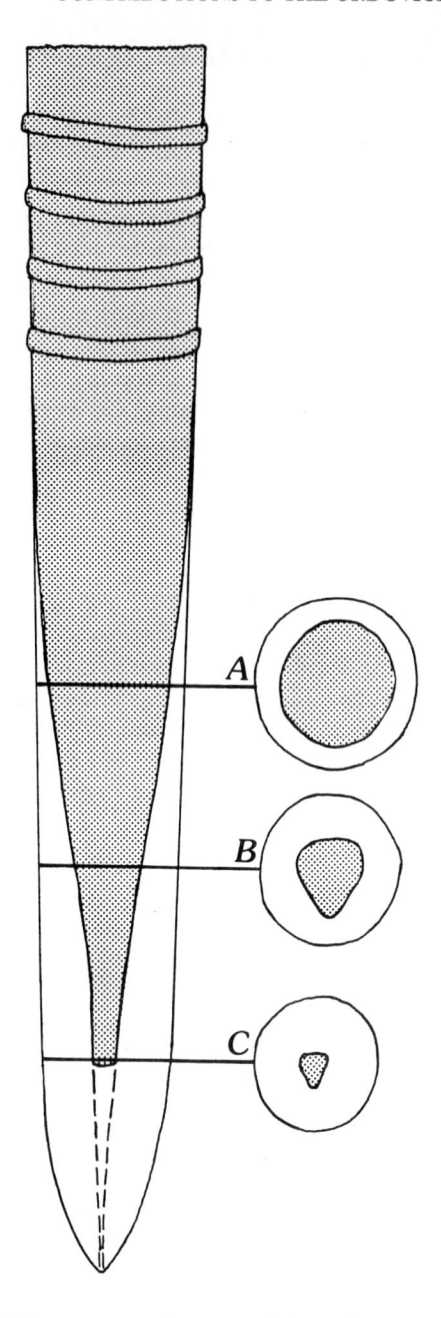

FIGURE 22.—Diagrammatic drawing of the siphuncular structure in *Triendoceras? davisi*. Based on holotype (MU 410T) from the Kope Formation, Ft. Mitchell, Ky. (see pl. 16, fig. 1). *A*, Section through adoral portion of siphuncle, circular in cross section; *B*, section through medial portion of siphuncle tube, initiation of more triangular cross section; *C*, adapical terminus of endosiphuncular tube with triangular cross section. ×0.65.

Covington area. It occurs primarily in thin packstone beds in association with crinoids, bryozoans, the brachiopods *Sowerbyella* and *Zygospira*, and the trilobite *Cryptolithus*.

The distinctive subtriangular cross section of the endosiphuncular tube in *Triendoceras? davisi* is similar

GENUS *ACTINOCERAS* BRONN

Type species.—*Actinoceras bigsbyi* Bronn, 1837, p. 97, by original designation; Black River Formation, Blackriveran Stage, Middle Ordovician, Thessalon Island, Lake Huron, Ontario.

Description.—Large (up to 1 m in length), robust, straight to slightly endogastrically curved, rapidly expanding fusiform shells; subcircular to moderately depressed in cross section with sutures that slope slightly in the adapical direction from the dorsum to the venter. Camerae short, cameral lengths decreasing with ontogeny. Phragmocone rather short, body chamber nearly one-third the length of the shell, contracting toward the aperture. Maximum shell diameter at the junction of the body chamber and the phragmocone. Aperture straight, transverse, open. Shell exterior smooth.

Siphuncle large having maximum diameters up to one-half to three-fourths the diameter of the shell. Segments subcentral to midventral in position, having the ventral connecting ring in contact with the ventral shell wall in many species. Siphuncle typically moving from the venter to a more central position with ontogeny. Segments strongly expanded within camerae, highly compressed in shape. Segments decompressing with ontogeny to form equiaxial or depressed globular segments. Septal necks long, recurved, but not recumbent, with relatively short brims. Massive endosiphuncular deposits consisting of fused annulosiphonate deposits that fill most of the siphuncle, leaving a narrow central canal and curved, radial, lateral canals in the adapical portions of large individuals. Well developed episeptal and hyposeptal cameral deposits.

Stratigraphic and geographic occurrence.—Middle Ordovician (Blackriveran) to Lower Silurian (Llandoverian), North America, Baltic Europe, and Siberia.

Remarks.—Flower (1957a) divided this genus into a number of generalized "species groups." Members of his "*Actinoceras ruedemanni* Group" were characterized as having straight fusiform shells with depressed cross sections and flattened venters. Sutures have broad ventral lobes, and siphuncle segments are large and broadly expanded in early stages, becoming smaller in diameter with ontogeny. Segments are sinuate in outline, the curve of the septal neck being as broadly rounded as the expanded portions of the segment (see fig. 23). In Flower's "*A. centrale* Group" were species with closely spaced septa, broad, short siphuncles, and the outline of the segment more broadly rounded than the contracted area at the septal neck. Species assigned to his "*A. paquettense* Group" are described as being essentially breviconic; short phragmocones with rarely more than 20 camerae, compared with the 30–40 present in species belonging to the two previous groups. Members of the "*A. winstoni* Group" have slender shells, circular in section, that are slightly curved adapically. Segments are also narrower in diameter for a much greater length of the siphuncle compared with other species groups.

The first three of Flower's species groups are characteristic of Middle Ordovician (Blackriveran and Rocklandian) carbonate facies exposed in portions of Eastern and Central North America. It is in these strata that *Actinoceras* reaches its apogee in terms of abundance and diversity. Members of the *A. winstoni* Group occur in younger, Middle Ordovician (Kirkfieldian and Shermanian) strata and are much less abundant than their Blackriveran progenitors. Flower also described two additional species groups from younger (Shermanian to Richmondian) carbonates in Western and Arctic North America. These are the "*A. simplicem* Group," characterized by species with broad, evenly rounded siphuncle segments in early stages having the siphuncle remaining large and ventral with ontogeny, but the expansion of the segments within the camerae becoming reduced. The largely Upper Ordovician "*A. anticostiense* Group" is distinguished by siphuncle segments that show a long, tubular adoral region and a short adapical region, in which segment expansion is abruptly concentrated. The septa have long septal necks that terminate in very short brims that are abruptly recurved. Flower (1968a) reviewed the early evolution of actinocerids.

Species of *Actinoceras* described from the Kentucky Blue Grass region are "*Leurorthoceras*" *altopontense* Foerste and Teichert and *Actinoceras libanum* Foerste and Teichert from the Camp Nelson Limestone (Middle Ordovician, Blackriveran); *A. altopontense* Foerste and Teichert, *A. arcanum* Foerste and Teichert, *A. bigsbyi* Bronn, *A. jessaminense* Foerste and Teichert, and *A. kentuckiense* Foerste and Teichert, all from the Tyrone Limestone (Middle Ordovician, Rocklandian); and *A. curdsvillense* Foerste and Teichert from the Curdsville Limestone Member of the Lexington Limestone (Middle Ordovician, Kirkfieldian). Kentucky species of *Actinoceras* from the Tyrone Limestone are all members of Flower's *A. ruedemanni* Group. *Actinoceras curdsvillense* from the Lexington Limestone appears to be a member of the *A. centrale* Group on the basis of its narrowly rounded, nearly recumbent septal necks and shallow camerae. Most of these Kentucky species of *Actinoceras* were described from fragmentary portions of the phragmocone. This, coupled with a lack of extensive collections of individuals, makes accurate species comparisons difficult, especially in light of the pronounced ontogenetic changes that are typical of this genus. Species of *Actinoceras* described below are those that are able to be recognized from additional material collected in the central Kentucky region by U.S. Geological Survey and Kentucky Geological Survey field crews.

ACTINOCERAS ALTOPONTENSE FOERSTE AND TEICHERT, 1930

Plate 16, figure 6; plate 17, figures 6–10

Actinoceras altopontense Foerste and Teichert, 1930, p. 225, pl. 35, figs. 1A–C; pl. 56, fig. 1.

Diagnosis.—Rapidly expanding species of *Actinoceras* with comparatively short camerae and broadly expanded, highly compressed siphuncle segments, in which the length of the connecting ring is nearly twice that of the septal necks; segments in contact with the ventral shell margin most of the length of the shell.

Description.—Medium-sized (phragmocone lengths in excess of 35 cm and shell diameters of up to 7.2 cm), rapidly expanding (apical angle 10.5–11.5 degrees) orthoconic longicones, depressed circular in cross section and having the venter more flattened than the dorsum, causing sutures to slope from the dorsum to the venter in the adapical direction. Sutures forming a broad ventral lobe. Cameral chambers short, cameral lengths ranging from one-fifth to one-sixth the diameter of the shell, decreasing with ontogeny. Body chamber unknown. Shell exterior smooth.

Siphuncle subcentral (SPR=0.28–0.33) for the length of the siphuncle and having ventral portions of the connecting rings in contact with the ventral shell margin. Siphuncle segments highly compressed (SCR=3.7–2.5), decompressing only slightly with ontogeny. Segments broad, having diameters ranging from two-thirds the diameter of the shell at diameters of 30 mm or less, to one-half the diameter of the shell adorally. In dorsal-ventral profile, connecting rings twice as long as the length of the septal necks. Septal necks long, cyrtochoanitic, recurved, with short free brims. Siphuncle segments with well-developed annulosiphonate deposits that delineate a broad, central canal having numerous fine, curved radial canals. Cameral deposits not observed in known portions of the shell, 30–50 mm in diameter.

Stratigraphic and geographic occurrence.—Upper part of the Rocklandian portion of the Tyrone Limestone (Middle Ordovician), in the vicinity of the Kentucky River, Jessamine County, Ky.

Studied material.—Holotype, USNM 82196, from the Tyrone Limestone at High Bridge, Ky.; USNM 158640 (pl. 16, fig. 6), USNM 158641 (pl. 17, figs. 8, 9), 158642 (pl. 17, fig. 10), 158646 and 158648, all silicified material from USGS collection 6034-CO, Tyrone Limestone, Little Hickman section A, Sulfur Wells, Ky.

Remarks.—*Actinoceras altopontense* is most similar to a group of Blackriveran and Rocklandian species of *Actinoceras* distinguished by their large, rapidly expanding shells and by their broadly expanded siphuncle segments that are typically in contact with the ventral shell margin and that consist of connecting rings whose lengths are nearly twice the lengths of the septal necks (fig. 23).

Actinoceras altopontense differs from the associated *A. kentuckiense* Foerste and Teichert in its consistently shorter camerae at comparable shell diameters, the more ventral position of the siphuncle, higher segment compression ratios at the diameters studied, and in the narrower diameter of segments relative to shell diameter at comparable diameters. *Actinoceras glenni* Foerste and Teichert from equivalent strata at Nashville, Tenn., is similar to *A. kentuckiense* and differs from *A. altopontense* in the same manner. *Actinoceras ruedemanni* Foerste and Teichert, from the Black River Formation at Watertown, N.Y., is a more slender form (apical angle 9 degrees), having longer camerae and more compressed, broader siphuncle segments at shell diameters less than 45 mm. *Actinoceras beloitense* (Whitfield) from the Platteville Formation and equivalents in the upper Mississippi valley is similar to *A. altopontense* in the ventral position of the siphuncle, but differs from the Kentucky species in its longer camerae and lower SCR values. Siphuncle segment diameters relative to shell diameter are similar in early stages of growth, but segment diameters in *A. beloitense* become narrower with ontogeny. A comparison of the morphological features of a number of these species of *Actinoceras* from Middle Ordovician strata in North America is presented in table 10.

Actinoceras altopontense and *A. kentuckiense* appear to be equally common within the USGS 6034-CO collection, but the sample size is too small to determine this relationship with any certainty. Identifiable phragmocone fragments of both species are uncommon in this collection, and disarticulated siphuncle segments referable to species of *Actinoceras* are abundant. These segments often preserve impressions of the fine radial canals typical of members of this genus (pl. 17, figs. 6, 7).

ACTINOCERAS KENTUCKIENSE FOERSTE AND TEICHERT, 1930

Plate 17, figures 1–5

Actinoceras kentuckiense Foerste and Teichert, 1930, p. 244, pl. 32, figs. 1A–1B; pl. 33, fig. 1; pl. 57, fig. 3.

Diagnosis.—Rapidly expanding species of *Actinoceras* with comparatively long camerae and broadly expanded, highly compressed siphuncle segments, in which the lengths of the connecting rings are nearly twice the length of the septal necks; segments not in contact with the ventral shell margin in adoral portions of shell.

Description.—Medium-sized (phragmocone lengths of at least 30 cm and shell diameters up to 6.0 cm), rapidly expanding (apical angle 10–11 degrees) orthoconic longicone; depressed in cross section such that sutures slope from the dorsum to the venter in the adapical direction, forming a shallow ventral lobe. Cameral chambers comparatively long, cameral lengths ranging from one third to one fourth the diameter of the shell, decreasing with

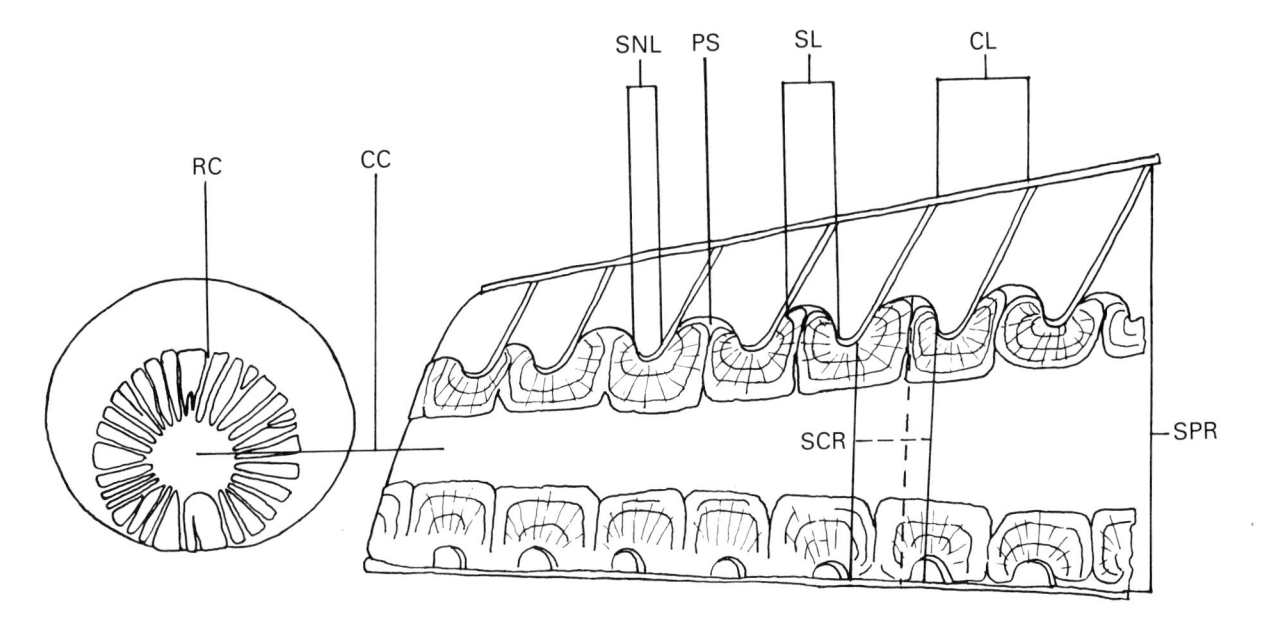

FIGURE 23.—Important morphological features used to distinguish species of the genus *Actinoceras* from Middle Ordovician strata in Eastern North America. CC, central canal; RC, radial canals; PS, perispatia; CL, cameral length; SL, segment length; SNL, length of septal neck; SPR, segment position ratio; and SCR, segment compression ratio.

ontogeny. Body chamber unknown. Shell exterior smooth.

Siphuncle subcentral the known length of the phragmocone (SPR=0.28–0.40). Siphuncle segments not in contact with ventral shell margin at diameters greater than 45 mm. Segments highly compressed (SCR=2.3–2.9) at shell diameters of 35–45 mm, compression ratio decreasing slightly with ontogeny. Segments broad, diameters three-fourths the diameter of the shell at shell diameters less than 40 mm, decreasing to one-half the diameter of the shell adorally, indicative of segment decompression with ontogeny. In dorsal-ventral section, length of connecting rings nearly twice the length of septal necks. Septal necks cyrtochoanitic, broadly recurved, long, with short, free brims. Well developed annulosiphonate deposits, having the radial vascular system typical of the genus. Hyposeptal cameral deposits weakly developed in phragmocone sections less than 35 mm in diameter.

Stratigraphic and geographic occurrence.—Upper part of the Rocklandian portion of the Tyrone Limestone (Middle Ordovician), in the vicinity of the Kentucky River, Jessamine County, Ky.

Studied material.—Holotype USNM 82237 from the Tyrone Limestone at High Bridge, Ky.; plus USNM 158670 (pl. 17, figs. 1, 2, 4, 5), 158647 (pl. 17, fig. 3), and 468720, all silicified material from the Tyrone Limestone, USGS 6034-CO collection, Little Hickman section A, Sulfur Wells, Ky.

Remarks.—The differences between *A. kentuckiense* and the associated *A. altopontense* were listed in the previous description of that species. Although general morphological features are similar, phragmocones can be distinguished by the longer camerae of *A. kentuckiense* at comparable shell diameters. *A. kentuckiense* is most similar to *A. glenni* Foerste and Teichert from the Tyrone Limestone in central Tennessee, in that both species have rapidly expanding shells and comparatively long camerae. *Actinoceras kentuckiense* differs from *A. glenni* in possessing broader, more highly compressed siphuncle segments at comparable shell diameters (table 10). *Actinoceras kentuckiense* differs from *A. ruedemanni* Foerste and Teichert from Blackriveran strata in New York in being a broader, more rapidly expanding shell having more highly compressed segments at diameters of 50 mm and less compressed segments at diameters less than 45 mm. *Actinoceras kentuckiense* is distinct from *A. beloitense* (Whitfield) from Blackriveran strata in Wisconsin and insular portions of Lake Huron in possessing more centrally located, more highly compressed siphuncle segments. *Actinoceras kentuckiense* is similar to the type species of *Actinoceras*, *A. bigsbyi* Bronn, in that adoral siphuncle segments are removed from the ventral shell margin. It differs from this species in being a more rapidly expanding form with longer camerae at comparable shell diameters. Ontogenetic siphuncle decompression also is much more extensive in *A. bigsbyi*, the segment diameters being slightly less than two-thirds the diameter of the shell at a diameter of 35 mm and less than one-third the diameter of the shell at diameter of 45 mm.

TABLE 10.—Comparisons of the morphological features of selected Middle Ordovician (Blackriveran and Rocklandian) species of *Actinoceras*, showing variation in rate of shell expansion, cameral length, siphuncle position, siphuncle segment shape, and siphuncle diameter relative to shell diameter

[Data from silicified material from Tyrone Limestone in Kentucky and specimens illustrated by Foerste and Teichert (1930). CL/D, ratio of cameral length to shell diameter; SPR, siphuncle segment position ratio; SCR, siphuncle segment compression ration; SD/D, ratio of siphuncle diameter to shell diameter; NA, not applicable; mm, millimeter]

Feature	A. altopontense	A. kentuckiense	A. glenni	A. ruedemanni	A. beloitense
Apical angle, degrees	10.5	11	12	9	10.5
Diameter 50 mm					
CL/D	.18	.24	.22	.22	.26
SPR	.28	.34	.30	.30	.26
SCR	2.88	2.33	2.27	1.80	1.46
SD/D	.52	.56	.50	.40	.38
Diameter 45 mm					
CL/D	.20	.24	.22	.20	.27
SPR	.31	.40	.28	.33	.29
SCR	3.125	2.70	2.65	3.00	1.66
SD/D	.55	.66	.58	.60	.44
Diameter 40 mm					
CL/D	.20	.25	.25	.21	.25
SPR	.325	.35	NA	.36	.275
SD/D	.625	.725	.50	.675	.55
Diameter 35 mm					
CL/D	.20	.285	.285	.228	.257
SPR	.285	.37	.34	.37	.257
SCR	2.85	2.60	2.30	3.25	2.20
SD/D	.57	.74	.65	.74	.57

ACTINOCERAS CURDSVILLENSE FOERSTE AND TEICHERT, 1930

Plate 17, figure 11

Actinoceras curdsvillense Foerste and Teichert, 1930, p. 240, pl. 39, figs. 2A–B; pl. 56, fig. 9.

Diagnosis.—Species of *Actinoceras* distinguished by its very short camerae and peculiar short, highly recurved, nearly recumbent septal necks.

Description.—Medium-sized (estimated phragmocone length of at least 26 cm and shell diameters of up to 4.0 cm), rapidly expanding (apical angle 10 degrees) species of *Actinoceras*; circular in cross section, with essentially straight, transverse sutures that form a weakly developed ventral lobe. Cameral chambers short, cameral lengths one-seventh to one-eighth the diameter of the shell, decreasing with ontogeny. Body chamber unknown. Shell exterior apparently smooth.

Siphuncle midventral in position at a diameter of 40 mm and having siphuncle segments in contact with the ventral shell margin. Segments for known portions of the phragmocone (from 32 to 40 mm in diameter) are very highly compressed (SCR=4.0) and broadly expanded, constituting from two-thirds to one-half the diameter of the shell, decreasing in relative diameter with ontogeny. Septal necks short, highly curved, nearly recumbent, with brims as long or longer than necks. Endosiphuncular vascular system unknown. Cameral deposits consisting of well developed hyposeptal deposits.

Stratigraphic and geographic range.—Curdsville Limestone Member of the Lexington Limestone (Middle Ordovician, Kirkfieldian Stage), Blue Grass region of north-central Kentucky.

Studied material.—Holotype USNM 48425 (pl. 17, fig. 11), a silicified fragment of the phragmocone; Curdsville, Ky.

Remarks.—This species is an apparently uncommon form in the Curdsville Limestone Member of the Lexington Limestone, known only from the holotype, a natural section of a fragment of the phragmocone 50 mm in length. Although this species is known only from this one specimen, the preserved features indicate a distinctive morphology characterized by the very short length of the camerae, highly compressed, broad siphuncle segments, and peculiar recurved, nearly recumbent nature of the short septal necks. These features distinguish this species from any other species of *Actinoceras* known to the writer. As indicated above, this species appears to belong in Flower's "*A. centrale* Group," as indicated by its short camerae and short, broad siphuncle segments. It represents a Blackriveran "relict" that survived the Rocklandian extinction event to persist into the Kirkfieldian. *Actinoceras* is not known from the overlying Ordovician section exposed in the Kentucky-Indiana-Ohio area.

FAMILY ORMOCERATIDAE SAEMAN

Generally straight, medium-sized (less than 1 m in length) orthoconic longicones with circular to slightly depressed cross sections and straight, transverse sutures. Camerae typically short. Body chambers poorly known. Shell exteriors smooth or ornamented with longitudinal ribs or lirae.

Siphuncles subcentral to midventral in position. Segments rather narrow in diameter for this order, making up one-third or less the diameter of the shell. Segments globular to spheroidal, compressed to slightly depressed or equiaxial in outline. Septal necks cyrtochoanitic, typically short, with short brims, recumbent or free. Endosiphuncular deposits consisting of coalescing annulosiphonate deposits and simple endosiphuncular vascular systems consisting of a central canal and radial canals that are straight, generally at right angles to the central canal. Cameral deposits well developed, consisting of both hyposeptal and episeptal deposits.

Stratigraphic range.—Middle Ordovician (Whiterockian) to Middle Devonian (Eifelian).

GENUS *ORMOCERAS* STOKES

Type species.—*Ormoceras bayfieldi* Stokes, 1840, p. 709; by original designation; Manistique Formation, Middle Silurian, Michigan.

Description.—Medium-sized, moderately to rapidly expanding orthoconic longicones; circular to slightly depressed in section with straight, transverse sutures. Cameral chambers short, with cameral lengths decreasing with ontogeny. Body chamber unknown. Shell exterior smooth.

Siphuncle subcentral in position having segments becoming more central in position with ontogeny. Segments broad, typically one-third the diameter of the shell, moderately to highly compressed; becoming more compressed with ontogeny. Siphuncle segments do not demonstrate segment decompression with ontogeny. Septal necks cyrtochoanitic, with short necks and brims; necks strongly recurved, but free, not recumbent. Connecting rings thin, homogeneous; adapical portions of rings typically adnate both dorsally and ventrally to adjacent septa having the adoral portions of rings free. Endosiphuncular deposits consisting of coalescing annulosiphonate deposits and a vascular system consisting of straight radial canals situated at right angles to the central canal. Well developed hyposeptal and episeptal cameral deposits.

Stratigraphic and geographic occurrence.—Middle Ordovician (Chazyan Champlainian Provincial Series) to Early Silurian (Llandoverian Series); worldwide.

Remarks.—This genus has become a receptacle for a variety of small to medium-sized Ordovician and Silurian longiconic actinocerids with rather narrow, globular to compressed siphuncle segments. A brief investigation of some of the better known species assigned to this genus indicates that more than one genus-level taxon is contained in *Ormoceras* as it is currently defined (Teichert and others, 1964). At least three genus-level taxa can be recognized within *Ormoceras* at this time.

Flower (1957a) erected *Adamsoceras* for Middle Ordovician (Whiterockian) forms with thick, layered connecting rings and a recticular vascular canal system. In addition, species assigned to *Adamsoceras* are gradually expanding longicones (apical angle 4–5 degrees) with siphuncle segments only one-fifth to one-fourth the diameter of the shell and SCR < 1.5. Cameral length appears to increase with ontogeny in species assigned to this genus. Younger, more typical Middle Ordovician (Blackriveran) to Early Silurian (Llandoverian) species of *Ormoceras*, including the type species *O. bayfieldi*, have more rapidly expanding shells (apical angle 7–9 degrees) and broader, more highly compressed siphuncle segments (SCR=1.75–2.0), diameters averaging one-third the diameter of the shell. Connecting rings are thin and homogeneous, consisting of simple, straight radial canals. Based on these criteria, *Adamsoceras* Flower appears to be distinct from *Ormoceras* based on *O. bayfieldi* Stokes.

A third taxon, *Sactoceras* Hyatt, based on *Orthoceras richteri* Barrande from the Upper Silurian (Ludlovian) of Czechoslovakia, also has been placed in synonymy with *Ormoceras* Stokes. *Sactoceras richteri* is a medium-sized (phragmocone lengths up to 45 cm and shell diameters up to 5.5 cm), moderately expanding (apical angle 7–7.5 degrees) orthoconic longicone that is slightly curved endogastrically adapically. Camerae are short, decreasing with ontogeny from one-fifth to one-seventh the diameter of the shell. The siphuncle is in a central position for the length of the shell. Siphuncle segments are globular, slightly compressed having SCR values less than 1.33, and diameters from one-fourth to one-seventh the diameter of the shell. Endosiphuncular deposits consist of annulosiphonate deposits that fuse ventrally to form parietal deposits adapically. There is little evidence of a well-developed actinocerid type of vascular system, and perispatia are lacking. Dzik (1984), citing the absence of radial structure in the endosiphuncular deposits of these species, went as far as to remove *Sactoceras* from the Actinocerida, placing it in the Sactoceratidae within the order Orthocerida. The type species of *Sactoceras* appears to be distinct from *Ormoceras* in terms of siphuncle diameter relative to shell diameter, in segment compression, and in the nature of the endosiphuncular deposits. As such, *Sactoceras* appears to be a valid genus distinct from *Ormoceras* and may not, as suggested by Dzik, belong in the Actinocerida. The Kentucky material described here belongs to the complex of species that includes *O. bayfieldi* and constitutes true *Ormoceras*.

ORMOCERAS FERECENTRICUM FOERSTE AND TEICHERT, 1930

Plate 18, figures 1–7

Ormoceras ferecentricum Foerste and Teichert, 1930, p. 289, pl. 49, figs. 2*A–B*; pl. 58, fig. 10.

Diagnosis.—Moderately expanding species of *Ormoceras* with short camerae and a subcentral siphuncle with broad, highly compressed segments that average one-third the diameter of the shell.

Description.—Small to medium-sized (known portions of phragmocone up to 30 cm in length and 3.5 cm in diameter), moderately expanding (apical angle 7 degrees) orthoconic longicones, circular in cross section with straight, transverse sutures. Cameral chambers short, cameral lengths decreasing with ontogeny from one-third to one-seventh the diameter of the shell. Body chamber unknown. Shell exterior smooth.

Siphuncle subcentral in position the length of the phragmocone, moving to a more central position with ontogeny. Siphuncle segments have SPR values of 0.30 at diameters of 15–20 mm and 0.40 at diameters in excess of 30 mm. Segments broad and highly compressed (SCR=1.5–2.0) the length of the siphuncle, segment diameters one-third the diameter of the shell. Segments becoming more compressed with ontogeny. Endosiphuncular deposits consisting of annulosiphonate deposits and a vascular system of radial canals at right angles to the central canal. Adoralmost cameral deposits consisting of episeptal deposits.

Stratigraphic and geographic occurrence.—Upper part of the Rocklandian portion of the Tyrone Limestone (Middle Ordovician), in the vicinity of the Kentucky River, Jessamine County, Ky.

Studied material.—Holotype USNM 82214 (pl. 18, figs. 1–3), High Bridge, Ky.; plus silicified portions of the phragmocone from the USGS 6034-CO collection: USNM 158639 (pl. 18, figs. 4–6) and 468721 (pl. 18, fig. 7), from the Little Hickman section A, south of Sulfur Wells, Ky.

Remarks.—As indicated above, *O. ferecentricum* belongs to *Ormoceras*, in a narrow sense, by virtue of its shallow camerae and broad, highly compressed, subcentral siphuncle segments. It is most similar to a group of Middle and Upper Ordovician species from North America. This group lists *O. allumettense* (Billings) from the "Leray-Rockland beds" (Middle Ordovician, Rocklandian) in the Ottawa region of Ontario and Quebec, *O. cannonense* Foerste and Teichert from the Shermanian part of the Cannon Limestone (Middle Ordovician) of central Tennessee, and *O. dartoni* Flower from the Upper Ordovician Second Value Dolomite of New Mexico.

Ormoceras ferecentricum is most similar to *O. allumettense* (Billings), differing from this species only in its slightly less rapidly expanding shell (apical angle 7 degrees compared with 8–8.5 degrees) and the more central position of the siphuncle in the Kentucky species at comparable diameters. Larger collections of both species may indicate that these species represent regional variants of the same species, the observed morphological differences falling within the range of intraspecific variation. A comparison of the morphological features of these two species of *Ormoceras* is presented in figure 24.

Ormoceras ferecentricum is distinct from *O. cannonense* Foerste and Teichert in the much shorter camerae typical of the latter species, the cameral lengths averaging one-eighth the diameter of the shell compared with average cameral lengths one-fifth the diameter of the shell in the Kentucky species. Comparisons between *O. ferecentricum* and *O. dartoni* Flower are difficult due to the fact that the holotype and only known specimen of *O. dartoni* is represented by a section of phragmocone 40–50 mm in diameter, larger than any known fragment of *O. ferecentricum*.

At both the High Bridge and Sulfur Wells localities, *O. ferecentricum* is an uncommon species associated with more abundant specimens of the actinocerids *Actinoceras altopontense* Foerste and Teichert (pl. 16, fig. 6) and *A. kentuckiense* (pl. 17, figs. 1–5). At both localities, *O. ferecentricum* is represented by incomplete portions of the phragmocone and isolated siphuncle segments. It occurs in the 6034-CO collection in association with the diverse nautiloid fauna that is listed in table 1.

GENUS *DEIROCERAS* HYATT

Type species.—*Orthoceras python* Billings, 1857, p. 335; designated by Hyatt, 1884; Trenton Limestone, upper Shermanian Stage, Middle Ordovician, Ontario.

Description.—Medium-sized, gradually expanding orthoconic longicones; circular in cross section with straight transverse sutures. Cameral chambers moderately long, cameral lengths ranging from one-third to two-fifths the diameter of the shell. Body chambers unknown. Shell exterior poorly known; some species apparently smooth, other forms with faint longitudinal fluting.

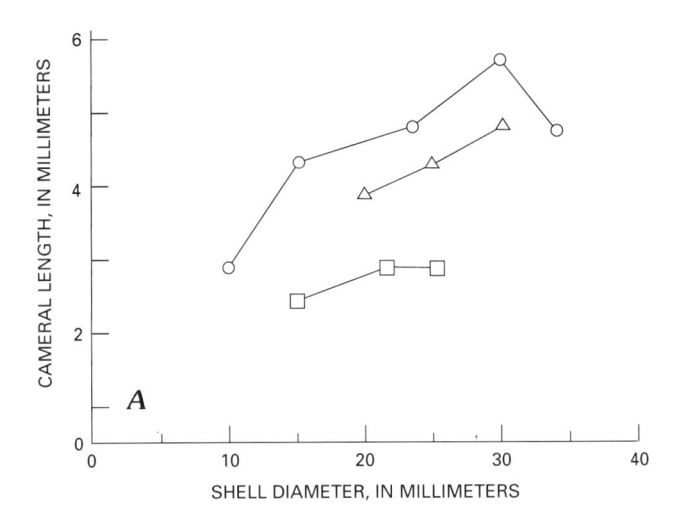

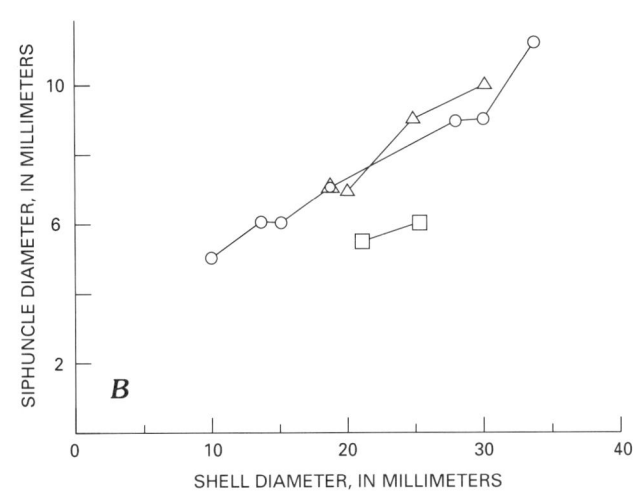

FIGURE 24.—Bivariate plots of cameral length versus shell diameter, *A*, and siphuncle diameter versus shell diameter, *B*, for Middle Ordovician species of *Ormoceras* from Eastern North America. Species are *O. ferecentricum* (USNM 82214, 158639, 468721), Tyrone Limestone, Kentucky, (circles); *O. allumettense* (GSC 1265) "Leray-Rockland beds," Quebec (triangles); and *O. cannonense* (USNM 82232), Cannon Limestone, central Tennessee (squares).

Siphuncles in midventral position for the length of the shell. Siphuncle segments typically from one-third to one-fourth the diameter of the shell; segments expanded ovate, globular, depressed in shape, longer than wide. Septal necks comparatively long, cyrtochoanitic, with necks gently curved, free, no areas of adnation. Endosiphuncular deposits consisting of annulosiphonate deposits and simple, straight radial canals set perpendicular to narrow central canal. Cameral deposits well developed, adorally consisting of episeptal and hyposeptal deposits.

Stratigraphic and geographic occurrence.—Middle Ordovician (Blackriveran) to Upper Ordovician (Edenian Stage), Eastern North America.

Remarks.—This genus is known primarily from poorly preserved fragmentary steinkerns and isolated siphuncles of late Middle Ordovician (Shermanian) to early Late Ordovician (Edenian) Age in Eastern and Central North America. *Deiroceras*, including the type species *D. python* Billings, is especially characteristic of the dark limestones of the Cobourg Formation in Ontario and adjacent portions of New York. Foerste and Teichert (1930) illustrated a number of species from coeval platform carbonates in Kentucky, Missouri, and Tennessee. Older Blackriveran species are *D. kindlei* Foerste, *D. paquettense* Foerste, *D. pertinax* (Billings), and *D. scofieldi* Foerste, all from Middle Ordovician carbonates exposed in the upper Mississippi valley and the Ottawa region of Ontario and Quebec.

Flower (1957a) pointed out the close similarity in terms of the siphuncle between the presumed smooth-shelled genus *Deiroceras* and the longitudinally fluted genus *Troedssonoceras* Foerste, both of which occur in the Catheys Formation of central Tennessee. Flower suggested that these two genera might prove to be one and the same. Shell exteriors are unknown for most described species of *Deiroceras*, as well as the type species. *Deiroceras pertinax* (Billings) and an unnamed species of the genus illustrated by Flower (1957, pl. 11, fig. 7) from upper Middle Ordovician limestones in Quebec both exhibit indications of a longitudinally fluted shell similar to that of *Troedssonoceras*. *Troedssonoceras* is currently known from upper Middle Ordovician (Shermanian) through upper Ordovician (Maysvillian) strata exposed in central Tennessee, the Cumberland River valley in south-central Kentucky, and the Cincinnati region of southwestern Ohio and adjacent portions of northern Kentucky. For the purposes of this paper, these taxa are retained as separate genera, pending the discovery of better preserved specimens of *Deiroceras* retaining the external shell. *Deiroceras* is interesting in that it was one of the few Blackriveran genera that survived the Rocklandian crisis to undergo a second radiation of species in the Shermanian and Edenian.

DEIROCERAS CURDSVILLENSE FOERSTE AND TEICHERT, 1930

Plate 18, figures 8, 9

Deiroceras curdsvillense Foerste and Teichert, 1930, p. 294, pl. 47, figs. 3*A–B*; pl. 58, fig. 6.

Diagnosis.—Slender, gradually expanding species of *Deiroceras*, cameral lengths averaging one-third the diameter of the shell and depressed siphuncle segments midventral to ventral in position.

Description.—Medium-sized (lengths of up to 30 cm and observed shell diameters of 3.0 cm), slender, gradually expanding (apical angle 6 degrees) orthoconic longicones; circular in section with straight transverse sutures. Cameral chambers comparatively long, having cameral lengths averaging one-third the diameter of the shell, decreasing slightly with ontogeny. Body chamber unknown. Shell exterior apparently smooth.

Siphuncle midventral to ventral in position, becoming more ventral with ontogeny. Siphuncle segments depressed, globular (SCR=0.70–0.88), showing no indication of siphuncle segment decompression with ontogeny. Segment diameters from slightly less than one-third to one-fourth the diameter of the shell. Connecting rings appear to have been thin and homogeneous. Septal necks cyrtochoanitic, long, with free brims. Endosiphuncular deposits unknown. Cameral deposits consisting of mural-episeptal deposits dorsally.

Stratigraphic and geographic occurrence.—Curdsville Limestone Member of the Lexington Limestone (Middle Ordovician, Kirkfieldian Stage), central Kentucky Blue Grass region.

Studied material.—Holotype USNM 48221 (pl. 18, fig. 8) from Curdsville, Ky. A number of silicified shell fragments including USNM 468723 (pl. 18, fig. 9) from the USGS 5100-CO collection, Little Hickman section B, Jessamine County, Ky.

Remarks.—This species is known only from a limited number of fragmentary specimens; collectively they make up a section of shell roughly 10 cm in length having shell diameters of 21 to 30 mm. All specimens studied were coarsely silicified internal molds that have been sectioned to expose the siphuncle. Siphuncle segments have been recrystallized, destroying any evidence of endosiphuncular structure. Comparison with other described species of *Deiroceras* is limited due to the incomplete and limited amount of material in collections.

In terms of general shell morphology, *D. curdsvillense* is most similar to the coeval species *D. paquettense* Foerste, *D. pertinax* (Billings), and *D. kindlei* Foerste, all from the Middle Ordovician (Blackriveran and Rocklandian) strata in the Ottawa, Ontario, region. *Deiroceras paquettense* is a more slender form having an apical angle of 4 degrees. *Deiroceras pertinax* is another slender form (apical angle 5 degrees), in which the outer shell is marked by a number of faint, raised longitudinal ribs. Its internal structure is poorly known, but cameral lengths approximate those of *D. curdsvillense*. The Kentucky species is most similar to *D. kindlei*, the specifics of the camerae and siphuncle segments in the holotype of *D. kindlei* matching up almost exactly with the Curdsville Limestone Member material. The holotype (GSC 14418) of *D. kindlei* differs from the holotype of *D. curdsvillense* only in its slightly more rapid rate of shell expansion: 7 degrees compared with 6 degrees in the Kentucky specimens. Further study may indicate that these species are the same, with *D. kindlei* Foerste having priority. *Deiroceras persiphonatum* (Flower), represented by a single known specimen (UC 22695) from the upper part of the Lexington Limestone at Cynthiana, Ky., differs from *D. curdsvillense* in being a more slender shell (apical angle 5 degrees), having somewhat shorter cameral lengths at comparable shell diameters. *Deiroceras curdsvillense* is evidently an uncommon form within the Curdsville Limestone Member, associated with a low-diversity nautiloid fauna that consists of *Actinoceras curdsvillense* Foerste and Teichert (pl. 17, fig. 11), *Anaspyroceras* cf. *A. cylindricum* (Foerste) (pl. 12, fig. 5), and poorly preserved specimens of the endocerids *Cameroceras* and *Vaginoceras*.

GENUS *ORTHONYBYOCERAS* SHIMIZU AND OBATA

Type species.—*Ormoceras? covingtonense* Foerste and Teichert, 1930, p. 288; designated by Shimizu and Obata, 1935, p. 7; Maysvillian Stage; Upper Ordovician, Kentucky.

Description.—Medium-sized, moderately to rapidly expanding orthoconic longicones; circular to slightly depressed in cross section with straight, transverse sutures. Camerae short, with cameral lengths typically from one-fourth to one-sixth the diameter of the shell, decreasing in length with ontogeny. Body chamber incompletely known. Shell exterior smooth.

Siphuncle subcentral to midventral in position, becoming more ventral with ontogeny. Siphuncle segments expanded, compressed in outline, segment diameters from one-third to one-fourth the diameter of the shell. Septal necks cyrtochoanitic, short, with recumbent brims. Connecting rings thin, adnate dorsally to both adoral and adapical septa; not adnate ventrally to adapical surface of adoral septum. Endosiphuncular deposits well developed, consisting of dorsal annuli developed more or less symmetrically around septal necks and more massive ventral annulosiphonate deposits that thicken adapical of the septal necks. Broad central canal having dorsal radial canals approaching connecting rings at the midpoint of the segment, while ventrally, canals enter in the adoral half of segment. Adoralmost cameral deposits consisting of discrete episeptal and hyposeptal deposits, which are more strongly developed ventrolaterally.

Stratigraphic and geographic occurrence.—Upper Ordovician (Maysvillian-Richmondian Stages), Eastern North America.

Remarks—The nomenclatural history of this genus is described in part in the preceding remarks concerning the orthocerid genus *Treptoceras*. As demonstrated by Aronoff (1979), *Orthonybyoceras* is an actinocerid and distinct from the associated orthocerid *Treptoceras*. Although the two genera are externally similar, *Orthonybyoceras* differs from *Treptoceras* in the possession of short, recumbent septal necks both dorsally and ventrally, in the development of true perispatia between annulosiphonate deposits, and in the possession of more compressed, ovate siphuncle segments that increase in height with ontogeny, in contrast to the heart-shaped segments in *Treptoceras*, which decompress during ontogeny (compare pl. 5, figs. 1–4 with pl. 18, figs. 10, 14, 17).

Orthonybyoceras does not fit neatly into any of the existing families of the Actinocerida. *Orthonybyoceras* differs from other members of the Ormoceratidae in possessing short, recumbent septal necks dorsally and ventrally and in the development of connecting rings that are adnate to septal surfaces. It differs from most members of the Armenoceratidae in possessing well developed cameral deposits, simple radial canal systems, and comparatively narrow diameter siphuncle segments. *Orthonybyoceras* is similar to members of the Wutinoceratidae in particulars of the siphuncle segments, their position, and the nature of the septal necks. It is distinct from these genera, however, in possessing thin, homogeneous connecting rings and a simple radial canal system. It is probable that *Orthonybyoceras* represents an "advanced" member of the Wutinoceratidae as suggested by Aronoff (1979). However, pending a more thorough restudy of the Actinocerida, *Orthonybyoceras* is retained here in the Ormoceratidae, following Teichert (*in* Teichert and others, 1964).

As currently recognized, *Orthonybyoceras* is known only from the type species, *O. dyeri* (Miller) from the Upper Ordovician (Maysvillian) of northern Kentucky and adjacent portions of Indiana and Ohio. Teichert and Glenister (1953) described a second species of *Orthonybyoceras*, *O. tasmanense*, from lower Middle Ordovician strata exposed in Tasmania. Aronoff (1979) removed this species from the genus due to the ontogenetic decrease in siphuncle diameter present in this species. He transferred it to "*Nybyoceras*" *paucicubiculatum* Teichert and Glenister, also from Middle Ordovician limestones exposed in Tasmania. Flower (1957a) and Stait (1984) placed this species in *Wutinoceras* Flower on the basis of the bifurcating radial canals. If these assignments are accepted, then the genus *Orthonybyoceras* is known only from the type species from the Upper Ordovician of the Cincinnati arch region.

ORTHONYBYOCERAS DYERI (MILLER), 1875

Plate 18, figures 10–17

Orthoceras dyeri Miller, 1875, p. 125, Fig. 11.
Orthoceras dyeri Miller. Miller, 1880, p. 236, pl. 7, fig. 7.
Ormoceras? covingtonense Foerste and Teichert, 1930, p. 288, pl. 45, figs. 2A–B; pl. 58, fig. 9.
Orthonybyoceras covingtonense (Foerste and Teichert). Shimizu and Obata, 1935, p. 7, 8.
Orthonybyoceras covingtonense (Foerste and Teichert). Teichert and Glenister, 1953, p. 223.
Orthonybyoceras covingtonense (Foerste and Teichert). Teichert, 1964, p. K214, Fig. 151,1a.
Orthonybyoceras covingtonense (Foerste and Teichert). Aronoff, 1979, p. 106–109, figs. 2D, 3A–C.

Diagnosis.—Small to medium-sized, rather rapidly expanding orthoconic longicone with compressed, subcentral siphuncle segments one-third to one-fourth the diameter of the shell; short recumbent septal necks, and thin connecting rings adnate dorsally to adjacent septa.

Description.—Small to medium-sized (phragmocones at least 25 cm in length and up to 4.3 cm in diameter), rapidly expanding (apical angle 8.5–10 degrees) orthoconic longicones; circular to slightly depressed in cross section with straight transverse sutures. Camerae short having cameral lengths ranging from one-fifth to one-seventh the diameter of the shell, decreasing in length with ontogeny relative to shell diameter. Body chamber poorly known, at least one-fifth the total diameter of the shell. Periphract dorsomyarian. Shell exterior smooth.

Siphuncle subcentral most of the length of the shell (SPR=0.40–0.285), becoming more ventral in position with ontogeny. Siphuncle segments expanded, compressed, the SCR values ranging from 1.3 at shell diameters of 25 mm to 1.7 at diameters of 35 mm. Segments from one-third to one-fourth the diameter of the shell, decreasing slightly in diameter with ontogeny. Septal necks short, with recumbent brims. Connecting rings thin, homogeneous, adnate to either adoral or adapical septa dorsally; ventrally adnate to adapical septa or not adnate at all. Dorsal endosiphuncular deposits consisting of annuli developed more or less symmetrically about the septal necks; deposits more asymmetrical ventrally, consisting of annulosiphonate deposits that are much thicker in development in the region adoral to septal necks. Perispatia well developed. Dorsal radial canals intersect the connecting ring wall at midpoint of segment. Ventrally, radial canals intersect connecting ring wall in adoral half of segment. Cameral deposits consisting of episeptal and hyposeptal deposits that, adorally, are more heavily developed ventrolaterally.

Stratigraphic and geographic occurrence.—Fairview Formation and Grant Lake Limestone and equivalents in northern Kentucky and adjacent portions of the Cincinnati, Ohio, region; lower part of the Bull Fork Formation, southeastern Indiana.

Studied material.—Holotype of *Orthoceras dyeri*, MCZ 3403 (pl. 18, figs. 13, 15, 16), Cincinnati, Ohio; holotype of *Ormoceras? covingtonense* Foerste and Teichert, USNM 48258 (pl. 18, figs. 14, 17), Covington, Ky.; hypotype of *O. dyeri*, UC 351, Corryville Member of the Grant Lake Limestone, Cincinnati, Ohio; specimens labeled "co-types" of *O. dyeri*, UC 352A (pl. 18, fig. 11), 352B (pl. 18, fig. 12), and 352C, labeled "Maysville-

Corryville, Cincinnati, Ohio; UC 31380" (pl. 18, fig. 10), "Waynesville biofacies" in the Bull Fork Formation, Weisburg, Ind.; and numerous uncataloged specimens.

Remarks.—Comparisons of the holotype of *Ormoceras*(?) *covingtonense* Foerste and Teichert (USNM 48258) with the holotype (MCZ 3403), hypotype (UC 351), and "co-types" of *Orthoceras dyeri* Miller (UC 352A–C) indicate that these specimens are all the same species. All of these specimens occur in the same strata in the greater Cincinnati region of southwestern Ohio and northern Kentucky and all are rapidly expanding longicones with relatively short camerae and compressed, subcentral siphuncle segments one-third to one-fourth the diameter of the shell. As indicated in the species synonymy, *Orthoceras dyeri* was the first published name given to this species (Miller, 1875) and has priority over *Ormoceras*? *covingtonense* (Foerste and Teichert, 1930).

Orthonybyoceras dyeri is the only species currently recognized as assignable to the genus *Orthonybyoceras* Shimizu and Obata. Species previously assigned to the genus (Teichert and Glenister, 1953) have been transferred to other genera by subsequent workers (Flower, 1957a; Aronoff, 1979; Stait, 1984). As indicated in the previous discussions of the genus *Treptoceras* Flower, these two taxa are internally distinct and are members of different orders. *Treptoceras* is by far the more abundant taxon in the Ordovician strata exposed in the Kentucky-Indiana-Ohio region. In this area, *Orthonybyoceras dyeri* is locally common in the Fairview Formation and in the overlying Corryville Member of the Grant Lake Formation.

A specimen (UC 31380, pl. 18, fig. 10) listed as coming from the "Waynesville Formation" (Upper Ordovician, Richmondian Stage) exposed at Weisburg, Ind., also appears to be assignable to *Orthonybyoceras*. Comparisons with other specimens of *Orthonybyoceras dyeri* are difficult due to the incomplete nature of the "Waynesville" specimen, but the sectioned phragmocone fragment is similar to *O. dyeri* in the rate of shell expansion and in the diameter and shape of siphuncle segments. It differs from Maysvillian specimens of *O. dyeri* in the shorter length of its camerae and in the more central position of its siphuncle at comparable shell diameters. The affinities of this specimen remain unclear pending the discovery of more complete specimens of this form from these younger strata.

SUBCLASS NAUTILOIDEA AGASSIZ

Breviconic, cyrtoconic, and coiled nautiloids with generally small diameter, primitively subtubular siphuncles having orthochoanitic or cyrtochoanitic septal necks and connecting rings of varying thickness. Periphract ventromyarian; hyponomic sinus variably developed, but typically present ventrally. Buoyancy regulation achieved through the development of short phragmocones coupled with large body chambers having the large volume of visceral mass and thick shell counteracting the small volume of gas-filled phragmocone; or by planispiral coiling of the shell, bringing the center of buoyancy directly over the center of mass, these centers being fixed in their relative position within the coil. This subclass is thought to represent a phylogenetic sequence from the primitive, evolutely coiled Tarphycerida through breviconic and cyrtoconic Oncocerida, to the involutely coiled Nautilida. This is the dominant nautiloid subclass in the post-Devonian.

Stratigraphic range.—Lower Ordovician to Holocene.

ORDER TARPHYCERIDA FLOWER

Primitive coiled nautiloids with evolutely coiled gyroconic or serpenticonic shells of variable whorl section and simple transverse or slightly curved sutures. Body chambers typically long, extending from one-half to one complete whorl in length, mature portions often slightly to greatly divergent from earlier coiled portions. Ventromyarian periphract and a well developed hyponomic sinus. Siphuncle varies in position, consisting of depressed, elongate tubular segments with orthochoanitic septal necks and thick connecting rings in primitive forms; cyrtochoanitic septal necks and thin, homogeneous rings in younger forms. Stability offered by the coiled shell morphology offset to some degree by length of body chamber, generating some rotational instability. This is countered by the divergence of the final one-quarter to one-half whorl of the body chamber. Modified apertures in some younger taxa.

Stratigraphic range.—Lower Ordovician to Middle Devonian.

SUBORDER TARPHYCERINA FLOWER

Gyrocones and serpenticones with simple septa, including both depressed and compressed forms. Siphuncles vary in position with orthochoanitic septal necks and thick, layered connecting rings.

Stratigraphic range.—Lower Ordovician (Ibexian Provincial Series) to Upper Silurian (Ludlovian Series).

FAMILY TROCHOLITIDAE CHAPMAN

Subglobular to discoidal nautilicones with three to five or more, gradually expanding, dorsally impressed whorls, typically depressed in cross section. Sutures simple transverse, or forming broad ventral lobes; camerae short. Shell sculpture varies in development, usually consisting of imbricating growth lamellae that form deep ventral sinuses. Strong ribs and periodic constrictions of the shell developed in some taxa.

Siphuncles dorsal in position, marginal or dorsal of the center except in initial volution, where they may be central or ventral in position. Siphuncle segments tubular, narrow in diameter, usually less than one-fourth the diameter of the shell. Septal necks short, orthochoanitic. Connecting rings typically thick, but becoming thin and homogeneous in some taxa.

Stratigraphic range.—Lower Ordovician (Ibexian Provincial Series) to Upper Silurian (Ludlovian Series).

GENUS *TROCHOLITES* CONRAD

Type species.—*Trocholites ammonius* Conrad, 1838, p. 118; by original designation; Trenton Limestone, Shermanian Stage, Middle Ordovician, New York.

Description.—Small, generally less than 10 cm in diameter, gradually expanding nautilicones, subdiscoidal in shape, having five or more volutions and a body chamber one-half to three-fourths of a volution in length that does not diverge from preceding whorls. Whorl section depressed, broadly reniform; dorsal part of whorl has shallow impressed zone. Umbilicus wide; umbilical perforation small.

Siphuncle tubular, narrow in diameter, segment diameters from one-fifth to one-sixth the diameter of the whorl; dorsal or subdorsal in position at all stages of growth. Septal necks short, orthochoanitic. Connecting rings thin, thickening adorally with ontogeny. Surface sculpture consisting of fasciculate growth lines or bands and raised ribs that form ventral sinuses.

Stratigraphic and geographic occurrence.—Middle Ordovician (Chazyan) to Upper Ordovician (Maysvillian Stage), North America, Baltic Europe, Iberian Peninsula.

TROCHOLITES FABERI FOERSTE, 1929C

Plate 19, figures 6–10

Trocholites faberi Foerste, 1929c, p. 323, pl. 63, figs. 1A–D.

Diagnosis.—Rather large species of *Trocholites* distinguished by its subdued external ornament of fine, oblique growth lines, strongly depressed whorl profile, and dorsal location of the siphuncle.

Description.—Gradually expanding, evolute nautilicones consisting of four to five whorls in close contact with one another; shell up to 9.0 cm in diameter and whorls up to 3.0 cm in diameter. Whorls reniform, depressed in cross section, nearly twice as wide as high. Shallow, concave, dorsal impressed zone, having venter convex, broadly rounded. Whorl profile becoming less depressed with ontogeny. Cameral chambers ranging from one-third to one-half the height of the whorl. Sutures essentially straight, transverse initially, becoming more sinuous with ontogeny marked by development of a shallow ventral lobe and ventrolateral saddles. Body chamber makes up one-half of a whorl, in contact with previous whorl its entire length and contracting at the aperture. Aperture marked by a deeply incised, narrow, U-shaped hyponomic sinus, 7 mm deep at an adoral diameter of 2.5 cm. Aperture sloping adorally from the venter, then curving back adapically at the dorsum. Shell exterior marked by fine, weakly developed, sinuous growth lines that slope obliquely from the dorsum to the venter in the adapical direction. Growth lines marking a broad sinus ventrally.

Siphuncle tubular, located nearly at the dorsal margin of the whorl. Siphuncle diameter slightly less than one-third the diameter of the shell at a shell height of 13 mm. Septal necks and nature of connecting rings unknown.

Stratigraphic and geographic occurrence.—Shermanian part of the Clays Ferry Formation, to Upper Ordovician (Edenian Stage), north-central Kentucky; Point Pleasant Tongue of the Clays Ferry Formation and the overlying Kope Formation; Covington, Ky., area and adjacent portions of Cincinnati, Ohio, region.

Studied material.—USNM 468724 (pl. 19, figs. 6, 7), 468725 (pl. 19, figs. 8, 9), and 468726 (pl. 19, fig. 10) from the Kope Formation at Orphanage Road roadcut, Ft. Mitchell, Kenton County, Ky.; plus additional material including UC 46011, an impression from the same locality as above; UC 8138, labeled *T. planorbiformis* from the Cincinnati, Ohio, area; and UC 24476, a large shelled specimen (7.7 cm in diameter) and an incomplete portion of a body chamber, listed from the Vaupel Collection, "Southgate Member [Kope Formation] exposed at the Big 4 RR cut at Sadieville, Scott County, Kentucky." Other specimens of *Trocholites* studied are UC 1483, labeled *T. ammonius* Conrad from the Trenton Group, Trenton Falls, N.Y.; UC 8139, labeled *T. circularis* from the "Hudson River Group," Cincinnati, Ohio; UC 24475, labeled *T. dyeri* from the Carley Collection, "Trenton Limestone," Cincinnati, Ohio; and MU 29869, labeled *T. ammonius*, Trenton Limestone, New York. The location of the holotype of *Trocholites faberi* (Foerste, 1929C, pl. 58, fig. 1a–c) is unknown at the present time.

Remarks.—Specimens of *Trocholites* from the greater Cincinnati region of Kentucky and Ohio have been placed in a number of species over the course of the past century. Examples are the type species, *T. ammonius* Conrad, *T. planorbiformis* Conrad, *T. circularis* Miller and Dyer, *T. minisculus* Miller and Dyer, *T. dyeri* Hyatt, and *T. faberi* Foerste. The specimens studied as part of this investigation, a majority of which are from the Edenian part of the Kope Formation (Upper Ordovician) in northern Kentucky, appear to be most similar to *T. faberi* Foerste. This is indicated by the large size of most specimens (in excess of 6.0 cm in diameter), the depressed whorl profile, and the weakly developed external ornament of fine, irregular, oblique growth lines preserved in most specimens. Unfortunately, none of the studied material preserves evidence of the structure of the siphuncle.

Trocholites ammonius Conrad was originally described (Conrad, 1838) from the middle part of the Trenton Limestone (Middle Ordovician, Shermanian) of New York. Emmons (1842) and Hall (1847) remarked that this species also occurs in the equivalent black shale facies (=Canajoharie Shale). Ruedemann (1926) indicated that the species ranges up into the overlying Utica Shale (Upper Ordovician) in the same area. Foerste (1929c) distinguished *T. ammonius* from *T. faberi* on the basis of the smaller size of *I. ammonius*, the more central position of the siphuncle, and its distinctive surface ornament of alternating coarse and fine, raised, transverse lamellose ribs and lirae that give the shell surface a peculiar netlike appearance. This distinctive external shell ornament has not been observed in specimens of *Trocholites* from the Cincinnati area.

Trocholites faberi is similar to a variety of *T. ammonius* described by Ruedemann (1926) as *T. ammonius* var. *major* from the Utica Shale at Holland Patent, N.Y. These specimens were described as consisting of five whorls and attaining a larger diameter than typical *T. ammonius*, as well as possessing deeper camerae and lacking evidence of the coarse external ornament (Ruedemann, 1926, pl. 17, figs. 1–3). The position of the siphuncle was unknown. These specimens appear to be more similar to the Kentucky material from the Kope Formation.

The second species of *Trocholites* described by Conrad from the Ordovician strata in New York, *T. planorbiformis* from the Pulaski Formation (Upper Ordovician, Maysvillian), was distinguished by Hall (1847) and Ruedemann (1926) from *T. ammonius* primarily on the basis of its surface ornament. This ornament was described as consisting of oblique, raised, transverse and revolving longitudinal lines, forming a distinct cancellate pattern. The prominent, raised, longitudinal lirae described as being characteristic of this species have not been observed in any of the studied specimens of *Trocholites* from the Cincinnati-Covington area. Specimens of *Trocholites* preserving the outer shell and surface ornament are rare in the Kope Formation and equivalents in the greater Cincinnati area. Specimens preserving the siphuncle are equally uncommon. Both of these factors make further comparisons with these New York species difficult.

The remainder of the species of *Trocholites* described from the Cincinnati region are poorly known, and the locations of the type specimens of *T. circularis*, *T. dyeri*, *T. faberi*, and *T. minisculus* are unknown. Comparisons of these various species are, as a result, difficult. Assignment of the Kope material to *T. faberi* is tentative, pending further study of these various species and discovery in the Kope Formation of specimens of *Trocholites* that preserve the siphuncle.

At the Orphanage Road locality (KY-1), specimens of *T. faberi* occur as uncommon matrix-filled internal molds in a meter-thick claystone bed. The species is associated with similarly preserved specimens of the orthocerid *Treptoceras transversum* (Miller) and a diverse fauna of mollusks, consisting of pleurotomariacean snails, the monoplacophorans *Cyrtolites* and *Sinuites*, rostroconchs, and pelecypods (Frey, 1987b). Additional associated fauna are the brachiopods *Onniellla* and *Zygospira*, twiglike bryozoans, and three-dimensional specimens of the graptolite *Geniculograptus typicalis*.

SUBORDER BARRANDEOCERINA FLOWER

Cyrtocones, gyrocones, serpenticones, and torticones with primitively tubular, empty, orthochoanitic, thin-walled siphuncles that vary in position but are usually central or ventral; many derived taxa have cyrtochoanitic septal necks and inflated siphuncle segments.

Stratigraphic range.—Middle Ordovician to Middle Devonian.

FAMILY PLECTOCERATIDAE HYATT

Coiled barrandeocerinids having comparatively broad, tubular, midventral siphuncles and short orthochoanitic septal necks and thin, structureless connecting rings. Differing from ancestral tarphyceratids primarily in the nature of the connecting rings: thin in the Plectoceratidae; thick, layered in the Tarphyceratidae.

Stratigraphic range.—Middle Ordovician (Chazyan Stage) to Middle Silurian (Wenlockian Series).

GENUS *PLECTOCERAS* HYATT

Type species.—*Nautilus jason* Billings, 1859, p. 464; designated by Hyatt, 1884, p. 268; Mingan Formation, Chazyan Stage, Middle Ordovician, Mingan Islands, Quebec.

Description.—Comparatively large (up to 20 cm in diameter), robust serpenticones consisting of three or four volutions with initial whorls in contact; fairly rapidly expanding, the dorsal-ventral diameter doubling in one volution. Umbilical perforation large. Initial whorls somewhat depressed, subcircular, the venter slightly flattened and dorsum with a shallow impressed zone, sides rounded. Later whorls becoming compressed, subquadrate. Camerae short, the sutures initially straight, developing shallow lateral lobes with ontogeny. Mature body chambers free, less than one-half a volution in length. Aperture has a deep hyponomic sinus ventrally. Surface of shell marked by strong costae that curve obliquely in an adapical direction from the dorsum to the venter and swing back ventrally, forming V-shaped elevations. Parallel to costae are coarse transverse lirae. Costae more conspicuous ventrolaterally, becoming obsolete dorsally.

Siphuncle near the ventral margin of the shell, removed from it by a distance equal to one-sixth the diameter of the shell. Siphuncle tubular, having diameters between one-eighth and one-fifth the diameter of the shell. Septal necks orthochoanitic, short, straight. Connecting rings thin, homogeneous.

Stratigraphic and geographic occurrence.—Middle Ordovician (Chazyan to Rocklandian Stages), Eastern and Arctic North America (Laurentia).

PLECTOCERAS CF. P. CARLETONENSE FOERSTE, 1933

Plate 19, figures 1–3

Plectoceras carletonense Foerste, 1933, p. 123, pl. 36, fig. 1.

Diagnosis.—Species of *Plectoceras* distinguished by its depressed whorl profile and an external ornament consisting of weakly developed ventrolateral costae.

Description.—Species incompletely known, represented by fragmentary specimens. These consist of a portion of the inner whorl expanding from 18 to 21 mm in diameter in a length of 50 mm. Whorl profile depressed, 18 mm in width and 15 mm high at the adapical end. Venter broadly rounded with a very shallow dorsal impressed zone, although much of the dorsal portion of the whorl has been lost. Camerae short, cameral lengths averaging one-sixth the diameter of the shell at diameters of 18–21 mm. Inner whorl in contact with outer whorl ventrally. Shell exterior marked by indistinct, raised costae spaced 3 mm apart, sloping obliquely in the adapical direction from the dorsum to the venter. Costae most distinct ventrolaterally. Costae forming broadly rounded V-shaped elevations ventrally. Siphuncle 2.5–3.0 mm in diameter, situated less than 1 mm from ventral shell margin.

Other fragments representing portions of the body chamber, including part of the aperture. Aperture disjunct from earlier whorls, the lateral portions curving sharply forward from venter and then curving sharply back at dorsum. Exterior marked by nearly flat, indistinct costae 4 mm wide, spaced 8 mm apart having intervening irregular, fine growth lines.

Stratigraphic and geographic occurrence.—Upper part of the Rocklandian portion of the Tyrone Limestone (Middle Ordovician), vicinity of the Kentucky River gorge, Jessamine County, Ky.

Studied material.—USNM 158651 (pl. 19, figs. 1–3) and additional fragmentary specimens, all silicified material from the USGS 6034-CO collection, Little Hickman section A, Sulfur Well, Ky.

Remarks.—The silicified Tyrone Limestone material described here, although fragmentary, preserves enough of the morphology of the shell to indicate placement of these specimens in the genus *Plectoceras* Hyatt. This is indicated by the ventral position of the siphuncle, the depressed whorl profile, the short camerae, and the raised costae forming a V-shaped pattern ventrally. A reconstruction of a complete specimen of this species based on this fragmentary material (fig. 25) indicates a specimen at a maximum diameter of 13.0 cm.

FIGURE 25.—Reconstruction of the entire shell of the coiled tarphycerid *Plectoceras* cf. *P. carletonense* from the upper part of the Tyrone Limestone, Jessamine County, Ky. Based on fragmentary silicified specimens from the USGS 6034-CO collection (see pl. 19, figs. 1–3). ×0.65.

The Tyrone specimens are most similar to *P. carletonense* Foerste from the "Black River Formation" (Leray beds) exposed in the Ottawa region of Ontario. Both species possess depressed, subrectangular whorl sections that are wider than high, have inner whorls that are in contact but not strongly impressed dorsally, and have siphuncles situated very close to the ventral shell margin. Costae in both forms are weakly developed, compared with other described species of *Plectoceras*, being most distinct ventrolaterally. More complete material from the Tyrone Limestone is needed to further compare these species.

The Tyrone material is less similar to *P. robertsoni* (Hall), another dorsally depressed species of *Plectoceras* from the Middle Ordovician (Blackriveran and Rocklandian) Platteville Formation in Wisconsin and Platteville Group in Illinois. It differs from this species in that the inner whorls of the Platteville form have a distinct dorsal impressed zone that appears to be lacking in the Kentucky material. The costae in specimens of *P. robertsoni* figured by Foerste (1932, pl. 36, figs. 2A–D, 3A–B) are also much more strongly developed than is evident in the Kentucky specimens.

In a similar manner, the Tyrone species is distinct from *P. halli* (Foord) described from Middle Ordovician strata exposed at Lorette, Quebec. This is another species of *Plectoceras* with a depressed whorl profile and strongly developed ventrolateral costae (pl. 19, figs. 4, 5). Further study may indicate that this species is a synonym of *P. robertsoni* (Hall).

The Tyrone species of *Plectoceras* is less similar to other described species of *Plectoceras*, such as the type species, *P. jasoni* (Billings) from Middle Ordovician (Chazyan) strata in New York and Quebec, *P. undatum* (Conrad) from the Watertown Limestone (Blackriveran and Rocklandian) in New York, and *P. occidentalis* (Hall) from the Platteville Group (Blackriveran and Rocklandian) in Illinois. This group of species is distinct from the preceding group, as well as the Tyrone form, in having whorls that are more circular in profile and that become disjunct at an earlier ontogenetic stage than the dorsal-ventrally depressed species described above. Additionally, *P. jasoni* and *P. occidentalis* possess costae that are much more strongly developed than is evident in the Tyrone species.

Study of the described species currently assigned to the genus *Plectoceras* Hyatt has indicated that the genus can be divided into two major groups. One group is of generally smaller diameter species that have the whorls in contact most of the length of the shell, giving the whorls a depressed cross section having a variably developed dorsal impressed zone. A second group, including the type species, consists of generally larger shells in which the whorls become disjunct at an earlier stage of development, giving the whorls a more circular cross section and typically lacking a dorsal impressed zone. Flower (1984) proposed splitting up *Plectoceras* along similar lines, although he noted there is such gradation in this respect that he regarded these groups as making up one genus. He differentiated *Plectoceras* instead on the basis of the more quadrate whorl and flattened venter typical of most younger (Blackriveran) species in contrast to the rounded venter typical of the type species, *P. jasoni* from older Chazyan strata in New York and Quebec. Pending further study of the genus, all of these species, as well as the Tyrone form, are retained here in *Plectoceras*.

ORDER ONCOCERIDA FLOWER

Primitively compressed, exogastric cyrtocones and brevicones with short camerae and simple, transverse or slightly curved sutures. Younger taxa are rapidly or gradually expanding longiconic cyrtocones, orthocones, torticones, gyrocones, and nautilicones, some depressed in section with endogastric curvature. Body chambers typically large; apertures primitively simple with a ventral hyponomic sinus; later forms developing bizarre contracted apertures. Periphract ventromyarian. Siphuncle ventral of center, segments generally narrow in diameter, but may become large and inflated. Septal necks suborthochoanitic or cyrtochoanitic; connecting rings typically thin but may thicken with ontogeny, producing actinosiphonate deposits. Major means of buoyancy regulation is the large volume of visceral mass and shell material relative to the phragmocone. Actinosiphonate deposits may have served a hydrostatic function, controlling the volume of cameral fluids in the phragmocone.

Stratigraphic range.—Middle Ordovician to lower Carboniferous.

FAMILY ONCOCERATIDAE HYATT

Rapidly to gradually expanding exogastric cyrtocones, brevicones, and longiconic cyrtocones; compressed in section and having straight transverse sutures that tend to become more curved with ontogeny, sloping adorally from the dorsum to the venter. Camerae typically short, cameral length decreasing with ontogeny. Body chamber large, contracting toward the aperture. Aperture simple, open, marked by a variously developed ventral hyponomic sinus. Shell exteriors typically smooth but may be marked by strong transverse costae or crenulate frills.

Siphuncle ventral of center, often located close to ventral shell margin. Segments empty, depressed, tubular in early stages, becoming expanded, scalariform to heart-shaped in later stages. Septal necks suborthochoanitic initially, becoming cyrtochoanitic with ontogeny. Connecting rings thin. Actinosiphonate deposits develop in some younger taxa.

Stratigraphic range.—Middle Ordovician (Chazyan Stage) to Upper Silurian (Ludlovian Series).

GENUS *ONCOCERAS* HALL

Type species.—*Oncoceras constrictum* Hall, 1847, p. 197; by original designation; Trenton Limestone, Shermanian Stage, Middle Ordovician, New York.

Description.—Small, exogastrically curved brevicones, compressed in cross section having the venter more narrowly rounded than dorsum. Greatest diameter of the shell typically located just adapical of the body chamber, so that the dorsal shell outline is initially concave, becoming broadly convex at junction with body chamber, then slightly concave at the aperture. Ventral shell outline remaining broadly convex the length of the shell. Cameral chambers short, cameral length decreasing with ontogeny. Sutures straight, transverse, sloping with ontogeny from the dorsum to the venter in the adoral direction. Body chamber short but broad, contracting toward the aperture. Aperture marked by a variously developed hyponomic sinus ventrally and a shallow sinus dorsally. Periphract ventromyarian. Shell exterior essentially smooth in most species, marked by fine transverse growth lines and a color pattern of transverse bands that completely encircle the shell.

Siphuncle small in diameter, located close to the ventral shell margin. Segments subtubular adapically, becoming more expanded with ontogeny to form ovoid to scalariform segments adorally. Septal necks short, recurved, cyrtochoanitic. Connecting rings thin, homogeneous. Segments devoid of deposits.

Stratigraphic and geographic occurrence.—Middle Ordovician (Chazyan Stage) to Upper Ordovician (Gamachian Stage), North America (Laurentia).

Remarks.—Flower (1946) gives an excellent synopsis of this genus and its stratigraphic and geographic distribution, and a discussion of its relationships with the genera *Beloitoceras* Hyatt and *Neumatoceras* Foerste. *Oncoceras* differs from *Beloitoceras* in being a less compressed, more rapidly expanding shell. *Beloitoceras* lacks the well-developed gibbosity characteristic of *Oncoceras*, namely, its evenly concave dorsal profile and a weak convexity in the lower half of the body chamber. These distinctions may, however, be blurred in early stages of growth and in the development of transitional species that share features common to both genera. *Beloitoceras* is the dominant form in earlier Chazyan and Blackriveran strata, *Oncoceras* becomes more prevalent in younger Shermanian to Maysvillian rocks. Flower (1946) described *Oncoceras* as a more advanced, "phylogeronitic" version of *Beloitoceras*. Curiously, both genera are equally represented in younger Richmondian strata in the Cincinnati region.

Flower (1946) described the closely related genus *Neumatoceras* Foerste as being a less well established genus, based on species that attain an extreme in the convexity of the shell just adapical to the body chamber. This genus is most typical of Shermanian to Richmondian strata that are part of the areally extensive carbonate platform facies that were deposited over much of Western and Arctic North America during this time. Species grouped in this genus were thought to represent the end product of the evolutionary sequence described above for *Beloitoceras* and *Oncoceras*. The differences between *Neumatoceras* and *Oncoceras* lie chiefly in the areas of size and the degree of shell convexity, vaguely defined characters to use to distinguish these genera. Flower further pointed out the possible effects of taphonomy. He replicated a cast of a typical specimen of *Oncoceras* and then, simulating the effects of sediment compaction, flattened the cast, resulting in a typical *Neumatoceras*. As a result of these observations, he suggested placing *Neumatoceras* in synonymy with *Oncoceras* as a junior synonym.

ONCOCERAS MAJOR NEW SPECIES

Plate 19, figures 11–15

Diagnosis.—Large species of *Oncoceras* distinguished by the lack of a constricted area adapical of the aperture and by its distinctive external ornament of coarse transverse costae.

Description.—Comparatively large (estimated lengths of 12.5 cm and a maximum observed shell diameter of 4.3 cm), broadly inflated, rapidly expanding (apical angle 25 degrees), exogastric brevicone. Lateral shell profile uniformly convex ventrally, dorsum slightly convex adorally such that the maximum shell diameter is at the juncture between the phragmocone and the body chamber. Compressed-ovate in cross section, height to width ratio 1.16 at a shell height of 30 mm and 1.22 at a shell height of 43 mm. In cross section the dorsum is broadly rounded, the venter narrowly rounded, almost angular. Sutures nearly straight, transverse, becoming slightly oblique in adoralmost portions of the phragmocone, sloping from the dorsum to the venter in adoral direction. Cameral chambers short, cameral lengths ranging from one-eighth to one-eleventh the diameter of the shell at shell diameters from 30 to 40 mm. Body chamber inflated but laterally compressed, contracting slightly towards the aperture. Aperture open, subtriangular, broadly rounded dorsally, the venter narrowly rounded, marked by a shallow V-shaped hyponomic sinus. Periphract ventromyarian. Shell surface marked by fine, incised, transverse striae between broad, 1–2 mm wide, raised convex costae of varying width and prominence. Costae transverse dorsally, curving ventrally to delineate a shallow V-shaped sinus.

Siphuncle situated at the ventral shell margin, the siphuncle 5 mm in diameter at a shell diameter of 30 mm. Opening for siphuncle in adoralmost septum broad, 9 mm in diameter. Segment shape unknown. Nature of septal necks unknown.

Stratigraphic and geographic occurrence.—Upper part of the Rocklandian portion of the Tyrone Limestone (Middle Ordovician), vicinity of the Kentucky River, Jessamine County, Ky.

Studied material.—Holotype USNM 158665 (pl. 19, figs. 11–15), USGS 6034-CO collection, Little Hickman section A, Sulfur Well, Ky.

Remarks.—This writer is hesitant to erect a new species on the basis of a single specimen, but this specimen is quite distinct from all other species of *Oncoceras*. Diagnostic of *O. major* is its large size, on which Flower (unpublished notes) remarked that the species is "The largest species of *Oncoceras* known to me." He stated further that "forms of comparable size have not been found in the eastern United States or Canada." Its coarse external ornament of broad, raised, transverse costae is equally distinctive, not being described or figured for any other species of *Oncoceras* known to this writer. Additional diagnostic features are the convex dorsal outline of the shell coupled with the lack of a strong constricted area adapical of the aperture.

Oncoceras major is most similar to a group of primarily Middle Ordovician species of *Oncoceras* characterized by a distinctly convex dorsal outline at the juncture of the phragmocone with the body chamber. These species are the type species *O. constrictum* Hall, from Rocklandian to Shermanian strata exposed in New York and adjacent portions of Canada; *O. collinsi* Foerste, *O. scalariforme* Wilson, and *O. tetreauvillense* Foerste, all from the "Leray beds" exposed in the Ottawa, Ontario region; and *O. douglassi* (Clarke) from the Shermanian Prosser Limestone in Minnesota. *Oncoceras major* differs from the first three species in its larger size, distinctive transverse external ornament, and lack of a well-developed constriction adapical of the aperture, the latter being a feature common to all three of these species of *Oncoceras*. *Oncoceras major* also is a more inflated, more dorsally convex species compared with *O. collinsi* and *O. scalariforme*. The Kentucky specimen is more similar to *O. tetreauvillense* and *O. douglassi* in inflation and dorsal convexity but differs from both in its coarse transverse surface markings.

Oncoceras major is less similar to Upper Ordovician species of *Oncoceras* described by Flower (1946) from the Cincinnati arch region. These Cincinnatian species, consisting of *O. arlandi* Flower from the Leipers Limestone in Kentucky, *O. deliculatum* Flower from the Bull Fork Formation in southwestern Ohio, and *O. madisonense* Flower from the Hitz Limestone Member of the Saluda Formation at Madison, Ind., all lack the obvious dorsal convexity and coarse transverse ornament characteristic of *O. major*.

As indicated above, *O. major* is apparently a rare element of the diverse, abundant nautiloid fauna present in the USGS 6034-CO collection from the upper part of the Tyrone Limestone in central Kentucky. There it is associated with abundant specimens of the baltoceratid *Cartersoceras shideleri* (pl. 1, figs. 1–6), the orthocerids *Pojetoceras floweri* (pl. 2, figs. 1–9) and *Proteoceras tyronensis* (pl. 4, figs. 1–7), the actinocerids *Actinoceras altopontense*, *A. kentuckiense* (pl. 17), and *Ormoceras ferecentricum* (pl. 18, figs. 1–7), the endocerid *Vaginoceras* sp. A (pl. 15, figs. 1–4), uncommon specimens of the tarphycerid *Plectoceras* cf. *P. carletonense* (pl. 19, figs. 1–3), and the oncocerids *Beloitoceras* cf. *B. huronense* (pl. 20, figs. 3–6), *Maelonoceras* cf. *M. praematurum* (pl. 20, figs. 7–12), and *Laphamoceras* cf. *L. scofieldi* (pl. 20, figs. 13–18).

ONCOCERAS SP.

Plate 20, figures 1, 2

Diagnosis.—Small, rapidly expanding species of *Oncoceras*, having exterior marked by fine, raised transverse lirae.

Description.—Incomplete, fragmentary specimens indicate a comparatively small (estimated lengths of 85 mm and a maximum observed shell diameter of 26 mm), rapidly expanding (apical angle 30 degrees), exogastrically curved brevicone having venter broadly convex and the dorsum concave adapically at shell heights of 10 mm, becoming straight to slightly convex adorally at shell heights of 15–25 mm. Shell compressed ovate in cross section, the height to width ratio being 1.11 at a shell height of 10 mm and 1.18 at a shell height of 20 mm. Camerae short, with cameral lengths increasing from one-eighth to one-sixth the diameter of the shell at shell heights from 15 to 23 mm, and then decreasing to one-ninth the diameter of the shell at a shell height of 25 mm. Body chamber incompletely known, inflated, making up roughly one-third the total length of the shell. Aperture unknown. Periphract ventromyarian. Shell surface marked by fine (0.5 mm wide), raised, transverse lirae that delineate a shallow V-shaped ventral sinus.

Siphuncle located at the ventral margin, narrow in diameter, one-ninth the diameter of the shell at a shell diameter of 25 mm. Segments scalariform, broader adorally than apically. Segments empty. Septal necks unknown.

Stratigraphic and geographic occurrence.—Grier Limestone Member of the Lexington Limestone (Middle Ordovician, Kirfieldian to Shermanian), Jessamine County, Ky.

Studied material.—USNM 468732 (pl. 20, fig. 1) and 468731 (pl. 20, fig. 2), both fragmentary silicified specimens from USGS 4879-CO and 4883-CO collections from exposures along Antioch Church Road, Valley View section C, Valley View, Ky.

Remarks.—The fragmentary nature of the material makes comparisons with other described species of *Oncoceras* difficult. Especially critical is the lack of any specimens preserving the dorsal outline of the adoral portion of the shell and adjacent portions of the body chamber. Additionally, no specimens available to this writer preserve the aperture of the body chamber. More complete material is necessary to further the affinities of these specimens.

In the Grier Limestone Member, *Oncoceras* sp. is an uncommon species found in association with a number of specimens of an unidentified large, smooth, orthoconic longicone (*Treptoceras*?). Other faunal associates are well-preserved specimens of the pleurotomariacean

GENUS *BELOITOCERAS* FOERSTE

Type species.—*Oncoceras pandion* Hall, 1861, p. 45; designated by Foerste, 1924b, p. 244; Platteville Limestone, Blackriveran Stage, Middle Ordovician, Wisconsin.

Description.—Small to moderate in size (rarely up to 20 cm in length and 7.5 cm in diameter), strongly curved, exogastric brevicones that are ovate in cross section, becoming more compressed with ontogeny. Maximum shell diameter situated within the adapical one-half to two-thirds of the body chamber; shell only slightly inflated. Dorsal outline of shell typically concave for its entire length. Ventral margin broadly convex. Sutures straight and transverse adapically, sloping from the dorsum to the venter in the adoral direction with ontogeny. Camera chambers short, cameral lengths ranging from one-fifth to one-eighth the diameter of the shell, increasing with ontogeny until onset of maturity. Body chamber short, laterally compressed; well-developed hyponomic sinus ventrally; shallow sinus dorsally. Periphract ventromyarian. Shell smooth or with fine, raised transverse ribs.

Siphuncle small in diameter, located close to the ventral shell margin. Segments subtubular adapically, becoming expanded, elongate ovate adorally. Septal necks, adapically short, suborthochoanitic, becoming more recurved, cyrtochoanitic, with ontogeny. Connecting rings thin, homogeneous. Segments empty. Cameral deposits described for some species.

Stratigraphic and geographic occurrence.—Middle Ordovician (Chazyan Stage) to Upper Ordovician (Gamachian Stage); North America and Baltic Europe.

Remarks.—As indicated in the previous discussion of *Oncoceras*, *Beloitoceras* forms the primitive stock from which *Oncoceras* and *Neumatoceras* were later derived in the Middle Ordovician. It differs in its typical form from these related genera in being a more laterally compressed, strongly curved shell that lacks the dorsal convexity in adoral portions of the shell that is typical of specimens of the latter two genera. *Beloitoceras* is the dominant oncocerid in Chazyan and Blackriveran strata in Eastern and Central North America, becoming less important in succeeding upper Middle Ordovician (Kirkfieldian to Shermanian) and lower Upper Ordovician (Edenian) strata. The genus has a resurgence of speciation in the Richmondian, being widespread across both Arctic and East-Central North America.

As noted by Flower (1946), the genus currently embraces more species than *Oncoceras* and *Neumatoceras* combined. Flower (1946) divided the genus into a number of "species groups," based primarily on the nature of the dorsal outline of the shell. He noted that these groups graded into one another, and therefore, he did not formally recognize these groups as distinct taxa.

BELOITOCERAS AMOENUM (MILLER), 1879

Plate 21, figure 3; plate 22, figure 2

Cyrtoceras amoenum Miller, 1879, p. 105, pl. 3, fig. 8.
Cyrtoceras amoenum Miller. James, 1886, p. 247.
Cyrtoceras amoenum Miller. Cumings, 1908, p. 1027, pl. 49, fig. 1.
Beloitoceras amoenum (Miller). Flower, 1946, p. 272, pl. 32, figs. 15, 17; pl. 34, figs. 4, 5; pl. 37, fig. 9; pl. 38, fig. 6; pl. 42, figs. 2, 5.

Diagnosis.—Relatively large, elongate, slender species of *Beloitoceras*, the greatest shell diameter just adapical of the body chamber, an essentially straight dorsum, and sutures that, with ontogeny, become strongly oblique, sloping adorally from the dorsum to the venter.

Description.—Size large for genus (estimated lengths up to 12.0 cm and observed shell diameters up to 3.8 cm); rapidly expanding (apical angle 22–26 degrees), exogastric brevicones whose dorsal outline is concave adapically, becoming nearly straight or slightly convex with ontogeny, having the maximum shell diameter just adapical of body chamber. Venter broadly convex for the length of the shell. Shell cross section ovate adapically, becoming more compressed adorally with ontogeny. Sutures straight, transverse adapically, becoming strongly oblique with ontogeny, sloping adorally from the dorsum to the venter. Cameral chambers short, cameral length decreasing with ontogeny from one-sixth to one-ninth the diameter of the shell at diameters of 15–25 mm, then increasing adorally from one-ninth to one-seventh the diameter of the shell at diameters greater than 25 mm. Adoralmost septa becoming approximate in larger specimens, cameral lengths reduced to one-twentieth the diameter of the shell. Body chamber short, one-fifth the total length of the shell, highly compressed, tapering toward aperture. Aperture marked by well-developed ventral hyponomic sinus and shallow dorsal sinus. Shell exterior marked by coarse tranverse lirae.

Siphuncle located at ventral margin. Segments narrow, diameters one-eighth to one-ninth the diameter of the shell; slightly expanded, depressed ovate in outline. Structure of septal necks and connecting rings unknown.

Stratigraphic and geographic occurrence.—Upper Ordovician (Richmondian Stage); Whitewater Formation in southeastern Indiana, upper part of the Bull Fork Formation in southwestern Ohio.

Studied material.—Holotype UC 106, Whitewater Formation, Richmond, Ind.; hypotypes Earlham College nos. 7744 and 14953, Whitewater Formation, Richmond, Ind.; USNM 468668 (pl. 21, fig. 3), upper part of the Bull Fork Formation, Dayton, Ohio; and USNM 468669 (pl. 22, fig. 2), Whitewater Formation, Richmond, Ind.

Remarks.—This species is a common form in the rubbly bedded, fossiliferous limestones of the Whitewater Formation exposed in the vicinity of Richmond, Wayne County, Ind. It usually occurs as poorly preserved internal molds associated with a diverse fauna of orthid and rhynchonellid brachiopods, ramose trepostome bryozoans, the coral *Grewingkia canadense*, pelecypods, and other mollusks. Figured specimen USNM 468668 (pl. 21, fig. 3) was collected from the interbedded, thin, fossiliferous packstone and shale in the upper part of the Bull Fork Formation exposed in north Dayton, Montgomery County, Ohio. The species is apparently a rare form in these strata, where it is associated with ramose trepostome bryozoans, the rugosan coral *Grewingkia canadense*, and the rhynchonellid brachiopod *Hiscobeccus capax*.

Flower (1946) commented that, on the basis of having the maximum shell diameter adapical of the body chamber, this species could be placed in the genus *Oncoceras*. The adorally compressed shell and the concave dorsal margin, however, indicate assignment of the species to *Beloitoceras*. *Beloitoceras amoenum* is most similar to *B. cumingsi* Flower, from equivalent fine-grained limestones of the Saluda Formation in southeastern Indiana and adjacent portions of western Ohio. Both species have adorally compressed shells that have their maximum diameter just adapical of the body chamber, straight dorsums, and sutures that slope adorally from the dorsum to the venter. *Beloitoceras cumingsi* differs from *B. amoenum* in being a smaller, more rapidly expanding shell. *Beloitoceras amoenum* is also similar to *B. magisterium* Foerste, a somewhat larger species from the Vaureal Formation (Gamachian) on Anticosti Island, Quebec. Similar in size to *B. amoenum* is *B. whitneyi* (Hall) from the Maquoketa Formation in Iowa. This species, however, differs from *B. amoenum* in possessing a more slender, longer phragmocone and longer body chamber. *Beloitoceras amoenum* differs from *B. bucheri* Flower (pl. 22, figs. 4, 5) from the Saluda Formation in being a larger, more inflated species with deeper camerae and sutures that slope from the dorsum to the venter adorally.

BELOITOCERAS BUCHERI FLOWER, 1946

Plate 22, figures 4, 5

Beloitoceras bucheri Flower, 1946, p. 283, pl. 39, fig. 6.

Diagnosis.—A slender, highly compressed species of *Beloitoceras* distinguished by its slight curvature, nearly straight dorsum, and very short camerae.

Description.—Medium-sized (estimated lengths up to 11.0 cm and observed diameters up to 3.0 cm), slender (apical angle 19–20 degrees), exogastric cyrtocone. Dorsal outline nearly straight adorally, the maximum shell diameter occurring at the junction of phragmocone and body chamber. Venter gently convex for the length of the shell. Shell cross section compressed ovate, the shell height exceeding the width for the entire length of the shell. Sutures straight, transverse. Cameral chambers very short, cameral length decreasing with ontogeny from one-eighth the diameter of the shell adapically to one-eleventh the diameter of the shell adorally. Body chamber long, in excess of one-fourth the length of the shell, compressed tubular, sides subparallel, contracting slightly at the aperture. Aperture has well-developed V-shaped ventral hyponomic sinus and very shallow dorsal sinus. Shell marked by closely spaced, raised transverse lirae.

Siphuncle located at ventral shell margin. Segment diameters narrow, one-seventh the diameter of the shell. Structure of septal necks and connecting rings unknown.

Stratigraphic and geographic occurrence.—Upper Ordovician (Richmondian Stage), Bull Fork Formation in southwestern Ohio; Saluda Formation in southeastern Indiana.

Studied material.—Holotype MU 437T, "lower Whitewater Beds," near Winchester, Adams County, Ohio; MU 433T (pl. 22, figs. 4, 5) and MU 29865, both from the "Concrete layer" (Bliss, 1984) at the top of the Saluda Formation, Hamburg, Franklin County, Ind.

Remarks.—Flower (1946) distinguished this species from all other Cincinnatian species of *Beloitoceras* on the basis of its "slender form, very short camerae, and relatively slight curvature." The species previously was known only from the holotype, a somewhat flattened internal mold preserving the body chamber and adoral portions of the phragmocone. Several better preserved, calcite-replaced specimens of a slender *Beloitoceras* (pl. 22, figs. 4, 5) have been collected from the "Concrete layer," a 1-ft (0.3-m) thick micritic limestone bed that occurs at the top of the Saluda Formation in portions of Franklin County, Ind. The slender, compressed, only slightly curved shell shape and short camerae in these specimens suggest placement of these specimens in Flower's species. In this massive limestone bed, *B. bucheri* occurs associated with a diverse nautiloid fauna that consists of *Gorbyoceras gorbyi* (Miller) (pl. 12, figs. 12, 13), *Cameroceras* sp., the ascoceroid *Schuchertoceras prolongatum* Flower, the large oncocerids *Diestoceras indianense* (Miller and Faber) (pl. 22, figs. 6–9) and *D. shideleri* (pl. 22, figs. 1, 3), and several species of the slender oncocerid *Manitoulinoceras*. This fauna is unique in terms of both its diversity and the complete nature of many of the preserved shells.

Beloitoceras bucheri is probably most similar to *B. ohioense* Flower, a slender, compressed species of *Beloitoceras* from the equivalent Whitewater Formation in Indiana and the upper part of the Bull Fork Formation in southwestern Ohio. It differs from this species in lacking the obliquity of the adoral sutures and deeper camerae characteristic of *B. ohioense* and similar species like *B. amoenum* and *B. cumingsi*. *Beloitoceras bucheri* differs from the associated slender, compressed oncocerid *Oonoceras fennemani* Flower in its greater shell height and more strongly curved shell form.

BELOITOCERAS CF. *B. HURONENSE* (BILLINGS), 1865

Plate 20, figures 3–6

Cyrtoceras huronense Billings, 1865, p. 176, figs. 158A–B.
Beloitoceras huronense (Billings). Foerste, 1932, 1933, p. 100, pl. 30, figs. 3A–C.
Beliotoceras huronense (Billings). Wilson, 1961, p. 84, pl. 32, figs. 3, 4.

Diagnosis.—Small, compressed species of *Beloitoceras* with a short body chamber whose dorsal outline is only slightly concave.

Description.—Small (estimated lengths of up to 6.0 cm and observed shell diameters up to 1.7 mm), slender (apical angle 15.5 degrees), exogastrically curved brevicone, the dorsum slightly concave in outline adorally, more strongly concave adapically at shell diameters less than 10 mm. Venter uniformly convex for the length of the shell. Shell circular in cross section at shell diameters less than 10 mm, becoming more compressed adorally with ontogeny. Sutures adapically straight, transverse; becoming slightly oblique with ontogeny, sloping from the dorsum to the venter in the adoral direction. Camerae short, cameral lengths ranging from one-fifth to one-sixth the diameter of the shell, decreasing slightly with ontogeny. Body chamber short, one-fifth the total length of the shell, laterally compressed, the maximum shell diameter just adoral to the juncture of the phragmocone and the body chamber. Aperture marked by a shallow dorsal sinus. Ventral portions of aperture not preserved. Shell exterior essentially smooth, marked by fine transverse growth lines.

Siphuncle located at the ventral shell margin. Segments small in diameter, shape unknown. Structure of septal necks and connecting rings unknown.

Stratigraphic and geographic occurrence.—Upper part of the Rocklandian part of the Tyrone Limestone (Middle Ordovician), vicinity of the Kentucky River, Jessamine County, Ky.

Studied material.—USNM 158666 (pl. 20, figs. 3, 4, 6) and 468733 (pl. 20, fig. 5), plus a suite of fragmentary silicified specimens (USNM 468729), all from the USGS 6034-CO collection, Little Hickman section A, Sulfur Well, Ky.

Remarks.—Specimens of *Beloitoceras* from the USGS 6034-CO collection in the Tyrone Limestone consist of incomplete, fragmentary silicified material, one specimen consisting of the adoral portion of the phragmocone and the attached body chamber (pl. 20, figs. 3, 4, 7) and a second specimen consisting of the adoral part of the phragmocone (pl. 20, fig. 5). The remainder of the specimens of *Beloitoceras* from this collection consist of numerous adapical portions of the phragmocone 5 mm or less in diameter. As no complete phragmocones are known for any of the various oncocerids present in the 6034-CO collection, the identity of these latter specimens is uncertain. The strong curvature of the small adapical portions of shells suggests placement in either *Beloitoceras huronense* or in *Laphamoceras* cf. *L. scofieldi* (pl. 20, figs. 13–18), as both *O. major* (pl. 19, figs. 11–15) and *Maelonoceras* cf. *M. praematurum* (pl. 20, figs. 7–12) are less strongly curved, rapidly expanding forms. These small adapical specimens may also be assignable, in part, to the genus *Loganoceras* Foerste. Larger diameter portions of the shell recognizable as belonging to *Loganoceras* have not, however, been observed in the 6034-CO collection.

The fragmentary specimens of *Beloitoceras* from the 6034-CO collection appear to be most similar to *B. huronense* (Billings) from the Middle Ordovician (Rocklandian) limestones exposed on St. Joseph Island in Lake Huron. The Kentucky material and the types of Billing's species (GSC 1297) are similar in the small size of the shell, the short body chamber, the laterally compressed adoral cross section, and the minimum concavity of the dorsal profile of the adoral portion of the shell. The Kentucky material appears to differ from *B. huronense* in the more dorsally produced shape of the body chamber and the lack of shell contraction at the aperture.

Beloitoceras cf. *B. huronense* differs from the type species of *Beloitoceras*, *B. pandion* (Hall), and from *B. janesvillense* Foerste, both from the Platteville Formation in Wisconsin, in being a smaller, more laterally compressed, less inflated species. It appears to be distinct from similar small, laterally compressed Middle Ordovician species like *B. clochense* Foerste, *B. houghtoni* (Clarke), and *B. isodorum* (Billings) in having a less strongly concave dorsal shell outline and possessing a shorter body chamber. More complete specimens of the Kentucky species are needed to more fully determine the affinities of the species and the identity of the small adapical shell fragments common in the 6034-CO collection.

GENUS *MAELONOCERAS* HYATT

Type species.—*Phragmoceras praematurum* Billings, 1860, p. 173; designated by Hyatt, 1884, p. 280; "Cloche Island beds," Rocklandian Stage, Middle Ordovician, Ontario.

Description.—Relatively small, slender, inflated exogastric cyrtocones with a distinctive subtriangular cross section, the dorsum wide, nearly flat, and the venter more narrowly rounded, subangular. Venter broadly convex in lateral profile, the dorsal outline subparallel, slightly concave. Sutures transverse, slightly oblique, sloping from dorsum to venter in the adoral direction. Cameral chambers short; septa thin, flat. Body chamber short, inflated, having the maximum diameter of the shell occurring just adapical of the aperture. Aperture distinctive, pear-shaped, constricted laterally by two rounded lateral, visorlike projections such that the ventral portion of the aperture is restricted to a narrow hyponomic sinus. Dorsal portion of aperture broad, open.

Siphuncle small in diameter, situated at the ventral shell margin. Segment shape unknown. Structure of septal necks and connecting rings unknown.

Stratigraphic and geographic occurrence.—Middle Ordovician (Rocklandian Stage), East-Central North America (Laurentia).

Remarks.—This poorly known genus was originally based on *Phragmoceras praematurum* Billings, known from a single body chamber from Rocklandian strata ("Cloche Island beds") exposed on Cloche Island, in Lake Huron, Ontario. This species was selected as the type species for the genus *Maelonoceras* by Hyatt. The genus was distinguished on the basis of its distinctive pear-shaped aperture and subparallel dorsal and ventral shell margins. Foerste initially used this name for the group of compressed cyrtocones he later (Foerste, 1924b) placed in his new genus *Beloitoceras*. *Maelonoceras* was then restricted in scope to include only the type species, *M. praematurum*. The Kentucky specimens described here are the only other examples of the genus known to this writer.

Maelonoceras is distinct from *Beloitoceras* in being a more inflated, less compressed form adorally that is also less strongly curved. In this respect, the genus is more similar to *Oncoceras*. It differs from *Oncoceras* in being concave in dorsal outline, in contrast to the pronounced convexity of this portion of the shell in *Oncoceras*. *Maelonoceras* differs from both *Beloitoceras* and *Oncoceras* in its distinctive, partially visored, pear-shaped aperture and in the development of the maximum shell diameter just adapical of the aperture.

MAELONOCERAS CF. M. PRAEMATURUM (BILLINGS), 1860

Plate 20, figures 7–12

Phragmoceras praematurum Billings, 1860, p. 173, fig. 19.
Maelonoceras praematurum (Billings). Hyatt, 1884, p. 280.
Maelonoceras billingsi Foerste, 1924a, p. 244, pl. 39, figs. 5a–c.
Maelonoceras praematurum (Billings). Flower, 1946, p. 288, figs. 11X–Y.

Diagnosis.—Small, slender, only slightly exogastrically curved cyrtocone, subtriangular in cross section, the maximum shell diameter just adapical of the distinctive, pear-shaped aperture.

Description.—Small (estimated lengths up to 8.0 cm observed diameters up to 2.1 cm), slender, inflated (apical angle 20 degrees), exogastric cyrtocones. Venter gently convex in outline the length of the shell; dorsum subparallel to venter, slightly concave. Shell cross section roughly heart-shaped; dorsum wide, flattened; venter narrowly rounded, nearly angular. Shell height equal to shell width. Sutures straight, transverse adapically, becoming slightly inclined adorally, sloping from the dorsum to the venter in the adoral direction at diameters of 20 mm. Cameral chambers short, cameral lengths averaging one-ninth the diameter of the shell at diameters of 15–17 mm. Body chamber short, inflated, about one-sixth the estimated length of the shell, the ventral and dorsal margins straight, subparallel. Maximum shell diameter is just adapical of the aperture. Aperture pear-shaped, having rounded visorlike lateral projections that constrict the aperture ventrolaterally, defining a narrowly rounded, distinct hyponomic sinus. Aperture broadening dorsally, subtriangular, open. Shell surface marked by fine, faint transverse growth lines.

Siphuncle small in diameter, situated at ventral shell margin. Segment shape unknown. Structure of septal necks and connecting rings unknown.

Stratigraphic and geographic occurrence.—Upper half of the Rocklandian part of the Tyrone Limestone (Middle Ordovician), vicinity of the Kentucky River, Jessamine County, Ky.

Studied material.—USNM 158649 (pl. 20, figs. 7, 8, 11, 12) and 468735 (pl. 20, figs. 9, 10), both silicified specimens from USGS 6034-CO collection, Little Hickman section A, Sulfur Well, Ky.

Remarks.—*Maelonoceras* cf. *M. praematurum* is an uncommon species within the diverse assemblage of nautiloids identified from the 6034-CO collection. Comparisons between the Kentucky specimens and the figures of the type specimen of *M. praematurum* (Foerste, 1924a, pl. 39, figs. 5a–c) are difficult as the Canadian specimen is known only by a single body chamber. This appears to be quite similar to the body chamber described and illustrated here (pl. 20, figs. 9, 10) from the Tyrone Limestone. Better, more complete specimens of both species are necessary to further determine the affinities of the Kentucky material. This material does suggest, however, that *Maelonoceras* is a valid genus, distinctive from similar Middle Ordovician oncocerids.

GENUS *LAPHAMOCERAS* FOERSTE

Type species.—*Cyrtoceras tenuistriatum* Hall, 1877, p. 243; designated by Foerste, 1933, p. 76; Platteville Limestone, Blackriveran Stage, Middle Ordovician, Wisconsin.

Description.—Small, rapidly expanding, strongly curved exogastric brevicones that are subcircular in cross section adapically, dorsum and venter equally rounded; becoming depressed in section with ontogeny, the venter more narrowly rounded than dorsum. Sutures straight, transverse adapically, becoming inclined, sloping from dorsum to venter adorally. Cameral chambers short. Body chamber short, open; venter marked by variably developed hyponomic sinus. Shell surface marked by numerous, closely spaced, raised, narrow, low transverse lamellae; occasionally marked by small nodes, but not crenulate or with extensive development of frills.

Siphuncle close to venter, consisting of narrow, subtubular, depressed segments that expand only slightly within camerae. Septal necks short, weakly recurved, suborthochoanitic. Segments empty.

Stratigraphic and geographic occurrence.—Middle Ordovician (Blackriveran Stage) to Late Ordovician (Richmondian Stage); East-Central North America (Laurentia).

Remarks.—Foerste (1933) originally erected this genus for species of *Zittelloceras* Hyatt that possessed transverse, linear ornament but lacked the prominent crenulate frills characteristic of more typical *Zittelloceras*. These species are *Laphamoceras tenuistriatum* (Hall) and *L. scofieldi* Foerste, both from the Middle Ordovician (Blackriveran) Platteville Formation of Wisconsin and Minnesota. Flower (1946) described *Zittelloceras russelli* from the Upper Ordovician (Richmondian) of southwest Ohio and noted that this species was very similar to described species of *Laphamoceras* in general shell shape and in the nature of the external ornament, but possessed faintly crenulate short frills. He suggested that, based on its external ornament, *Z. russelli* was gradational between *Laphamoceras* and *Zittelloceras* and that such subdivision of these species was artificial. Sweet (1964) followed suit and placed *Laphamoceras* as a synonym of *Zittelloceras*. Upon study of the various species placed in *Zittelloceras*, it was observed that species placed in *Laphamoceras*, as well as "*Z.*" *russelli*, are distinct from other members of *Zittelloceras*, including the type species *Z. hallianum* (D'Orbigny), in possessing a more adorally depressed, more rapidly expanding, and more strongly curved shell. These morphological differences, independent of the nature of the shell ornament, seem to be significant enough to warrant resurrecting *Laphamoceras* as a separate genus distinct from *Zittelloceras*.

Both *Laphamoceras* and *Zittelloceras* are most abundant in the Middle Ordovician (Blackriveran to Rocklandian) Platteville Formation and its equivalents in the upper Mississippi valley and the Ottawa–St. Lawrence Lowland in Ontario and Quebec, represented by 8 of the 15 described species. Younger "Trenton" species consist of only the type species of *Zittelloceras*, *Z. hallianum*, from the lower part of the Trenton Group in New York. No members of either genus are known from lower Upper Ordovician (Edenian and Maysvillian) strata in North America. In Upper Ordovician strata, *Laphamoceras* and *Zittelloceras* are known only from Richmondian rocks in the Cincinnati area, represented by *Laphamoceras russelli* (Flower) and five additional species of *Zittelloceras* (Flower, 1946). The other described Upper Ordovician species is *Z. costatum* Teichert from the Drummock Group (Ashgillian) of Girvan, Scotland.

LAPHAMOCERAS CF. *L. SCOFIELDI* FOERSTE, 1932

Plate 20, figures 13–18

Cyrtoceras billingsi Clarke, 1897, p. 806, pl. 60, fig. 10.
Laphamoceras scofieldi Foerste, 1932, pl. 27, fig. 5; 1933, p. 77.

Diagnosis.—Rather small, rapidly expanding, strongly curved species of *Laphamoceras* with an external ornament of transverse lamellae.

Description.—Rather small (lengths estimated to be up to 8.0 cm and observed shell diameters up to 2.5 cm), rapidly expanding (apical angle 28 degrees), strongly curved exogastric cyrtocones; shell remaining strongly curved the length of the shell. Cross section adapically circular, becoming depressed ovate with ontogeny such that the shell has a height to width ratio of 0.88 at a shell height of 15 mm. Dorsum and venter evenly rounded. Sutures sloping obliquely from the dorsum to the venter in the adoral direction. Cameral chambers short, cameral lengths averaging between one-seventh and one-eighth the diameter of the shell. Body chamber incompletely known, aperture unknown. Shell exterior marked by fine, transverse, raised lamellae, not crenulated, spaced 1 mm apart at a shell diameter of 18 mm. Preserved surface markings indicate a broad, shallow, V-shaped, ventral hyponomic sinus.

Siphuncle diameter small, one-seventh the diameter of the shell at a shell diameter of 13 mm, located at the ventral shell margin. Segments short, only slightly expanded, subtubular. Septal necks and structure of connecting rings unknown.

Stratigraphic and geographic occurrence.—Upper part of the Rocklandian part of the Tyrone Limestone (Middle Ordovician), vicinity of the Kentucky River, Jessamine County, Ky.

Studied material.—USNM 158654 (pl. 20, figs. 16–18) and 158663 (pl. 20, figs. 13–15), plus USNM 158655, 158664, and four uncataloged specimens, all from the USGS 6034-CO collection, Little Hickman section A, Sulfur Well, Ky.

Remarks.—The Kentucky specimens from the 6034-CO collection, in terms of shell size, extent of shell curvature, and rate of shell expansion, are similar to the holotype of *L. scofieldi* Foerste (University of Minnesota, no.207) illustrated by Foerste (1932, pl. 27, fig. 5). Flower (unpublished notes) stated that the Kentucky specimens differed from this specimen in being much more rapidly expanding shells distinguished by the development of paired ventrolateral nodes adapically. Comparisons of the

Kentucky material with Foerste's illustration indicate similar rates of shell expansion. The nodes described by Flower appear to be adventitious, associated with irregular sponge encrustations on these shells. The surface ornament is evident in only two of the Kentucky specimens, USNM 158655 and 158663 (pl. 20, figs. 13–15) and appears similar to that of the holotype. As is the case with many of these Tyrone species, a more definitive assignment of these specimens awaits the discovery of more complete material preserving the distinctive surface markings of this genus.

Laphamoceras scofieldi is distinct from the type species, *L. tenuistriatum* (Hall) from the Platteville Formation in Wisconsin, in being a much more rapidly expanding, more strongly curved shell. Similarly, *L. scofieldi* is a more strongly curved form compared with *L. russelli* (Flower) from the "Waynesville Formation" (Late Ordovician, Richmondian) in southwestern Ohio. The apparent curvature of the "Waynesville" specimen may, however, be the result of postdepositional compaction of the enclosing shale rather than being a true representation of the shell's original shape. There is evidence of dorsal-ventral crushing of the holotype and only known specimen (Flower, 1946, pl. 27, figs. 1–3).

Laphamoceras cf. *L. scofieldi*, while being an uncommon species in the 6034-CO collection, is the second most abundant oncocerid species identified from this collection, second to specimens of *Beloitoceras* cf. *B. huronense*. It is represented by 10 incomplete portions of the phragmocone, all greater than 10 mm in diameter. A number of these specimens are encrusted by nodose holoperipheral encrustations of the problematic sponge *Dermatostroma*.

FAMILY TRIPTEROCERATIDAE FLOWER

Straight to faintly exogastric longicones with depressed cross sections; flattened venter with dorsum broadly rounded to slightly keeled. Cross sections vary from depressed ovate to subtriangular. Sutures forming broad lobes dorsally and ventrally. Body chambers broad, tubular, apertures open. Paired weak sinuses dorsallaterally and a broad, shallow sinus ventrally. Shells smooth, marked with transverse growth lines, or less commonly, longitudinally fluted.

Siphuncles close to venter, separated from it by a distance equal to diameter of the siphuncle. Siphuncles narrow in diameter, from less than one-tenth to nearly one-quarter the diameter of the shell. Segments typically subtubular, depressed, twice as long as high. Septal necks short, orthochoanitic to cyrtochoanitic. Segments empty.

Stratigraphic range.—Middle Ordovician (Chazyan Stage) to Upper Ordovician (Richmondian Stage).

Remarks.—Flower erected this family for a group of rather poorly known, typically rapidly expanding orthoconic longicones distinguished by the development of a depressed, planoconvex shell cross section. In this family are the genera *Allumettoceras* Foerste, *Tripteroceras* Hyatt, *Tripterocerina* Foerste, and *Rasmussenoceras* Foerste. *Allumettoceras* is the oldest member of the family, occurring in Chazyan and Blackriveran strata in eastern North America, and is distinct from the other genera in this group in being a less rapidly expanding form, and having a less depressed cross section and globular, expanded siphuncle segments. *Tripteroceras* is a more rapidly expanding, highly depressed form with a strongly flattened venter and overlaps with *Allumettoceras* in the Blackriveran and then becomes most abundant in younger Kirkfieldian and Shermanian carbonates in the same region. *Tripterocerina* from the Bighorn Dolomite in Wyoming (Upper Ordovician) is similar in morphology to *Tripteroceras* except for the development of longitudinal fluting on the shell's dorsal exterior. *Rasmussenoceras*, a large, rapidly expanding, highly depressed shell, is known from Upper Ordovician carbonates of the Cape Calhoun Formation in Greenland (Troedsson, 1926) and from the rubbly limestones of the Whitewater Formation (Richmondian) in Indiana and Ohio (Flower, 1946). The latter genera are distinct from *Alumettoceras* in the possession of narrow, tubelike siphuncle segments and orthochoanitic septal necks. The variation in shell morphology and, especially, in the nature of the siphuncle segments suggests that this family is an artificial grouping of independently derived, ventrally flattened Middle and Upper Ordovician "flatfish" nautiloids.

The Tripteroceratidae have traditionally been placed in the Oncocerida on the basis of the ventral position of the siphuncle and the lack of cameral or endosiphuncular deposits. A logical origin for this family can be found in the primitive, longiconic Graciloceratidae, the probable ancestral stock for all succeeding Ordovician oncocerid families. Dzik (1984) removed the Tripteroceratidae from the Oncocerida and placed it in the Orthocerida, presumably on the basis of the orthoconic shell form of its members. However, an orthoconic shell morphology occurs several times within the Oncocerida, namely within the Tripteroceratidae in the Ordovician, the Jovellanidae in the Silurian, and the Tripleuroceratidae in the Devonian. The possession of an orthoconic shell morphology alone does not seem be sufficient cause to remove the Tripteroceratidae to the Orthocerida, and here they are retained, in the Oncocerida.

GENUS *ALLUMETTOCERAS* FOERSTE

Type species.—*Tripteroceras paquettense* Foerste, 1924a, p. 233; designated by Foerste, 1926, p. 311; "Leray-Rockland beds," Rocklandian Stage, Middle Ordovician, Ontario.

Description.—Small, rather rapidly expanding, orthoconic to slightly exogastric longicones, depressed ovate in cross section, with the dorsum broadly convex, not keeled; venter only slightly flattened, lateral margins narrowly rounded. Sutures sinuate, forming broad dorsal and deeper ventral lobes and lateral saddles. Cameral chambers short. Body chamber long; aperture open, not constricted. Shell exterior basically smooth, marked by fine transverse growth lines.

Siphuncle situated at or near the ventral shell margin. Siphuncle segments expanded, globose, nearly equiaxial with height equal to the length. Septal necks short, slightly recurved, cyrtochoanitic. Cameral and endosiphuncular deposits apparently absent.

Stratigraphic and geographic occurrence.—Middle Ordovician (Chazyan to Kirkfieldian Stages), East-Central North America (Laurentia).

Remarks.—As indicated above, *Allumettoceras* differs from other members of the Tripteroceratidae in possessing a less depressed, less angular cross section and in the development of expanded, equiaxial siphuncle segments whose diameters average one-fourth the diameter of the shell. The nature of the siphuncle segments in many taxa assigned to these genera is unknown. Specimens of *Tripteroceras* cf. *T. planocovexum* (Hall) and *Rasmussenoceras variable* Flower studied by the writer indicate the presence of narrow, tubular siphuncles less than one-tenth the diameter of the shell. The morphological differences between *Allumettoceras* and the other members of the Tripteroceratidae are probably sufficient to remove this genus from the family. The affinities of this genus, however, remain unknown pending discovery of more complete material preserving key internal structures of adapical portions of the shell. This writer has elected to retain it in the Tripteroceratidae at this time.

Dzik (1984) made *Allumettoceras* a junior synonym of *Tripteroceras*. For the reasons discussed above, this is totally erroneous and indicates a lack of knowledge of the morphological features of these two taxa. They are distinct genera and probably do not even belong in the same family.

ALLUMETTOCERAS CF. *A. TENERUM* (BILLINGS), 1860

Plate 20, figures 19–22

Orthoceras tener Billings, 1860, p. 174.
Allumettoceras tenerum (Billings). Foerste, 1932, p. 124, pl. 13, figs. 1A–B.
Allumettoceras tenerum (Billings). Wilson, 1961, p. 77, pl. 30, figs. 13–16.

Diagnosis.—A rather slender, moderately expanding species of *Allumettoceras*, cameral lengths averaging one-ninth the diameter of the shell.

Description.—Small (estimated lengths of up to 16.0 cm with shell diameters of up to 2.4 cm), moderately expanding (apical angle 7.5–8 degrees), orthoconic longicones; depressed ovate in cross section, the dorsum more convex than venter. Venter strongly flattened adapically, becoming more convex with ontogeny. Shell becoming slightly more depressed with ontogeny, the height to width ratios ranging from 0.71 at a height of 7 mm and 0.65 at a height of 20 mm. Sutures somewhat sinuous, having shallow dorsal and ventral lobes and lateral saddles. Cameral chambers short, cameral lengths one-ninth the diameter of the shell at a shell diameter of 22 mm. Body chamber incompletely known, at least one-sixth the total length of the shell; smooth, continuous with the phragmocone. Aperture open, no evidence of a ventral hyponomic sinus, consisting of ventral and dorsal lobes and lateral sinuses. Shell exterior smooth.

Siphuncle situated at ventral shell margin. Segment diameter nearly one-third the height of the shell at a shell height of 11 mm. Segments expanded, spherical in shape, equiaxial. Structure of septal necks and connecting rings unknown.

Stratigraphic and geographic occurrence.—Lower part of the Logana Member of the Lexington Limestone (Middle Ordovician, Kirkfieldian), exposed in Jessamine and Woodford Counties, Ky.

Studied material.—Incomplete silicified specimens USNM 468738 and 468670 (pl. 20, figs. 19, 21, 22) from the USGS 6419-CO collection, Wilmore quadrangle, Jessamine County, Ky., and USNM 468739 (pl. 20, fig. 20) from the USGS 4865-CO collection, Tyrone quadrangle, Woodford County, Ky.; plus additional uncataloged shell fragments from both localities.

Remarks.—The Kentucky material described here consists of fragmentary silicified portions of the phragmocone that, collectively, indicate a depressed longicone 16.0 cm in total length. Comparisons with other described species of *Allumettoceras* are complicated by the incomplete, fragmentary nature of most of the figured type specimens, often consisting of an internal mold of the body chamber and a few attached adoral camerae. The Kentucky specimens are less rapidly expanding compared with the type species of the genus, *A. paquettense* Foerste, as well as *A. josephium* Foerste, both from Blackriveran and Rocklandian strata in Ontario. Both of these latter species have apical angles of 13 degrees compared with values of 7.5–8 degrees for the Kentucky specimens. Similarly, the Logana Member specimens are less expanded than *A. huronense* Foerste from the "Black River Formation" on St. Joseph Island, Lake Huron (apical angle 11 degrees). In terms of the rate of shell expansion, the Logana material is most similar to *A. carletonense* Foerste and *A. tenerum* (Billings), both from Blackriveran and Rocklandian strata in Ontario. Both of these species have apical angles of 8 degrees. *Allumettoceras carletonense* is based on an isolated body chamber. *Allumettoceras tenerum* is known from the adapical portion of the phragmocone. The incomplete specimens from the Logana Member compare more favorably with *A. tenerum* in cameral depth and cross sectional shape, the ventral portion of the shell increasing in convexity with ontogeny. A complete restudy of this genus is needed to better define the known species of *Allumettoceras* and to more accurately determine the affinities of the Kentucky specimens. These specimens from the Logana Member of the Lexington Limestone are the youngest members of the genus (Middle Ordovician, Kirkfieldian) currently known to this writer.

Within the Logana Member, the species is a rare form found associated with the more abundant phragmocones of the small orthocerid *Isorthoceras albersi* (Miller and Faber). Other associated fauna consists of rare fragments of a small species of *Trocholites*, and abundant, well-preserved specimens of the univalved mollusks *Lophospira*, *Liospira*, and *Sinuites*, and the small pelecypod *Deceptrix*.

FAMILY DIESTOCERATIDAE FOERSTE

Straight to endogastrically curved, compressed brevicones with inflated body chambers that more or less contract toward the aperture, where a ventral hyponomic sinus is well developed. Sutures transverse or slightly sloping adorally from the venter to the dorsum. Periphract ventromyarian.

Siphuncle situated close to the ventral shell margin. Segments typically expanded, scalariform to subquadrate, with cyrtochoanitic septal necks. Segments possessing discrete, rather irregular actinosiphonate deposits.

Stratigraphic range.—Middle Ordovician (Blackriveran Stage) to Upper Ordovician (Gamachian Stage).

GENUS *DIESTOCERAS* FOERSTE

Type species.—*Gomphoceras indianense* Miller and Faber, 1894b, p. 137; designated by Foerste, 1924b, p. 262; Saluda Formation, Richmondian Stage, Upper Ordovician, Indiana.

Description.—Typically large (up to 30 cm in length), straight to faintly endogastrically curved, rapidly expanding brevicones that commonly reach their greatest diameter close to the base of the body chamber. Slightly compressed in cross section, the venter more narrowly rounded than the dorsum. Sutures typically transverse or sloping slightly adorally from the venter to the dorsum. Cameral chambers short, cameral lengths typically decreasing with ontogeny. Body chamber large, inflated, contracting toward the aperture. Aperture marked by a well-developed, broad, ventral hyponomic sinus. Periphract ventromyarian, the well-developed basal zone marked by short, longitudinal buttresses. Shell exterior marked by variably developed, raised, transverse growth lines.

Siphuncle situated close to the ventral shell margin. Segments broadly expanded adorally, the siphuncle segments one-fifth to one-fourth the diameter of the shell. Segments compressed ovate to subquadrate in outline. Septal necks short and abruptly recurved, cyrtochoanitic. Connecting rings thin and homogeneous except near septal foramen. Rings broadly adnate to adjacent adoral septa dorsally, but free ventrally. Rings adnate to adapical septa both dorsally and ventrally. Adoral segments of large individuals have irregularly developed deposits consisting initially of annulosiphonate deposits that then give rise to irregular calcareous processes that extend adapically from septal necks.

Stratigraphic and geographic occurrence.—Middle Ordovician (Blackriveran Stage) to Upper Ordovician (Gamachian Stage), North America and Baltic Europe.

Remarks.—Flower (1946) reviewed the morphological features and history of this genus and pointed out that it is probably overly broad in its scope. The critical features of the siphuncle are unknown for the majority of species currently assigned to the genus. This is particularly true of Middle Ordovician taxa placed in the genus. These consist of small, slender, atypical species compared with younger members of the genus, and include the type species, *D. indianense* (Miller and Faber) from Richmondian strata in the Cincinnati arch region.

Diestoceras is a prominent member of the so-called "Arctic" fauna, for which a number of large, characteristic species are described from Upper Ordovician carbonate facies in Wyoming, Manitoba, Quebec, the Canadian Arctic archipelago, and northern Greenland. The genus occurs in the Kentucky-Indiana-Ohio region in middle and upper Richmondian strata, being most abundant along with other "Arctic" nautiloid taxa in the Saluda and Whitewater Formations in southeastern Indiana and adjacent portions of western Ohio. The bulk of the figured specimens of *Diestoceras* from these strata come from a single locality in the "Lower Whitewater Formation" along Little Four-Mile Creek, near Oxford, Preble County, Ohio. Here, several species of *Diestoceras* have been identified associated with hundreds of specimens of the coiled nautiloid *Characteroceras baeri* (Meek) and many other, less common oncocerids (Flower, 1946, 1957b). The genus is also known from the equivalent, more clastic-rich Bull Fork Formation exposed in Montgomery and Clinton Counties in southwestern Ohio.

Foerste (1924b) and Flower (1946) identified five species of *Diestoceras* from these strata in Indiana and Ohio; all being large, essentially straight brevicones differentiated primarily on the basis of the shape and size of the body chamber and the relative length of cameral chambers. Unfortunately, most Cincinnatian species of

Diestoceras are preserved as matrix-filled internal molds in argillaceous limestones, many of these specimens showing evidence of postdepositional deformation due to sediment compaction. This makes comparisons of body chamber shape and cross section difficult. Curiously, specimens of *Diestoceras* in the Bull Fork and Saluda Formations often consist of just the body chamber with a few attached camerae. This complicates comparisons of specimens based on length of individual cameral chambers. Larger suites of more complete, uncrushed specimens are necessary to more fully determine the distinctiveness of these different species of *Diestoceras* described from the Cincinnati arch region.

DIESTOCERAS INDIANENSE (MILLER AND FABER), 1894B

Plate 22, figures 6–9

Gomphoceras indianense Miller and Faber, 1894b, p. 137, pl. 7, figs. 3–5.
Gomphoceras indianense Miller and Faber. Cumings, 1908, p. 1030, pl. 19, figs. 4–4B.
Diestoceras indianense (Miller and Faber). Foerste, 1924b, p. 263, pl. 25, figs. 1A–B; pl. 26, figs. 1A–B, 2A–B.
Diestoceras indianense (Miller and Faber). Flower, 1946, p. 482, pl. 36, figs. 1–3; pl. 41, figs. 10, 11.

Diagnosis.—Large, slightly endogastrically curved species of *Diestoceras* having the maximum shell diameter at the base of the body chamber, cameral lengths one-seventh to one-ninth the diameter of the shell at shell diameters greater than 35 mm, and compressed, subquadrate siphuncle segments adorally.

Description.—Large (estimated lengths of up to 14.0 cm and shell diameters up to 6.0 cm), rapidly expanding (apical angle 34–38 degrees), exogastrically curved, inflated brevicone; compressed ovate in cross section, the venter more narrowly rounded than the dorsum. Venter slightly concave adapically, becoming slightly convex adorally. Dorsum convex the length of the shell. Maximum shell diameter at base of body chamber. Adapically, sutures inclined slightly from the venter to dorsum; straight transverse adorally. Cameral chambers short, cameral lengths from one-seventh to one-ninth the diameter of the shell at diameters greater than 35 mm. Adoralmost two to three septa approximate at shell diameters of 55 mm or more. Body chamber inflated, the dorsal and ventral margins subparallel, straight, contracting adorally toward aperture. Body chamber length roughly three-fourths its maximum diameter. Aperture subtriangular, open and broadly rounded dorsally; constricted and narrowly rounded ventrally, forming a distinct hyponomic sinus. Shell exterior marked by closely spaced, narrow, raised transverse growth lines that slope ventrally to define a shallow, V-shaped sinus.

Siphuncle slightly removed from ventral shell margin. Siphuncle segments broad, expanded, subquadrate adorally, segment diameters one-fifth the diameter of the shell. Segments compressed, SCR values of 1.40 at shell diameters between 35 and 50 mm. Septal necks short, strongly recurved, cyrtochoanitic. Connecting rings not in contact with ventral shell margin, thin and homogeneous. Rings adnate to adoral septum both dorsally and ventrally. Endosiphuncular deposits developed adorally, consisting of flat, longitudinal radial plates projecting adapically from septal necks into siphuncle segments.

Stratigraphic and geographic occurrence.—Saluda Formation (Upper Ordovician, Richmondian Stage), southeastern Indiana; and equivalent portions of the "Liberty" and "Lower Whitewater biofacies" within the Bull Fork Formation in southwestern Ohio.

Studied material.—MU 431T (pl. 22, fig. 8) and 432T (pl. 22, figs. 6, 7, 9), both from the Saluda Formation at Hamburg, Franklin County, Ind.; MU 29866 from the "Saluda tongue" exposed north of Camden, Preble County, Ohio; MU 29867 from the "Saluda tongue" exposed on Talawanda Creek, just north of Oxford, Butler County, Ohio; Hypotype UC 24480 of Flower (1946, pl. 36, fig. 6), and hypotype of Foerste (1924b, pl. 25, figs. 1A–B), both from the "Saluda tongue" at Little Four-Mile Creek, McDills Mills, Preble County, Ohio.

Also studied was the holotype of *Gomphoceras indianense* (University of Chicago 8818), which Miller and Faber (1894B, pl. 7, figs. 3–5) reported as coming from "Upper part of the Hudson River Group, Versailles, Indiana."

Remarks.—Contrary to Flower's (1946) remarks, the holotype of *Gomphoceras indianense* resides in the Faber collection at the Field Museum of Natural History, Chicago, Illinois. It is cataloged under the University of Chicago, Walker Museum number 8818. Flower noted that the specimen came from the Saluda Formation. Foerste (1924b) and Flower (1946) described and illustrated a number of specimens they referred to this species, all of which were from the *Charactoceras baeri* bed in the "Lower Whitewater Formation" along Little Four-Mile Creek, just north of Oxford, in Preble County, Ohio (Flower, 1957b). This unit is now believed to be a tongue of the Saluda Formation that extends eastward into the Bull Fork Formation from southeastern Indiana into adjacent portions of western Ohio (J. Marak, Miami University, oral commun.). The two specimens of *D. indianense* illustrated here (pl. 22, figs. 6–9) are from the "Concrete layer," a 12-in (0.3-m) thick, resistant micritic limestone bed that overlies the coralliferous zone of the Saluda Formation in portions of Franklin County, Ind. These specimens consist of a fairly complete, uncrushed internal mold (pl. 22, figs. 6, 7, 9) and a similar specimen preserving some of the external shell that is crushed dorsally (pl. 22, fig. 8). These specimens are also of interest in that they preserve a blackened pigment around the aperture of

TABLE 11.—Comparisons of the morphological features of species of the oncocerid *Diestoceras* from Upper Ordovician strata in the Cincinnati arch region

[Data from specimens listed in text. mm, millimeter]

Feature	*D. indianense*	*D. shideleri*	*D. eos*
Apical angle from venter, degrees.	34–40	30–35	45–48
Location of maximum diameter.	At base of body chamber.	Midlength of body chamber.	1–2 cm adoral of base of body chamber.
Aperture shape	Triangular	Triangular	Compressed ovate.
Cameral length at 55-mm diameter, mm.	5.5–7	6–7.5	5.5–6
Cameral length at 40-mm diameter, mm.	4–7	6–8	5.5–6
Siphuncle segment shape.	Subquadrate	Compressed ovate.	Compressed ovate.
Adapical endogastric curvature.	Yes	Yes	No

the shell. Flower (1964b, pl. 1, figs. 6–11) observed a similar phenomenon in oncocerids from the Middle Ordovician of Quebec. This pigmentation is thought to represent the "black layer," a blackened area located at the aperture border and inside the shell margin all around the aperture that has been observed in specimens of *Nautilus* (Stenzel, 1964).

Foerste (1924b) distinguished *D. indianense* from associated species of *Diestoceras* primarily on the basis of the position of the area of maximum shell diameter within the shell. This was described as being located at the base of the body chamber in *D. indianense* and "at a point distinctly above the base of the living chamber" in *D. eos* (Hall and Whitfield) and "at the mid-height of the living chamber" in the holotype of *D. shideleri* Foerste.

Flower (1946) described *D. indianense* as a shell with "a rather short living chamber; little contracted adorally." He also noted that the phragmocone, adapically, was distinctly concave ventrally and strongly convex dorsally, its sutures spaced "moderately close" together. *Diestoceras eos* was described as "a larger form with living chamber nearly symmetrical, contracting more strongly towards the aperture," the shell being most inflated just adoral of the base of the body chamber. *Diestoceras shideleri* was described as being a large form with a body chamber comparable in length to *D. eos* but being less inflated and contracting only slightly toward the aperture.

Flower also noted that *D. shideleri* had a "phragmocone with exceptionally deep camerae; early part curved as in *D. indianense*."

Study of the few more or less complete specimens of *Diestoceras* available to this writer (N=15) indicate significant variation within individual specimens assigned to these species in terms of body chamber dimensions and cameral lengths. Specimens currently assigned to *D. indianense* have the maximum shell diameter at the base of the body chamber and body chambers whose lengths are roughly three-fourths or less their adapical diameters. Length to diameter ratios for body chambers of specimens of *D. indianense* studied range from 0.58 to 0.86, averaging 0.78. Cameral lengths in specimens assigned to *D. indianense* were 4–7 mm (average 5.25) at a diameter of 40 mm and 5.5–7 mm (average 6.2 mm) at a diameter of 55 mm.

Specimens currently assigned to *D. shideleri* have the maximum shell diameter roughly at a point midlength on the body chamber, and body chamber lengths are roughly equal to the adapical diameter of the body chamber. Adapically, cameral lengths in specimens of *D. shideleri* tend to be greater than those of typical *D. indianense* at the same diameter. There is, however, some overlap between specimens assigned to these two species.

Diestoceras eos is less well represented in collections, and the holotype (OSU 38082) is flattened due to compaction of the enclosing shaly limestone, making the dimensions of its original shell uncertain. *Diestoceras eos* appears to be distinguishable from *D. indianense* in being a more rapidly expanding species, the maximum shell diameter some distance adoral to the base of the body chamber and lacking any indications of endogastric curvature. *Diestoceras eos* also has a compressed ovate aperture compared to the broader, strongly triangular apertures present in *D. indianense* and *D. shideleri*. Comparison of the apertures of these three species of *Diestoceras* is illustrated in figure 26. Both *D. eos* and *D. shideleri* have more rounded, compressed ovate siphuncle segments compared with the subquadrate, scalariform segments in *D. indianense*. Comparisons of some of the morphological features used to distinguish these three species of *Diestoceras* are presented in table 11.

Diestoceras attenuatum Flower is known only from the holotype (UC 24524) from the "Lower Whitewater Formation" at Little Four-Mile Creek, just north of Oxford, in Preble County, Ohio. This species is a much more gradually expanding, slender form compared with *D. indianense*. Likewise, *D. waynesvillense* Flower is known only from the holotype (MU 446T) from the middle of the "Waynesville Formation" at Oldenburg, Franklin County, Ind. This species appears to be a smaller version of *D. eos* (Hall and Whitfield) and differs from *D. indianense* in the same manner.

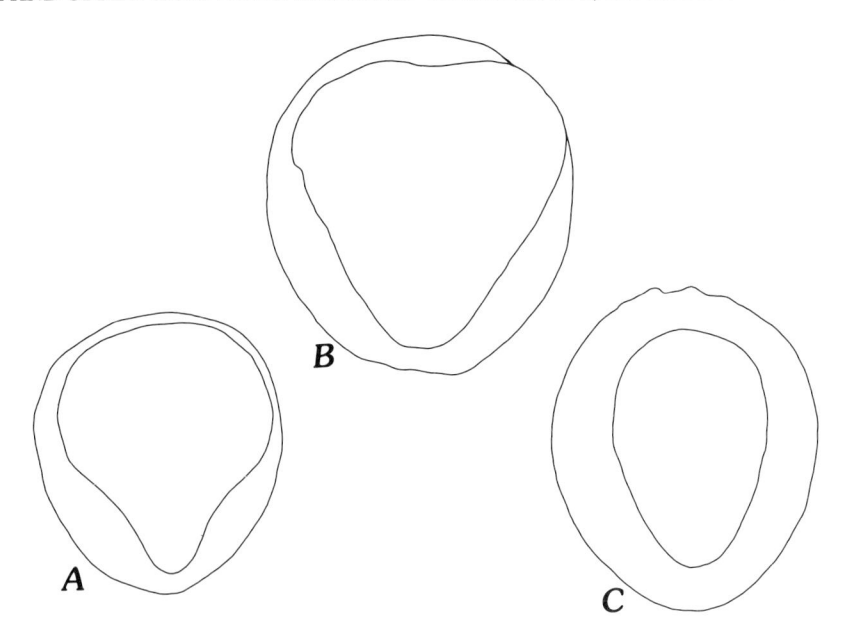

FIGURE 26.—Drawings of the apertures of species of *Diestoceras* from Richmondian strata in the Cincinnati arch region. The drawings are based on internal molds. The outer line in each figure is the shape of the filling of the body chamber. The inner line is the shape of the aperture. All figures ×1. *A*, Apertural view of *Diestoceras indianense* (Miller and Faber), drawing based on specimen MU432T, Saluda Formation, Hamburg, Indiana. *B*, Apertural view of *D. shideleri* Foerste, drawing based on specimen MU430T, Saluda Formation, Hamburg, Indiana. *C*, Apertural view of *D. eos* (Hall and Whitfield), drawing based on specimen MU382T, lower Whitewater Formation, Talawanda Creek, Oxford, Ohio. All ×0.65.

Flower (1946) describes *D. indianense* as being similar, especially in the shape of the siphuncle segments, to *D. scalare* Foerste from the Richmondian Vaureal Formation on Anticosti Island, Quebec. The other Cincinnati arch species of *Diestoceras*, as well as *D. indianense*, are more laterally compressed adorally compared with a number of large, highly inflated, bulbous species of *Diestoceras* described from coeval or slightly older (Edenian) carbonate facies in northern Greenland (Cape Calhoun Formation), the Hudson Bay region (Shammattawa and Nelson Limestones), the lakes region of southern Manitoba (Red River Formation), and Wyoming (Bighorn Formation and its Lander Sandstone Member).

DIESTOCERAS SHIDELERI FOERSTE, 1924B

Plate 21, figures 1, 2, 5; plate 22, figures 1, 3

Diestoceras shideleri Foerste, 1924b, p. 266, pl. 27, figs. 1A–B.
Diestoceras shideleri Foerste. Flower, 1946, p. 407, pl. 38, figs. 1, 8.

Diagnosis.—Large, slightly exogastrically curved species of *Diestoceras* having the maximum shell diameter at the midpoint of the body chamber, cameral lengths one-fifth to one-ninth the diameter of the shell at shell diameters greater than 35 mm, and compressed ovate siphuncle segments adorally.

Description.—Large (estimated lengths up to 18.0 cm and shell diameters of up to 7.7 cm), exogastrically curved brevicone, compressed ovate in cross section, the venter more narrowly rounded than the dorsum. Venter slightly concave adapically, becoming strongly convex adorally. Dorsum convex the length of the shell. Maximum shell diameter near the midpoint of the body chamber. Sutures adapically slightly inclined from the venter to the dorsum in the adapical direction; becoming straight, tranverse adorally. Cameral chambers of moderate length, the cameral lengths ranging from one-fifth to one-ninth the diameter of the shell at shell diameters in excess of 35 mm, decreasing with ontogeny. Adoralmost two or three septa approximate in specimens having phragmocone diameters in excess of 60 mm. Body chamber inflated, not laterally compressed. Venter and dorsum both gently convex; chamber expanding to its midlength, then contracting toward the aperture. Body chamber length roughly equal to maximum diameter. Aperture subtriangular, open, broadly rounded dorsally; more narrowly rounded ventrally, forming a distinct hyponomic sinus. Shell exterior marked by low, indistinct, rounded transverse costae that slope towards the venter, marking a shallow V-shaped sinus.

Siphuncle located at ventral shell margin. Siphuncle segments broad, expanded, compressed ovate, with diameters nearly one-third the diameter of the shell adorally at shell diameters in excess of 55 mm. Segments highly compressed adorally, SCR values of 3.2 at a shell diameter of 55 mm. Septal necks strongly recurved, cyrtochoanitic. Connecting rings in contact with ventral shell margin, thin and homogeneous. Rings adnate dorsally to adoral septum and ventrally to adapical septum. Endosiphuncular deposits developed adorally, consisting of flat, longitudinal radial plates projecting from septal necks adapically into siphuncle segments.

Stratigraphic and geographic occurrence.—Saluda Formation (Upper Ordovician, Richmondian Stage), southeastern Indiana; and Saluda Tongue within the Bull Fork Formation in portions of Butler and Preble Counties, Ohio.

Studied material.—MU 429T (pl. 21, figs. 1, 2) from the "Concrete layer" just above the *Tetradium* zone within the Saluda Formation, Route 101, just north of Brookville, Franklin County, Ind.; MU 430T (pl. 21, fig. 5; pl. 22, figs. 1, 3), same horizon exposed just north of Hamburg, Franklin County, Ind.; hypotypes MU 442T (Flower, 1946, pl. 38, fig. 1) and UC 24508 (Flower, 1946, pl. 38, fig. 8), both from the "Saluda tongue," Little Four-Mile Creek, Preble County, Ohio; and MU 14078 from the rubbly facies in the Whitewater Formation at Route 725, just north of Camden, Preble County, Ohio.

Also studied was the holotype of *Diestoceras shideleri* (USNM 81286) Foerste (1924B, pl. 27, Figs. 1A–B), which he listed as coming from the "basal part of the Whitewater Member of the Richmond near Oxford, Ohio."

Remarks.—Comparisons between *D. shideleri* and the associated *D. indianense* (Miller and Faber) are presented in the previous description of that species. The important distinctions between these species are in the relative shape and dimensions of the body chamber. These distinctions and the common association of both species together in the same beds exposed at a number of localities suggests that these morphological differences could be the result of sexual dimorphism. This feature has been recognized in several species of living *Nautilus* (Stenzel, 1964; Saunders and Spinoza, 1978; Ward, 1987), the mature males having a broader aperture and body chamber to accommodate the male sexual organs (spadix). Oncocerids are believed to belong to the same evolutionary group as the Nautilida (Flower, 1955d), and sexual dimorphism had been suggested previously for other members of the order (Ruedemann, 1921). The smaller, less inflated body chambers currently assigned to *D. indianense* may represent female individuals, the larger, more inflated body chambers of *D. shideleri* being the males. Large collections of relatively well-preserved, undistorted specimens of both species need to be studied in order to further validate sexual dimorphism in this genus.

Comparisons between *D. shideleri* and *D. eos* (Hall and Whitfield) are more difficult due to the deformed nature of the holotype of *D. eos* and the lack of well-preserved museum material referable to this species. These taxa are similar, both in being large species and in having the maximum diameter of the body chamber at a point adoral of the base of the body chamber. Siphuncle segments in both species also appear to be compressed ovate rather than subquadrate as in *D. indianense*. Flower (1946) indicated that *D. eos* is a more rapidly expanding species that lacks the endogastric curvature characteristic of *D. shideleri* and has a body chamber that contracts more strongly toward the aperture. The limited number of specimens of *D. eos* available to this writer also indicate that *D. eos* has a shorter body chamber relative to chamber diameter, more comparable with that characteristic of *D. indianense*.

DIESTOCERAS EOS (HALL AND WHITFIELD), 1875

Plate 21, figure 4

Gomphoceras eos Hall and Whitfield, 1875, p. 100, pl. 3, fig. 5.
Diestoceras eos (Hall and Whitfield). Foerste, 1924b, p. 265, pl. 28, figs. 1A–B.
Diestoceras eos (Hall and Whitfield). Flower, 1946, p. 403, pl. 34, figs. 1, 6; pl. 35, fig. 9; pl. 37, fig. 10; pl. 38, fig. 7; pl. 39, fig. 10.

Diagnosis.—Large, inflated, somewhat bulbous species of *Diestoceras* having the maximum shell diameter just adoral of the base of the body chamber and possessing a short, rapidly expanding phragmocone that lacks any clear indication of endogastric curvature.

Description.—Large (estimated lengths of up to 12.0 cm and shell diameters up to 7.4 cm), rapidly expanding (apical angle 45–48 degrees), orthoconic brevicones, the venter and dorsum both concave adapically, nearly symmetrical; becoming more inflated with ontogeny such that the venter adorally becomes more convex than dorsum. Cross section compressed ovate adapically, the venter more narrowly rounded than the dorsum; becoming inflated, nearly circular in cross section adorally. Maximum shell diameter just adoral of the base of the body chamber. Sutures straight, transverse. Cameral chambers short, cameral lengths ranging from one-sixth to one-tenth the diameter of the shell at diameters in excess of 30 mm, decreasing with ontogeny. Adoralmost two or three septa become approximate at diameters in excess of 60 mm. Body chamber length roughly three-fourths the maximum diameter of the shell; chamber inflated, roughly circular in cross section, the maximum shell diameter situated just adoral of the base, then contracting evenly toward the aperture. Aperture compressed ovate, the venter more narrowly rounded than the dorsum, delineating a ventral hyponomic sinus. External ornament of raised, fine, closely spaced transverse growth lines.

Siphuncle located at ventral shell margin, the segments adorally expanded, their diameters nearly one-fourth the diameter of the shell. Segments highly compressed ovate, the SCR values being 2.85 at a shell diameter of 54 mm. Septal necks strongly recurved, cyrtochoanitic. Connecting rings thin, in contact with ventral shell wall. Rings adnate to adoral septum dorsally and to adapical septum ventrally. Endosiphuncular deposits developed adorally, consisting of flat, radial plates projecting adapically from the septal necks into the segment.

Stratigraphic and geographic occurrence.—Upper part of the Bull Fork Formation (Upper Ordovician, Richmondian Stage) in southwestern Ohio and equivalent portions of the Saluda Formation, southeastern Indiana.

Studied material.—Holotype OSU 3082 (Hall and Whitfield, 1875, pl. 3, fig. 5; Foerste, 1924b, pl. 28, figs. 1A–B; Flower, 1946, pl. 34, fig. 6), "Whitewater beds" at Dayton, Montgomery County, Ohio; hypotype MU 382T (Foerste, 1924b, pl. 26, figs. 1A–B; Flower, 1946, pl. 34, fig. 6), "Lower Whitewater Formation" at Talawanda Creek, near Oxford, in Preble County, Ohio; hypotype MU 440T (Foerste, 1924b, pl. 27, fig. 2; Flower, 1946, pl. 39, fig. 10), Saluda Formation, Madison, Jefferson County, Ind.; hypotype UC 24500 (Flower, 1946, pl. 37, fig. 10), "Lower Whitewater beds," Little Four-Mile Creek, near Oxford, Ohio; and USNM 468730 (pl. 21, fig. 4) and MU 15905, both from the *Drepanella richardsoni* zone in the upper part of the Bull Fork Formation, Cowan Lake spillway, near Wilmington, Clinton County, Ohio.

Remarks.—As indicated in the preceding descriptions of *D. indianense* and *D. shideleri*, comparisons between these species and *D. eos* are difficult due to the flattened, distorted condition of Hall and Whitfield's type specimen. However, study of hypotype MU 382T, a well-preserved, undistorted internal mold preserving all of the body chamber and most of the phragmocone, indicates that *D. eos* is a valid taxon, distinct from these associated species of *Diestoceras*. *Diestoceras eos* differs from *D. indianense* and *D. shideleri* in lacking any indication of endogastric curvature of the shell and in possessing a laterally compressed ovate aperture, compared with the strongly triangular apertures typical of these other two species (fig. 26). Additionally, *D. eos* differs from *D. indianense* in having the maximum shell diameter some distance adoral of the base of the body chamber and having broader, more highly compressed siphuncle segments compared with the narrower, subquadrate segments present in *D. indianense* at the same shell diameters. *Diestoceras eos* differs from *D. shideleri* in possessing a shorter body chamber relative to shell width and having shorter cameral lengths at comparable diameters. The differences between these Cincinnati arch species of *Diestoceras* are summarized in table 11.

Diestoceras eos is most similar to *D. waynesvillense* Flower from the underlying Waynesville "biofacies" in the Bull Fork Formation at Oldenburg, Ind. Detailed comparison between these species is difficult due to the poor preservation of the holotype (MU 446T) and only known specimen of *D. waynesvillense*. The general outline of the shell is, however, very similar to that of *D. eos* and additional specimens of this species are needed to further determine the affinities.

Diestoceras eos is more similar to the large species of *Diestoceras* described from slightly older and coeval carbonates in the Western United States and Canada than either *D. indianense* or *D. shideleri*. The "Arctic" species, like *D. eos*, are more inflated, nearly hemispherical forms, the flanks evenly convex, lacking any indication of endogastric curvature. These species, consisting of *D. gibbosum* Foerste, *D. nobile* (Whiteaves), and *D. whiteaves* Foerste from the Red River Formation in Manitoba, and *D. staufferi* Foerste from the Maquoketa Formation in Minnesota, all tend to be more inflated than *D. eos*, possessing greater apical angles and shorter body chambers.

In the Cincinnati arch region, specimens identified as *D. eos* have been collected from the Saluda Formation in southeastern Indiana, a tongue of the Saluda Formation extending into the Bull Fork Formation in Preble and Butler Counties, Ohio, and in the upper portions of the Bull Fork Formation at Dayton and near Wilmington, in southwestern Ohio. Well-preserved, undistorted specimens are, unfortunately, rather rare, the species being typically represented by distorted matrix-filled molds or isolated body chambers.

FAMILY VALCOUROCERATIDAE FLOWER

Typically small, slender, gradually to rapidly expanding exogastric cyrtocones. Cross sections varying with ontogeny from compressed ovate adapically to depressed ovate to subtriangular adorally, the venter more narrowly rounded than the dorsum, often being subangular. Sutures sloping adorally from the dorsum to the venter. Broad lobe developed dorsally in some taxa. Cameral chambers typically very short, cameral lengths decreasing with ontogeny. Body chambers short relative to phragmocone length. Apertures open, variably developed ventral hyponomic sinus. Shells smooth or marked by fine, raised growth lines.

Siphuncle located near ventral shell margin, consisting of expanded ovate segments that typically become larger in diameter and more compressed with ontogeny. Septal necks cyrtochoanitic; recurved with short free brims. Connecting rings becoming thickened with ontogeny, generating complex actinosiphonate deposits consisting of radial, platelike processes that converge inward but fail to meet in the center of the segment.

Stratigraphic range.—Middle Ordovician (Chazyan Stage) to Upper Ordovician (Richmondian Stage); currently known only from North America (Laurentia).

Remarks.—The Valcouroceratidae were defined by Flower (1946) as a group of typically slender, exogastric cyrtocones with ventral cyrtochoanitic siphuncles. They were differentiated from the Oncoceratidae by the presence of actinosiphonate deposits within the siphuncle. The oldest known genus, *Valcouroceras* Flower from Chazyan strata in New York, is remarkable for the dramatic changes in morphology that the shell undergoes with ontogeny (Flower, 1946, fig. 14). Shell cross sectional shape changes from highly compressed ovate adapically to broad, depressed subtriangular adorally, then contracting to compressed ovate at the aperture. The siphuncle segments initially are slender, subtubular with thin connecting rings; becoming more expanded, compressed heart-shaped adorally, the connecting rings thickening with ontogeny, leading to the development of thick actinosiphonate deposits in adoralmost segments of large specimens. Early slender, subtubular adapical segments with suborthochoanitic septal necks are similar to those developed at the same stage of growth in the Oncoceratidae and in some members of the Tripteroceratidae, suggesting to Flower (1946) a common origin for all three families in the early Middle Ordovician.

Younger members of the family retain the same sequence of morphologic changes with ontogeny that are characteristic of *Valcouroceras*, the shells being compressed ovate adapically, becoming depressed, subtriangular with venter subangular with ontogeny. These younger taxa, consisting of *Augustoceras* Flower, *Kindleoceras* Foerste, and *Manitoulinoceras* Foerste, tend to be more depressed, slender forms compared with the adorally inflated shells typical of *Valcouroceras*.

Small, fragmentary silicified specimens assignable to the family occur in the USGS 6034-CO collection from the upper part of the Tyrone Limestone (Middle Ordovician, Rocklandian) in central Kentucky. These specimens may be assignable to either *Valcouroceras* or *Manitoulinoceras*, but not enough of the shell is preserved to be diagnostic of either genus. The family is best represented in Upper Ordovician strata in the Cincinnati arch region where Flower (1946) has described four species of *Augustoceras*, four species of *Kindleoceras*, and seven species of *Manitoulinoceras*. Within these Cincinnatian strata, members of this family are most abundant in the Leipers Limestone (Edenian and Maysvillian) in south-central Kentucky and in younger carbonates in the Saluda and Whitewater Formations (Richmondian) in southeastern Indiana.

GENUS *AUGUSTOCERAS* FLOWER

Type species.—*Augustoceras shideleri* Flower, 1946, p. 343; by original designation; Leipers Limestone, (Maysvillian part), south-central Kentucky.

Description.—Small, slender, somewhat fusiform cyrtocones; shell initially rapidly expanding, then becoming more gradually expanding, tubular with ontogeny. Venter convex and dorsum concave the length of the shell. Shell cross section rounded subtriangular, the dorsum flattened and venter subangular. Sutures oblique, sloping adorally from the dorsum to the venter. Cameral chambers very short, length decreasing with ontogeny. Body chamber short, slightly inflated adapically, contracting slightly toward aperture. Aperture open, inclined adorally from dorsum to venter, the ventral hyponomic sinus weakly developed. Shell exterior marked by closely spaced, fine, raised growth lines.

Siphuncle situated near ventral shell margin; segments small in diameter (averaging one-eighth the diameter of the shell), slightly expanded within camerae, scalariform. Segments becoming more compressed, broader in diameter with ontogeny. Septal necks short, strongly recurved, cyrtochoanitic. Connecting rings slightly thickened. Segments occupied by actinosiphonate deposits originating from connecting ring wall and consisting of longitudinal plates that, in cross section, consist of numerous radially disposed rays extending inward from the connecting ring wall. Deposits thickening adorally with ontogeny.

Stratigraphic and geographic occurrence.—Middle Ordovician (Shermanian Stage) to Upper Ordovician (Maysvillian Stage); Ohio valley, central Kentucky, and central basin of Tennessee.

Remarks.—*Augustoceras* is distinct from other members of the Valcouroceratidae in its slender, fusiform shell shape and the restriction of the ontogenetic changes in the siphuncle segments to the earliest adapical portions of the shell. The latter results in the presence of expanded, cyrtochoanitic siphuncle segments and actinosiphonate deposits over much of the length of the phragmocone. The genus also differs from the more primitive *Minganoceras* and *Valcouroceras* in the development of a shallow hyponomic sinus, a feature absent in these older Middle Ordovician taxa.

Cincinnatian species of *Kindleoceras* Foerste and *Manitoulinoceras* Foerste differ from *Augustoceras* in having the development of actinosiphonate deposits restricted to adoral portions of the shell adjacent to the body chamber. Additionally, *Kindleoceras* has a more pronounced subtriangular cross section and both *Kindleoceras* and *Manitoulinoceras* have a more slender, gradually expanding shell compared with known species of *Augustoceras*. As indicated above, in the Cincinnati arch region, *Augustoceras* is best known from the

Maysvillian Leipers Limestone in south-central Kentucky, whereas species of *Kindleoceras* and *Manitoulinoceras* are known from younger Richmondian strata in southeastern Indiana and adjacent portions of southwestern Ohio.

Known occurrences of *Augustoceras* consist of fragmentary specimens from the Catheys Formation in central Tennessee and the Strodes Creek Member of the Lexington Limestone in northern Kentucky (Middle Ordovician, Shermanian). Poorly preserved specimens from the Edenian part of the Kope Formation (Upper Ordovician) in the Cincinnati, Ohio, region may also belong in this genus. *Augustoceras* is represented by abundant, well-preserved, calcite-replaced specimens from the Maysvillian part of the Leipers Limestone (Upper Ordovician, Maysvillian) in the Cumberland River valley in south-central Kentucky (pl. 20, figs. 23–27). Flower (1946) differentiated four species from these strata. *Cyrtoceras vallandighami* Miller, a poorly known species from the Maysvillian part of the Fairview Formation (Late Ordovician) at Cincinnati, Ohio, might also be a species of *Augustoceras*.

Augustoceras, along with the actinocerid *Troedsonoceras* Foerste and the large discosorids *Faberoceras* Flower and *Antiphragmoceras* Foerste, was endemic to the southern Cincinnati arch region in the Middle and Late Ordovician and is characteristic of Flower's "Austral" fauna. This indigenous nautiloid fauna, as well as the orthocerid *Treptoceras* Flower and the endocerid *Cameroceras* Conrad, alternates with immigrant "Boreal" taxa derived from coeval carbonate facies in Western and Arctic North America throughout the Upper Ordovician strata exposed in the Kentucky-Indiana-Ohio area (Flower, 1946, 1976).

AUGUSTOCERAS SHIDELERI FLOWER, 1946

Plate 20, figures 23–27

Augustoceras shideleri Flower, 1946, p. 432, pl. 13, figs. 1–10; pl. 19, fig. 14; pl. 20, figs. 1–7.

Diagnosis.—Large species of *Augustoceras* distinguished by the nearly tubular nature of the adoral portion of the shell and its comparatively deep camerae and rapid rate of shell expansion adapically.

Description.—Small (estimated lengths of up to 11.6 cm and shell diameters of up to 2.7 cm), slender (apical angle 18–22 degrees), exogastrically curved, fusiform cyrtocones. Shell initially rapidly expanding up to diameters of 20 mm, then becoming more gradually expanding with ontogeny. Venter strongly convex the length of the shell; dorsum concave. Shell cross section compressed ovate at diameters of 15 mm or less, venter more narrowly rounded than dorsum; section gradually becoming more depressed, the dorsum slightly flattened and venter more angular at diameters greater than 15 mm. Shell height roughly equal to width much of the length of the shell. Sutures sloping adorally from dorsum to venter; very shallow dorsal lobe developing adorally. Cameral chambers short, cameral lengths ranging from one-fifth to one-eleventh the diameter of the shell, decreasing with ontogeny. Body chamber comparatively short, slightly less than one-fifth the total length of the shell; slightly inflated adapically, then contracting adorally toward aperture. Aperture open, the lateral flanks oblique, sloping forward from dorsum to venter. Dorsum broadly rounded, marked by shallow V-shaped ventral hyponomic sinus. Shell surface having crowded, fine, raised transverse growth lines that slope dorsally and ventrally, delineating a shallow lobe and ventral sinus.

Siphuncle located close to the ventral shell margin, separated from it by a distance equal to the diameter of the siphuncle segment. Segments narrow in diameter, averaging one-seventh the diameter of the shell. Segments slightly expanded, scalariform, wider adorally than adapically, SCR values being 0.90–1.10 (nearly equiaxial to slightly compressed) at shell diameters of 13.5–17 mm. Segments adnate adorally to adjacent septum. Septal necks short, cyrtochoanitic, more strongly recurved dorsally than ventrally. Connecting rings thickened much of the length of the phragmocone, producing well developed actinosiphonate deposits that thicken adorally. Deposits consisting of longitudinal plates that in cross section show 12–17 variably developed radial rays that project inward from the connecting ring wall but do not meet in the center.

Stratigraphic and geographic occurrence.—Leipers Limestone (Upper Ordovician, Maysvillian part); vicinity of the Cumberland River valley, south-central Ky.

Studied material.—Holotype UC 24290 (Flower, 1946, pl. 18, figs. 2–4), paratype UC 24292 (Flower, 1946, pl. 18, figs. 5–9), paratype MU 435T, and topotype MU 428T (pl. 20, figs. 23–27), all from the *Tetradium* biostrome in the Leipers Limestone exposed along the banks of the Cumberland River at Belk Island, near Rowena, Ky.

Remarks.—Flower (1946) distinguished *A. shideleri* from associated species of *Augustoceras* on the basis of its larger size and the deeper camerae and more rapid rate of shell expansion evident in adapical portions of the phragmocone. Comparisons of studied specimens of *A. shideleri* with Flower's types of *A. medium* Flower (UC 28284) and *A. minor* Flower (UC 24287) indicate that these differences can be used to differentiate these species. Additionally, overlays of the shell outlines of these specimens indicate that both of these species are much more gently curved compared with typical specimens of *A. shideleri*.

Augustoceras shideleri is more similar to *A. commune* Flower in the curvature of the shell and in the rate of shell expansion of the adapical portion of the shell (both 18–23 degrees). Cameral lengths in portions of the shell 15 mm in diameter tend to be shorter in specimens of *A. commune*. The holotype of *A. commune* (MU 448T) indicates that the maximum size attained by this species was approximately 10.0 cm, a maximum observed diameter being 2.0 cm; somewhat less than the maximum size observed in *A. shideleri* (up to 11.6 cm in length and 2.5 cm in diameter). Further study of large suites of more complete specimens is needed to determine the relation between these two similar species, associated in taxa. Flower (1946) indicated that in the Leipers Limestone, *A. commune* was the most abundant species of *Augustoceras* and, in fact, the most abundant nautiloid present in the *Tetradium* biostrome.

GENUS *MANITOULINOCERAS* FOERSTE

Type species.—*Cyrtoceras lysander* Billings, 1865, p. 161; designated by Foerste, 1924a, p. 230; Meaford Formation, Richmondian Stage, Upper Ordovician, Manitoulin Island, Ontario.

Description.—Small to medium-sized, slender exogastric cyrtocones, the shell moderately to rapidly expanding adapically, becoming more gradually expanding, almost tubular adorally with ontogeny. Shell nearly straight to strongly curved; venter typically convex the length of the shell, dorsum concave. Cross section depressed ovate, the dorsum more flattened than the venter. Venter rounded or subangular. Sutures weakly crenulate; straight, transverse adapically, becoming oblique with ontogeny, sloping from dorsum to venter. Dorsum with shallow lobe adorally. Cameral chambers very short, cameral lengths decreasing with ontogeny. Body chambers short, nearly tubular, not contracting towards aperture. Aperture open, marked by a broad shallow sinus dorsally and a shallow V-shaped hyponomic sinus ventrally. Shell surface marked by closely spaced, fine, transverse growth lines.

Siphuncles located at or near ventral shell margin; narrow in diameter, averaging one-fifth the diameter of the shell. Segments expanded, spherical to compressed ovate in shape; broad at the septal foramen relative to segment length. Septal necks short, recurved, cyrtochoanitic. Connecting rings thickening adorally, producing irregularly developed actinosiphonate deposits.

Stratigraphic and geographic occurrence.—Middle Ordovician (Blackriveran? Stage) to Upper Ordovician (Richmondian Stage); Eastern and Central North America (Laurentia).

Remarks.—*Manitoulinoceras* is evidently a long-ranging genus that has been reported from strata as old as Blackriveran in age by Foerste (1933). Flower (1946) noted, however, that older Middle Ordovician species assigned to the genus were atypical, consisting of forms that were too narrow in section and (or) more rapidly expanding than is typical of younger members of the genus, including the type species. He indicated that the oldest well-established species assignable to the genus was *M. wykoffense* Foerste from the Prosser Limestone (Middle Ordovician, Shermanian) of Minnesota. Undescribed typical specimens of *Manitoulinoceras* from Middle Ordovician (Rocklandian and Kirkfieldian) strata in Quebec indicate, however, that the genus goes back at least to the Rocklandian. The bulk of the described species of *Manitoulinoceras* are from younger Late Ordovician (Richmondian) strata in the Cincinnati arch region of Kentucky-Indiana-Ohio and adjacent portions of southern Ontario.

In contrast to the preceding genus, specimens of *Manitoulinoceras* from the Cincinnati region typically consist of matrix-filled internal molds in which the septal necks and connecting ring walls are not preserved. As a result, with the exception of several molds from the Saluda Formation in southeastern Indiana (Flower, 1946, pl. 23, figs. 7–9), actinosiphonate deposits are not known for most of these species of *Manitoulinoceras* from the Cincinnati arch region. Curiously, well-preserved, calcite-replaced specimens of *M. tenuiseptum* (Faber) and *M. williamsae* Flower from the *Treptoceras duseri* shale unit sectioned by the author lacked any evidence of actinosiphonate deposits (Frey, 1983, pl. 19, fig. 3).

Within the Cincinnati arch region, the earliest occurrence of *Manitoulinoceras* is *M. tenuiseptum* (Faber) and *M. williamsae* Flower in the "*Treptoceras duseri* Shale" within the Bull Fork Formation (lower Richmondian) in southwestern Ohio (Flower, 1946; Frey, 1983, 1989). I have found a number of incomplete internal molds of *Manitoulinoceras* in several other trilobite-bearing claystone beds higher up in the Bull Fork section in the same area. The genus, however, is most abundant in younger Richmondian carbonate facies, where *M. moderatum* Flower and *M. gyroforme* Flower occur in the micritic limestones of the Saluda Formation in southeastern Indiana and western Ohio and *M. moderatum*, *M. erraticum* Flower, and *M. ultimum* Flower occur in the various facies of the overlying Whitewater Formation in same area. *Manitoulinoceras gyroforme* Flower is described here from the uppermost Bull Fork Formation (Richmondian) exposed near Fairborn, Greene County, Ohio.

MANITOULINOCERAS GYROFORME FLOWER, 1946

Plate 20, figures 28, 29

Manitoulinoceras gyroforme Flower, 1946, p. 382, pl. 21, figs. 12–14.

Diagnosis.—Small species of *Manitoulinoceras* distinguished by its strongly curved shell and highly depressed cross section.

Description.—Small (estimated lengths of up to 11.0 cm and shell diameters up to 2.0 cm), moderately expanding (apical angle of 10 degrees), strongly exogastrically curved cyrtocones with a strongly convex venter and concave dorsum. Shell cross section depressed ovate, becoming more depressed in section with ontogeny at a height to width ratio averaging 0.86 for shell heights of 12–20 mm. Dorsum only slightly more flattened than venter. Sutures slightly crenulate, sloping adorally from dorsum to venter with the development of a broad, shallow dorsal lobe. Cameral chambers short, cameral lengths ranging from one-seventh to one-tenth the diameter of the shell, decreasing with ontogeny. Body chamber depressed, tubular, slightly less than one-fifth the total length of the shell. Aperture poorly known; open with a shallow dorsal lobe and sloping obliquely from dorsum to venter. Shell exterior marked by closely spaced, fine, raised growth lines.

Siphuncle located at ventral shell margin; segments expanded, compressed ovate, slightly greater than one-seventh the diameter of the shell; SCR values of 1.16–1.50 (compressed) at shell widths of 19–22 mm. Structure of septal necks and connecting rings unknown.

Stratigraphic and geographic occurrence.—"Saluda tongue" within the upper part of the Bull Fork Formation (Upper Ordovician, Richmondian Stage), Butler County, Ohio; uppermost Bull Fork Formation (late Richmondian), Greene County, Ohio.

Studied material.—Holotype MU 436T (Flower, 1946, pl. 21, figs. 12–14), Saluda Formation at Indian Creek, 5 mi west of Oxford, Ohio; USNM 468719 (pl. 20, figs. 28, 29), Bull Fork Formation, railroad cut at Wright Memorial, Fairborn, Greene County, Ohio.

Remarks.—*Manitoulinoceras gyroforme* is apparently a rare species in Richmondian strata exposed in the Cincinnati arch region, currently known from only two specimens: the holotype from the "Saluda" tongue near Oxford, Butler County, Ohio, and the specimen illustrated here from the uppermost Bull Fork Formation in Greene County, Ohio. This species is readily distinguished from all other species of *Manitoulinoceras* on the basis of the strong curvature of the shell and the more or less equally depressed venter and dorsum over the length of the shell. The most abundant species of *Manitoulinoceras* in these strata, *M. moderatum* Flower, is a larger, more gently curved, slender cyrtocone that is most common in the micritic limestones of the Saluda Formation in adjacent portions of southeastern Indiana.

REFERENCES CITED

Alberstadt, L.P., 1979, The brachiopod genus *Platystrophia*: U.S. Geological Survey Professional Paper 1066-B, 20 p.

Aronoff, S.M., 1979, Orthoconic nautiloid morphology and the case of *Treptoceras* and *Orthonybyoceras*: Neu Jahrbuch Geologie und Palaontologie, v. 158, no. 1, p. 100–122.

Baird, G.C., Brett, C.E., and Frey, R.C., 1989, "Hitchhiking" epizoans on orthoconic cephalopods—preliminary review of evidence and its implications: Senckenbergiana Lethaea, v. 69, p. 501–516.

Barrande, Joachim, 1865–1877, Systeme silurien du centre de la Boheme, Premiere Partie: Recherches Paleontologiques, v. 2, Classe des Mollusques, Order des Cephalopodes, Parts 1–5, Prague.

Bassler, R.S., 1915, Bibliographic index of American Ordovician and Silurian fossils: U.S. National Museum Bulletin 92, 2 v., 1250 p.

———1932, The stratigraphy of the central basin of Tennessee: Tennessee Division of Geology Bulletin, v. 38, 268 p.

Bell, B.M., 1979, Edrioasteroids (Echinodermata): U.S. Geological Survey Professional Paper 1066-E, 7 p.

Berdan, J.M., 1984, Leperditicopid ostracodes from Ordovician rocks of Kentucky and adjacent states and characteristic features of the order Leperditicopida: U.S. Geological Survey Professional Paper 1066-J, 40 p.

Bergstrom, S.M., Kolata, D.R., and Huff, W.D., 1991, Geologic significance of gigantic ash falls in the Ordovician of North America and Europe: Ohio Journal of Science, v. 91, no. 2, p. 34.

Bergstrom, S.M., and Mitchell, C., 1990, Trans-Pacific graptolite faunal relations: the biostratigraphic position of the base of the Cincinnatian Series in the standard Australian graptolite succession: Journal of Paleontology, v. 64, no. 6, p. 992–997.

Berry, W.B.N., and Boucot, A.J., 1973, Glacio-eustatic control of Late Ordovician-Early Silurian platform sedimentation and faunal change: Geological Society of America Bulletin, v. 84, no. 1, p. 275–284.

Billings, Elkhanah, 1857, Geological Survey of Canada, Reports of Progress for 1853–1856, p. 321–338.

———1859, *Nautilus jason*: Canadian Naturalist and Geologist, v. 4, p. 464.

———1860, New species of fossils from the Lower Silurian rocks of Canada: Canadian Naturalist, v. 5, p. 161–177.

———1865, Paleozoic fossils, v. 1: Geological Survey of Canada, 426 p.

Bird, J.M., and Dewey, J.F., 1970, Lithosphere plate-continental margin tectonics and the evolution of the Appalachian orogen: Geological Society of America Bulletin, v. 81, p. 1031–1060.

Black, D.F.B., and Cuppels, N.P., 1973, Strodes Creek Member (Upper Ordovician)—a new map unit in the Lexington Limestone of north-central Kentucky: U.S. Geological Survey Bulletin 1372-C, 16 p.

Bliss, Frank, 1984, The "concrete layer" of the Saluda Formation (Upper Ordovician) in Franklin County, Indiana: Oxford, Ohio, Miami University, M.S. thesis, 121 p.

Branstrator, J.W., 1979, Asteroidea (Echinodermata): U.S. Geological Survey Professional Paper 1066-F, 7 p.

Brenchley, P.J., 1984, Late Ordovician extinctions and their relationship to the Gondwana glaciation, *in* Brenchley, P.J., ed., Fossils and climate: Chichester, England, Wiley and Sons, p. 291–315.

——1989, Late Ordovician events and the terminal Ordovician extinction, *in* Donavan, S.K., ed., Mass extinctions—processes and evidence: New York, Columbia University Press, p. 104–132.

Brenchley, P.J., and Newall, G., 1980, A facies analysis of Upper Ordovician regressive sequences in the Oslo region, Norway—A record of glacio-eustatic change: Palaeogeography, Palaeoclimatology, Palaeoecology, v. 31, no. 1, p. 1–38.

Bronn, H.G., 1837, Lethaca Geognostica, Band 1: Stuttgart, E. Schweizerbart, 672 p.

Cameron, B., and Mangion S., 1977, Depositional environments and revised stratigraphy along the Black River–Trenton boundary in New York and Ontario: American Journal of Science, v. 277, p. 486–502.

Catalani, J.A., 1987, Biostratigraphy of the Middle and Upper Ordovician cephalopods of the upper Mississippi valley area, *in* Sloan, R.E., ed., Middle and Late Ordovician lithostratigraphy and biostratigraphy of the upper Mississippi valley: Minnesota Geological Survey Report of Investigations 35, p. 187–189.

Chamberlain, J.A., Ward, P.D., and Weaver, J.S., 1981, Postmortem ascent of *Nautilus* shells—implications for cephalopod paleobiogeography: Paleobiology, v. 7, p. 494–509.

Chen, Yun-Juan, and Teichert, Curt, 1983, Cambrian Cephalopoda of China: Palaeontographica, sect. B, v. 181, p. 1–102

Clarke, J.J., 1897, The Lower Silurian cephalopods of Minnesota: Minnesota Geological Survey, Final Report 3, pt. 2, p. 761–812.

Conrad, T.A., 1838, Report of the paleontological department of the Survey: New York Geological Survey Annual Report 2, p. 107–119.

——1842, Observations on the Silurian and Devonian Systems of the United States, with descriptions of new organic remains: Journal of the Academy of Natural Sciences, Philadelphia, v. 8, p. 228–280.

——1843, *Orthoceras anellus*. Proceedings of the Academy of Natural Sciences, Philadelphia, v. 1, p. 334.

Cooper, G.A., 1956, Chazyan and related brachiopods: Smithsonian Miscellaneous Collections, v. 127, Pt. 1, p. 1–1024, Pt. 2, p. 1025–1245.

Copper, P., 1978, Paleoenvironments and paleocommunities in the Ordovician-Silurian Sequence of Manitoulin Island, *in* Sanford, J.T., and Mosher, R.E., eds., Geology of the Manitoulin Island area: Michigan Basin Geological Society Special Papers 3, p. 47–61.

Cressman, E.R., 1973, Lithostratigraphy and depositional environments of the Lexington Limestone (Ordovician) of central Kentucky: U.S. Geological Survey Professional Paper 768, 61 p.

Cressman, E.R., and Noger, M.C., 1976, Tidal-flat carbonate environments in the High Bridge Group (Middle Ordovician) of central Kentucky: Kentucky Geological Survey Report of Investigations 18, 15 p.

Crick, R.E., 1980, Integration of paleobiogeography and paleogeography—evidence from Arenigian nautiloid biogeography: Journal of Paleontology, v. 54, p. 1218–1236.

——1988, Buoyancy regulation and macroevolution in nautiloid cephalopods: Senckenbergiana Lethaea, v. 69, p. 13–42.

——1990, Cambro-Devonian biogeography of nautiloid cephalopods, *in* McKerrow, W.S., and Scotese, C.R., eds., Palaeozoic paleogeography and biogeography: Geological Society Memoir 12, p. 147–161.

Cumings, E.R., 1908, The stratigraphy and paleontology of the Ordovician rocks of Indiana: Indiana Department of Natural Resources 32nd Annual Report, p. 605–1188.

Denison, J.M., 1976, Appalachian Queenston delta related to eustatic sea-level drop accompanying Late Ordovician glaciation centered in Africa, *in* Bassett, M.G., ed., The Ordovician System: Cardiff, University of Wales Press and the National Museum of Wales, p. 107–120.

Dunham, R.J., 1962, Classification of carbonate rocks according to depositional texture, *in* Ham, W.H., ed., Classification of carbonate rocks: American Association of Petroleum Geologists Memoir 1, p. 108–122.

Dzik, J., 1984, Phylogeny of the Nautiloidea: Palaeontologica Polonica, v. 45, 203 p.

Elias, R.J., 1983, Middle and Late Ordovician solitary rugose corals of the Cincinnati arch region: U.S. Geological Survey Professional Paper 1066-N, 13 p.

——1991, Environmental cycles and bioevents in the Upper Ordovician Red River–Stony Mountain solitary rugose coral province of North America, *in* Barnes, C.R., and Williams, S.W., eds., Advances in Ordovician geology: Geological Survey of Canada Paper 90–9, p. 205–211.

Emmons, Ebenezer, 1842, Geology of New York, Pt. 2, Comprising the Survey of the second geological district: Albany, New York Geological Survey, 437 p.

Ettensohn, F.R., 1991, Flexural interpretation of relationships between Ordovician tectonism and stratigraphic sequences, central and southern Appalachians, USA, *in* Barnes, C.R., and Williams, S.W., eds., Advances in Ordovician geology: Geological Survey of Canada Paper 90–9, p. 213–224.

Flower, R.H., 1939, Study of the Pseudorthoceratidae: Paleontographica Americana, v. 2, no. 10, p. 1–214.

——1942, An Arctic cephalopod faunule from the Cynthiana of Kentucky: Bulletins of American Paleontology, v. 27, no. 103, p. 5–90.

——1943, Annulated orthoceraconic genera of paleozoic nautiloids: Bulletins of American Paleontology, v. 28, no. 109, p. 102–128.

——1946, Ordovician cephalopods of the Cincinnati region, Pt. 1: Bulletins of American Paleontology, v. 29, no. 116, 738 p.

——1952, New Ordovician cephalopods from Eastern North America: Journal of Paleontology, v. 26, no. 1, p. 24–59.

——1955a, Status of endoceroid classification: Journal of Paleontology, v. 29, no. 3, p. 329–371.

——1955b, New Chazyan orthocones: Journal of Paleontology, v. 29, no. 5, p. 806–830.

——1955c, Trails and tentacular impressions of Orthoconic cephalopods: Journal of Paleontology, v. 29, no. 5, p. 857–867.

——1955d, Saltations in nautiloid coiling: Evolution, v. 9, no. 3, p. 244–260.

——1957a, Studies of the Actinoceratida: New Mexico Bureau of Mines and Mineral Resources Memoir 2, 59 p.

———1957b, Nautiloids of the Paleozoic, *in* Ladd, H., ed., treatise of marine ecology and paleoecology: Geological Society of America Memoir 67, v. 2, p. 829–852.

———1958, Some Chazyan and Mohawkian Endoceratida: Journal of Paleontology, v. 32, no. 3, p. 433–458.

———1962, Notes on the Michelinoceratida: New Mexico Bureau of Mines and Mineral Resources Memoir 10, pt. 2, p. 21–42.

———1964a, The nautiloid order ellesmerocerida: New Mexico Bureau of Mines and Mineral Resources Memoir 12, 234 p.

———1964b, Nautiloid shell morphology: New Mexico Bureau of Mines and Mineral Resources Memoir 13, 62 p.

———1965, Cephalopods—as illustrated by early Paleozoic forms, *in* Kummel, B., and Raup, D., eds., Handbook of paleontological techniques: San Francisco, Calif., W.H. Freeman, p. 53–57.

———1968a, The first great expansion of the Actinocerida: New Mexico Bureau of Mines and Mineral Resources Memoir 19, pt. 1, p. 1–16.

———1968b, Silurian cephalopods of the James Bay Lowland with a revision of the family Narthecoceratidae: Geological Survey of Canada Bulletin 164, 88 p.

———1970, Early Paleozoic of New Mexico and the El Paso region: New Mexico Bureau of Mines and Mineral Resources Reprint Series, 44 p.

———1976, Ordovician cephalopod faunas and their role in correlation, *in* Bassett, M.G., ed., The Ordovician System: Cardiff, University of Wales Press and the National Museum of Wales, p. 523–552.

———1984, *Bodeiceras*, a new Mohawkian oxycone, with revision of the order Barrandeocerida and discussion of the status of the order: Journal of Paleontology, v. 58, no. 6, p. 1372–1379.

Flower, R.H., and Kummel, B., 1950, A classification of the Nautiloidea: Journal of Paleontology, v. 24, no. 5, p.604–616.

Flower, R.H., and Teichert, Curt, 1957, The cephalopod order Discosorida—Mullusca, Article 6: University of Kansas Paleontological Contributions, 144 p.

Fluegeman, R., and Pope, J.K., 1983, Brainard Shale outliers (Upper Ordovician), Maquoketa Group in southeastern Indiana: Geological Society of America, Abstracts with Programs, v. 15, p. 574.

Foerste, A.F., 1893, Fossils of the Clinton Group in Ohio and Indiana: Report of the Geological Survey of Ohio, v. 7, p. 516–601.

———1910, Preliminary notes on Cincinnatian and Lexington fossils: Journal of the Science Laboratories, Denison University, v. 16, p. 17–87.

———1912, *Strophomena* and other fossils from Cincinnatian and Mohawkian horizons, chiefly in Ohio, Indiana, and Kentucky: Journal of the Science Laboratories, Denison University, v. 17, p. 17–174.

———1914a, The Rogers Gap fauna of central Kentucky: Journal of the Cincinnati Society of Natural History, v. 21, p. 109–156.

———1914b, Notes on the Lorraine faunas of New York and the Province of Quebec: Journal of the Science Laboratories, Denison University, v. 17, p. 247–328.

———1924a, Upper Ordovician faunas of Ontario and Quebec: Geological Survey of Canada Memoir 138, 255 p.

———1924b, Notes on American Paleozoic cephalopods: Journal of the Science Laboratories, Denison University, v. 20, p. 193–268.

———1926, Actinosiphonate, trochocerid, and other cephalopods: Journal of the Science Laboratories, Denison University, v. 21, p. 285–384.

———1928a, The Cephalopod fauna of Anticosti Island, *in* Twenhofel, W.H., ed., The geology of Anticosti Island: Geological Survey of Canada Memoir 154, p. 257–321.

———1928b, A restudy of some of the Ordovician and Silurian cephalopods described by Hall: Journal of the Science Laboratories, Denison University, v. 23, p. 173–230.

———1928c, Restudy of American orthoconic Silurian cephalopods: Journal of the Science Laboratories, Denison University, v. 23, p. 236–320.

———1928d, Contributions to the geology of Foxe Land, Baffin Island, pt. 2—The cephalopods of the Putnam Highland: University of Michigan Museum of Paleontology Contributions, v. 3, no. 3, p. 25–70.

———1929a, The Ordovician and Silurian of American Arctic and Subarctic regions: Journal of the Science Laboratories, Denison University, v. 24, p. 27–79.

———1929b, The cephalopods of the Red River Formation of southern Manitoba: Journal of the Science Laboratories, Denison University, v. 24, p. 129–235.

———1932, Black River and other cephalopods from Minnesota, Wisconsin, Michigan, and Ontario: Journal of the Science Laboratories, Denison University, v. 27, p. 47–136.

———1933, Black River and other cephalopods from Minnesota, Wisconsin, Michigan, and Ontario, pt. 2: Journal of the Science Laboratories, Denison University, v. 28, p. 1–164.

———1935, Bighorn and related cephalopods: Journal of the Science Laboratories, Denison University, v. 30, p. 1–96.

———1936a, The cephalopods of the Maquoketa Shale of Iowa: Journal of the Science Laboratories, Denison University, v. 30, p. 231–260.

———1936b, Cephalopods from the Upper Ordovician of Perce, Quebec: Journal of Paleontology, v. 10, no. 5, p. 373–384.

———1938, Cephalopoda, *in* Twenhofel, W.H., ed., Geology and paleontology of the Mingan Islands, Quebec: Geological Society of America Special Paper 11, 107 p.

Foerste, A.F., and Savage, T.E., 1927, Ordovician and Silurian cephalopods of the Hudson Bay area: Journal of the Science Laboratories, Denison University, v. 22, p. 1–108.

Foerste, A.F., and Teichert, Curt, 1930, The actinoceroids of East-Central North America: Journal of the Science Laboratories, Denison University, v. 25, p. 201–296.

Folk, R., 1962, Spectral subdivision of limestone types, *in* Ham, W.H., ed., Classification of carbonate rocks: American Association of Petroleum Geologists Memoir 1, p. 108–122.

Foord, A.H., 1891, Catalogue of the fossil cephalopods of the British Museum (Natural History), part II, 407 p.

Frey, R.C., 1981, *Narthecoceras* (Cephalopoda) from the Upper Ordovician (Richmondian) of southwest Ohio: Journal of Paleontology, v. 55, no. 6, p. 1217–1224.

———1983, The paleontology and paleoecology of the *Treptoceras duseri* shale unit (Late Ordovician, Richmondian) of southwest Ohio: Oxford, Ohio, Miami University, Ph.D. dissertation, 719 p.

———1987a, The paleoecology of a Late Ordovician shale unit from southwest Ohio and southeastern Indiana: Journal of Paleontology, v. 61, no. 2, p. 242–267.

———1987b, The occurrence of pelecypods in early Paleozoic epeiric sea environments, Late Ordovician of the Cincinnati, Ohio, area: Palaios, v. 2, no. 1, p. 3–23.

———1988, The paleoecology of *Treptoceras duseri* (Michelinoceratida-Proteoceratidae) from the Late Ordovician of southwest Ohio, *in* Wolberg, D.L., ed., Contributions to Paleozoic paleontology and stratigraphy in honor of Rousseau H. Flower: New Mexico Bureau of Mines and Mineral Resources Memoir 44, p. 79–101.

———1989, Paleoecology of a well-preserved nautiloid assemblage from a Late Ordovician shale unit, southwest Ohio: Journal of Paleontology, v. 63, no. 5, p. 604–620.

Furnish, W.M., and Glenister, B.F., 1964, *in* Moore, R.C., ed., Treatise on invertebrate paleontology, pt. K, Mollusca 3, Cephalopods: Lawrence, Geological Society of America and University of Kansas Press, p. K129–K159.

Gil, A.V., 1988, Whiterock (lower Middle Ordovician) cephalopodfauna from the Ibex area, Millard County, Utah, *in* Wolberg, D.L., ed., Contributions to Paleozoic paleontology and stratigraphy in honor of Rousseau H. Flower: New Mexico Bureau of Mines and Mineral Resources Memoir 44, p. 27–59.

Grabau, A.W., and Shimer, H.W., 1910, North American index of fossils, invertebrates, v. 1: Seiler and Company, New York, 853 p.

Gray, J., and Boucot, A.J., 1972, Palynological evidence bearing on the Ordovician-Silurian paraconformity in Ohio: Geological Society of America Bulletin, v. 83, p. 1299–1314.

Hall, James, 1847, Paleontology of New York, v. 1 of Organic remains of the lower divisions of the New York system: Albany, N.Y., C. Van Beurthuysen, 338 p.

———1861, Geological survey of Wisconsin: Report of the Superintendent, Madison, 52 p.

———1877, Addenda, *in* Miller, S.A.: The American Paleozoic Fossils, Cincinnati, p. 243–245.

Hall, James, and Whitfield, R.P., 1875, Fossils of the Hudson River Group: Ohio Geological Survey Report, v. 2, p. 67–110.

Hatfield, C.B., 1968, Stratigraphy and paleoecology of the Saluda Formation (Cincinnatian) in Indiana, Ohio, and Kentucky: Geological Society of America Special Paper 95, 34 p.

Hay, H.B., 1981, Lithofacies and formations of the Cincinnatian Series (Upper Ordovician) in southeastern Indiana and southwestern Ohio: Oxford, Ohio, Miami University, Ph.D. dissertation, 236 p.

Hay, H.B., Pope, J.K., and Frey, R.C., 1981, Lithostratigraphy cyclic sedimentation, and paleoecology of the Cincinnatian Series in southwestern Ohio and southeastern Indiana, *in* Roberts, T.G., ed., Annual meeting, Cincinnati, 1981: Geological Society of America Field Trip Guidebook, v. 1, p. 73–86.

Hook, S.C., and Flower, R.H., 1977, Late Canadian (Zones J, K) cephalopod faunas from the southwestern United States: New Mexico Bureau of Mines and Mineral Resources Memoir 32, 102 p.

House, M.R., 1988, Extinction and survival in the Cephalopoda, *in* Larwood, G.P., ed., Extinction and the fossil record: Oxford, Clarendon Press, Systematics Association Special Volume 34, p. 121–138.

Howe, H.J., 1979, Middle and Late Ordovician Plectambonitacean, Rhynchonellacean, Syntrophiacean, Trimerellacean, and Atrypacean Brachiopods: U.S. Geological Survey Professional Paper 1066–C, 18 p.

Hrabar, S.V., Cressman, E.R., and Potter, P.E., 1971, Crossbedding of the Tanglewood Limestone Member of the Lexington Limestone (Ordovician) of the Blue Grass region of Kentucky: Brigham Young University Geological Studies, v. 18, p. 99–114.

Huff, W.D., Bergstrom, S.M., and Kolata, D.R., 1992, Gigantic Ordovician volcanic ash fall in North America and Europe—biological, tectomagmatic, and event-stratigraphic significance: Geology, v. 20, p. 875–878.

Huff, W.D., and Kolata, D.R., 1990, Correlation of the Ordovician Deicke and Millbrig K-bentonites between the Mississippi valley and the southern Appalachians: American Association of Petroleum Geologists Bulletin, v. 74, no. 11, p. 1736–1747.

Hyatt, Alpheus, 1884, Genera of fossil cephalopods: Boston Society of Natural History, Proceedings, p. 237–338.

———1900, Cephalopoda, *in* Zittel, K.A., ed. (translated by Eastman, C.R.), Textbook of palaeontology, 1st ed.: London, McMillan, v. 1, p. 502–592.

James, J.F., 1878, *Orthoceras* Hindei: The Paleontologist, No. 1, p. 6.

———1886, Cephalopods of the Cincinnati Group: Journal of the Cincinnati Society of Natural History, v. 8, no. 4, p. 236–253.

Karklins, O.L., 1984, Trepostome and cystoporate bryozoans from the Lexington Limestone and the Clays Ferry Formation of Kentucky: U.S. Geological Survey Professional Paper 1066–I, 104 p.

Kay, G.M., 1937, Stratigraphy of the Trenton Group: Geological Society of America Bulletin, v. 48, p. 252–255.

———1951, North American geosynclines: Geological Society of America Memoir 48, 143 p.

Kepferle, R.C., 1976, Geologic map of the Fisherville quadrangle, north-central Kentucky: U.S. Geological Survey, Geological Quadrangle Map GQ–1321.

Kepferle, R.C., Noger, M.C., Meyer, D.L., and Schumacher, G.A., 1987, Stratigraphy, sedimentology, and paleontology (Upper Ordovician), and glacial and engineering geology of northern Kentucky and southern Ohio: Kentucky Geological Survey, 18 p.

Kobayashi, T., 1927, Ordovician fossils from Corea and south Manchuria: Japanese Journal of Geology and Geography, v. 5, no. 4, p. 173–212.

Kolata, D.R., Frost, J.K., and Huff, W.D., 1986, K-bentonites of the Ordovician Decorah Subgroup, upper Mississippi valley—correlation by chemical finger-printing: Illinois State Geological Survey Circular 537, 30 p.

Kuhnhenn, G.L., Grabowski, G.J., and Dever, G.R., 1981, Paleoenvironmental interpretation of the Middle Ordovician High Bridge Group in central Kentucky, *in* Roberts, T.G., ed., Annual meeting, Cincinnati, 1981: Geological Society of America Field Trip Guidebook, v. 1, p. 1–30.

Kunk, M.J., and Sutter, J.F., 1984, ^{40}Ar/^{39}Ar age spectrum dating of biotite from Middle Ordovician bentonites, Eastern North America, in Bruton, D.L., ed., Aspects of the Ordovician System: University of Oslo Palaeontological Contribution 295, p. 11–22.

Landman, N.H., Rye, D.M., and Shelton, K.L., 1983, Early ontogeny of *Eutrephoceras* compared to Recent *Nautilus* and Mesozoic ammonites—evidence from shell morphology and light stable isotopes: Paleobiology, v. 9, p. 269–279.

Liberty, B.A., 1969, Paleozoic geology of the Lake Simcoe area, Ontario: Geological Survey of Canada Memoir 355, 201 p.

MacArthur, R.H., and Wilson, E.O., 1963, An equilibrium theory of insular zoogeography: Evolution, v. 17, p. 373–387.

McKerrow, W.S., 1979, Ordovician and Silurian changes in sea level: Journal of Geological Society of London, v. 136, p. 137–145.

Meyer, D.L., and others, 1981, Stratigraphy, sedimentology, and paleoecology of the Cincinnatian Series (Upper Ordovician) in the vicinity of Cincinnati, Ohio, in Roberts, T.G., ed., Annual meeting, Cincinnati, 1981: Geological Society of America Field Trip Guidebook, v. 1, p. 31–71.

Miller, A.K., and Carrier, J.B., 1942, Ordovician cephalopods from the Big Horn Mountains of Wyoming: Journal of Paleontology, v. 16, no. 5, p. 531–548.

Miller, A.K., and Furnish, W.M., 1937, Paleoecology of Paleozoic cephalopods: National Research Council, Committee on Paleoecology Report, 1936-37, p. 1129–1130.

Miller, A.K., Youngquist, W.L., and Collinson, C.W., 1954, Ordovician cephalopod fauna from Baffin Island: Geological Society of America Memoir 62, 234 p.

Miller, S.A., 1875, Class Cephalopoda (Cuvier) as represented in the Cincinnati Group: Cincinnati Quarterly Journal of Science, v. 2, no. 2, p. 121–134.

———1879, Description of twelve new fossil species: Journal of the Cincinnati Society of Natural History, v. 2, p. 104–118.

———1880, *Orthoceras dyeri*: Journal of the Cincinnati Society of Natural History, v. 3, p. 236.

———1881, New species of fossils from the Hudson River Group: Journal of the Cincinnati Society of Natural History, v. 4, no. 4, p. 315–323.

———1889, North American geology and paleontology for use of amateurs, students, and scientists: Cincinnati, Ohio, Western Methodist Book Concern, 664 p.

———1894, *Orthoceras gorbyi*: 18th Annual Report: Department of Geology and Natural Resources of Indiana, p. 322.

Miller, S.A., and Faber, C.L., 1894a, New species of fossils from the Hudson River Group, and remarks upon others: Journal of the Cincinnati Society of Natural History, v. 17, no. 1, p. 22–33.

———1894b, Description of some Cincinnati fossils: Journal of the Cincinnati Society of Natural History, v. 17, p. 137–158.

Mitchell, C.E., and Bergstrom, S.M., 1991, New graptolite and litho-stratigraphic evidence from the Cincinnati region, USA, for the definition and correlation of the base of the Cincinnatian Series (Upper Ordovician), in Barnes, C.R., and Williams, S.W., eds., Advances in Ordovician geology: Geological Survey of Canada Paper 90-9, p. 59–77.

Nelson, S.J., 1963, Ordovician paleontology of the northern Hudson Bay Lowland: Geological Society of America Memoir 90, 152 p.

Neuman, R.B., 1955, Middle Ordovician rocks of the Tellico-Sevier belt, eastern Tennessee: U.S. Geological Survey Professional Paper 274–F, p. 141–175.

———1967, Some silicified Middle Ordovician brachiopods from Kentucky: U.S. Geological Survey Professional Paper 583–A, 14 p.

Nickles, J.M., 1902, The geology of Cincinnati, Ohio: Journal of the Cincinnati Society of Natural History, v. 20, art. 3, 53 p.

Orth, C.J., Gilmore, J.S., Quintana, L.R., and Sheehan, P.M., 1986, The terminal Ordovician extinction—geochemical analysis of the Ordovician-Silurian boundary, Anticosti Island, Quebec: Geology, v. 14, p. 433–436.

Parks, W.A., and Fritz, Madeline, 1922, The stratigraphy and paleontology of Toronto and vicinity, pt. 9, no. 4: Ontario Department of Mines Annual Report 32, p. 16–28.

Parsley, R.L., 1981, Echinoderms from Middle and Upper Ordovician rocks of Kentucky: U.S. Geological Survey Professional Paper 1066–K, 9 p.

Pojeta, John, 1979, The Ordovician paleontology of Kentucky and nearby states—introduction: U.S. Geological Survey Professional Paper 1066–A, 48 p.

Pope, J.K., 1982, Some silicified strophomenacean brachiopods from the Ordovician of Kentucky, with comments on *Pionomena*: U.S. Geological Survey Professional Paper 1966–L, 30 p.

Raymond, P.E., 1903, Faunas of the Trenton at the type section and at Newport, New York: Bulletins of American Paleontology, v. 17, 18 p.

Reyment, R.A., 1958, Some factors in the distribution of fossil cephalopods: Stockholm Contributions to Geology, v. 1, no. 6, p. 97–184.

———1968, Orthoconic nautiloids as indicators of shoreline surface currents: Journal of Sedimentary Petrology, v. 38, p. 1387-1389.

Ross, R.J., 1967, Calymenid and other Ordovician trilobites from Kentucky and Ohio: U.S. Geological Survey Professional Paper 583–B, 19 p.

———1979, Additional trilobites from the Ordovician of Kentucky: U.S. Geological Survey Professional Paper 1066–D, 13 p.

Ross, R.J., and others, 1982, The Ordovician System in the United States—correlation chart and explanation: International Union of Geological Sciences Publication 12, 73 p.

Ruedemann, Rudolf, 1906, Cephalopods of the Beekmantown and Chazy in the Champlain basin: New York State Museum Bulletin 90, p. 395–611.

———1921, Paleontological contributions from the New York State Museum, pt. 2, on sex distinction in fossil cephalopods: New York State Museum Bulletin 227, p. 63–130.

———1926, Utica and Lorraine Formations of New York, pt. 2 (Mollusca, Crustacea, Miscellania): New York State Museum Bulletin 272, 227 p.

Samson, S.D., 1986, Chemistry, mineralogy, and correlation of Ordovician bentonites: Minneapolis, University of Minnesota, M.S. thesis, 128 p.

Sardeson, F.W., 1930, *Cameroceras* and its allies: Pan-American Geologist, v. 53, p. 175–182.

Saunders, W.B., and Spinosa, Claude, 1978, Sexual dimorphism in *Nautilus* from Palau: Paleobiology, v. 4, p. 349–358.

———1979, *Nautilus* movements and distribution in Palau, West Caroline Islands: Science, v. 204, p. 1199–1201.

Schumacher, G.A., Swinford, E.M., and Shrake, D.L., 1991, Lithostratigraphy of the Grant Lake Limestone and Grant Lake Formation (Upper Ordovician) in southwestern Ohio: Ohio Journal of Science, v. 9, no. 1, p. 56–68.

Scotese, C.R., and McKerrow, W.S., 1990, Revised world maps and introduction, *in* McKerrow, W.S., and Scotese, C.R., eds., Palaeozoic palaeogeography and biogeography: London, Geological Society Memoir 12, p. 1–21.

Self, S.M., Rampino, R., Newton, S., and Wolff, J.A., 1984, Volcanological study of the great Tamboro eruption of 1815: Geology, v. 12, p. 659–663.

Sepkoski, J.J., 1982, Mass extinctions in the Phanerozoic oceans: A review, *in* Silver, L.T., Schultz, P.H., Burke, K., and Raup, D.M., eds., Geological interpretations of impacts of large asteroids and comets on the Earth: Geological Society of America Special Paper 190, p. 283–290.

Serpagli, Enrico, and Gnoli, Maurizio, 1977, Upper Silurian cephalopods from southwestern Sardinia: Bollettino della Societa Paleontologica Italica, v. 16, p. 153–196.

Sharpf, C.D., 1990, Stratigraphy and associated faunas of the Middle Ordovician Millbrig K-bentonite in central Kentucky: Cincinnati, Ohio, University of Cincinnati, M.S. thesis, 280 p.

Sharpton, V.I., and Ward, P.D., eds., 1990, Global catastrophes in Earth history: Geological Society of America Special Paper 247, 631 p.

Sheehan, P.M., 1988, Late Ordovician events and the terminal Ordovician extinction, *in* Wolberg, D.L., ed., Contributions to Paleozoic paleontology and stratigraphy in honor of Rousseau H. Flower: New Mexico Bureau of Mines and Mineral Resources Memoir 44, p. 405–415.

Shimizu, S., and Obata, T., 1935, New genera of Gotlandian and Ordovician nautiloids: Journal of the Shanghai Science Institute, v. 2, p. 1–10.

Shrake, D.L., Schumacher, G.A., and Swinford, E.M., 1988, Field guide to the stratigraphy, sedimentology, and paleontology of the Upper Ordovician rocks in southwestern Ohio—Fifth midyear meeting: Columbus, Ohio, Society of Economic Paleontologists and Mineralogists, 81 p.

Sloan, R.E., 1987, Tectonostratigraphy, *in* Sloan, R.E., ed., Middle and Late Ordovician lithostratigraphy and biostratigraphy of the Mississippi valley: Minnesota Geological Survey Report of Investigations 35, p. 7–20.

Stait, B.A., 1984, Re-examination and description of the Tasmanian species of *Wutinoceras* and *Adamsoceras* (Nautiloidea-Ordovician): Geologica et Palaeontologica, v. 18, p. 53–57.

———1988, Nautiloids of the Lourdes Formation (Middle Ordovician), Port au Port Peninsula, western Newfoundland, *in* Wolberg, D.L., ed., Contributions to paleozoic paleontology and stratigraphy in honor of Rousseau H. Flower: New Mexico Bureau of Mines and Mineral Resources Memoir 44, p. 61–77.

Stanley, S.M., 1984a, Temperature and biotic crises in the marine realm: Geology, v. 12, p. 205–208.

———1984b, Marine mass extinctions—a dominant role for temperature, *in* Nitecki, M.H., ed., Extinctions: Chicago, Ill., University of Chicago Press, p. 71–117.

Steele, H.M., and Sinclair, G.W., 1971, A Middle Ordovician fauna from Braeside, Ottawa valley, Ontario: Geological Survey of Canada Bulletin 211, 96 p.

Stenzel, H.B., 1964, Living *Nautilus*, *in* Moore, R.C., ed., Treatise of invertebrate paleontology, pt. K, Mollusca 3, Cephalopoda: Lawrence, University of Kansas Press, p. K59–K93.

Stokes, XXX, 1940, *Ormoceras*: Transactions of the Geological Society of London, Series 2, v. 5, p. 709.

Swadley, W.C., Luft, S.J., and Gibbons, A.B., 1975, The Point Pleasant Tongue of the Clays Ferry Formation, northern Kentucky: U.S. Geological Survey Bulletin 1405-A, p. A30–A31.

Sweet, W.C., 1958, The Middle Ordovician of the Oslo region, Norway, pt. 10, Nautiloid cephalopods: Norsk Geologisk Tidsskrift, v. 18, no. 1, 178 p.

———1964, Nautiloidea-Orthocerida, *in* Moore, R.C., ed., Treatise on invertebrate paleontology, pt. K, Mollusca 3, Cephalopoda: Lawrence, Geological Society of America and University of Kansas Press, p. K216–K261.

———1979, Conodonts and conodont biostratigraphy of post-Tyrone Ordovician rocks of the Cincinnati region: U.S. Geological Survey Professional Paper 1066-G, 26 p.

———1984, Graphic correlation of upper Middle and Upper Ordovician rocks, North American Midcontinent Province, USA, *in* Bruton, D.L., ed., Aspects of the Ordovician System: Palaeontological Contributions from the University of Oslo, v. 295, p. 23–25.

Teichert, Curt, 1933, Der Bau der Actinoceroiden Cephalopoden: Palaeontolographica, sect. A, v. 78, p. 111–230.

———1940, Contributions to nautiloid nomenclature: Journal of Paleontology, v. 14, no. 6, p. 590–597.

———1964, Actinoceratoidra, *in* Moore, R.C., ed., Treatise on invertebrate paleontology, pt. K. Mollusca 3, Cephalopoda: Lawrence, Geological Society of America and University of Kansas Press, p. K190–K216.

———1970, Drifted *Nautilus* shells in the Bay of Bengal: Journal of Paleontology, v. 44, no. 6, p. 1129–1130.

———1988, Main features of cephalopod evolution, *in* Clarke, M.R., and Trueman, E.R., eds., The mollusca, v. 12 of Paleontology and neontology of cephalopods: San Diego, Calif., Academic, p. 11–79.

Teichert, Curt, and Glenister, B.F., 1953, Ordovician and Silurian cephalopods from Tasmania, Australia: Bulletins of American Paleontology, v. 34, no. 144, p. 187–248.

Teichert, Curt, and others, 1964, Descriptions of subclasses and orders, *in* Moore, R.C., ed., Treatise of invertebrate paleontology, pt. K, Mollusca 3, Cephalopoda: Lawrence, Geological Society of American and University of Kansas Press, p. K127–K466.

Titus, R., 1982, Fossil communities of the middle Trenton Group (Ordovician) of New York State: Journal of Paleontology, v. 54, no. 2, p. 477–485.

Titus, R., and Cameron, B., 1976, Fossil communities of the lower Trenton Group (Middle Ordovician) of central and northwestern New York State: Journal of Paleontology, v. 50, no. 6, p. 1209–1225.

Tobin, R.C., 1982, A model for cyclic deposition in the Cincinnatian Series of southwestern Ohio, northern Kentucky, and southeastern Indiana: Cincinnati, Ohio, University of Cincinnati, Ph.D. dissertation, 483 p.

Toriyama, R., Sada, T., and Komalarjun, P., 1965, *Nautilus pompilius* drifts on the west coast of Thailand: Japanese Journal of Geology and Geography, v. 36, p. 149–161.

Troedsson, G.T., 1926, On the Middle and Upper Ordovician faunas of northern Greenland, pt. I, Cephalopods: Meddelelser om Gronland, v. 71, p. 1–147.

———1932, Studies of Baltic fossil cephalopods, pt. II, Vertically striated or fluted orthoceracones in the *Orthoceras* limestone: Lund Universitets Arsskrift, v. 28, no. 6, p. 1–38.

Votaw, R.B., 1980, Middle and Upper Ordovician conodonts from the Upper Peninsula of Michigan, *in* Votaw, R.B., ed., Ordovician and Silurian geology of the northern Peninsula of Michigan: Michigan Basin Geological Society Field Excursion, 1980, p. 18–20.

Wade, Mary, 1988, Nautiloids and their descendants—cephalopod classification in 1986, *in* Wolberg, D.L., ed., Contributions to Paleozoic paleontology and stratigraphy in honor of Rousseau H. Flower: New Mexico Bureau of Mines and Mineral Resources Memoir 44, p. 15–25.

Wahlman, G.P., 1992, Middle and Upper Ordovician symmetrical univalved mollusks (Monoplacophora and Bellerophontina) of the Cincinnati arch region: U.S. Geological Survey Professional Paper 1066–O, 213 p.

Walker, L.G., 1982, The brachiopod genera *Hebertella*, *Dalmanella*, and *Heterorthina* from the Ordovician of Kentucky: U.S. Geological Survey Professional Paper 1066–M, 30 p.

Ward, P.D., 1987, The natural history of Nautilus: Boston, Mass., Allen & Unwin, 267 p.

Warshauer, S.M., and Berdan, J.M., 1982, Palaeocopid and podocopid Ostracoda from the Lexington Limestone and Clays Ferry Formation (Middle and Upper Ordovician) of central Kentucky: U.S. Geological Survey Professional Paper 1066–H, 80 p.

Weir, G.W., Peterson, W.L., and Swadley, W.C., 1984, Lithostratigraphy of the Upper Ordovician strata exposed in Kentucky, *with a section on* biostratigraphy by John Pojeta: U.S. Geological Survey Professional Paper 1151–E, 121 p. [1985]

Westermann, G.E.G., 1973, Strength of concave septa and depth limits of fossil cephalopods: Lethaia, v. 6, p. 383–403.

Willman, H.B., and Kolata, D.R., 1978, The Platteville and Galena Groups in northern Illinois: Illinois State Geological Survey Circular 502, 75 p.

Wilson, A.E., 1961, Cephalopoda of the Ottawa Formation of the Ottawa–St. Lawrence Lowland: Geological Survey of Canada Bulletin 67, 106 p.

Wilson, C.W., 1949, Pre-Chattanooga stratigraphy in central Tennessee: Tennessee Division of Geology Bulletin 56, 407 p.

Winchell, N.H., and Schuchert, Charles, 1895, The Lower Silurian Brachiopoda of Minnesota: Minnesota Geological and Natural History Survey Final Report, v. 3, pt. 1, p. 333–474.

Witzke, B.J., 1980, Middle and Upper Ordovician paleogeography of the region bordering the Transcontinental arch, *in* Fouch, T.D., and Magathan, E.R., eds., Paleozoic paleogeography of the West-Central United States—Paleogeography Symposium 1: Society of Economic Paleontologists and Mineralogists, Rocky Mountain Section, p. 1–18.

———1990, Climatic constraints for Palaeozoic palaeolatitudes of Laurentia and Euramerica, *in* McKerrow, W.S., and Scotese, C.R., eds., Palaeozoic palaeogeography and biogeography: Geological Society Memoir 12, p. 57–73.

Young, F.P., 1943, Black River stratigraphy and faunas: American Journal of Science, v. 241, p. 141–166, 209–240.

Zhuravleva, F.A., 1978, Devonian Orthocerids, superorder Orthoceratoidea: Trudy Paleontologii Instytut Akademiya Nauk SSSR, v. 168, 223 p.

Zinsmeister, W.J., 1987, Unusual nautilid occurrence in the upper Eocene La Meseta Formation, Seymour Island, Antarctica: Journal of Paleontology, v. 61, p. 724–726.

APPENDIX: LOCALITY REGISTER

During the geologic mapping of the State of Kentucky by the Kentucky and U.S. Geological Surveys, Ordovician fossils were collected from nearly 1,100 localities. Pojeta (1979) listed 317 of these localities from which Ordovician fossils were collected and studied. Below is a list of 18 USGS localities from which some of the fossils studied for this report were collected. The localities are arranged sequentially by collection number assigned to each in the USGS register of Cambrian and Ordovician (CO) localities. The following abbreviations are used in this locality listing: C.n., collection number; G.l., geographic locality; F., formation; S.p., stratigraphic position; Q.n., 7.5 min quadrangle name; S.n., section name. Not all localities have information in all categories.

U.S. Geological Survey Localities

C.n. - **4073-CO.**
G.l. - 0.75 mi east of Milner, Ky., turn right (southwest) on Shyrock Ferry Road. Continue down road about 1 mi to top of section.
F. - Grier Limestone Member, Lexington Limestone.
S.p. - 148 ft above the base of the Curdsville Limestone Member, Lexington Limestone.
Q.n. - Tyrone, Ky.
S.n. - Tyrone A.

C.n. - **4865-CO.**
G.l. - U.S. Route 62 at crossing of Kentucky River; from east side of Blackburn Memorial Bridge, take side road north 0.1 mi to railroad cut exposure.
F. - Logana Member, Lexington Limestone.
S.p. - Lower 12 ft of Logana Member.
Q.n. - Tyrone, Ky.
S.n. - Tyrone C, or Ky. Utilities Plant.

C.n. - **4879-CO.**
G.l. - Road exposures between Trinity Church and Antioch Church, southwest quadrant, Valley View quadrangle, Kentucky, 1 mi southwest of northern termination of Kentucky Route 595 at Kentucky River.
F. - Grier Limestone Member, Lexington Limestone.
S.p. - 43–45 ft above base of section.
Q.n. - Valley View, Ky.
S.n. - Valley View C (Antioch Church Road).

C.n. - **4883-CO.**
G.l. - Same as for 4879-CO.
F. - Grier Limestone Member, Lexington Limestone.
S.p. - 60.5 ft above base of section.
Q.n. - Valley View, Ky.
S.n. - Valley View C (Antioch Church Road).

C.n. - **5067-CO.**
G.l. - Hillside pasture exposure on west side of U.S. Route 227, 0.4 mi south of intersection of Ford-Hampton Road and U.S. 227, north of Ford, Ky.
F. - Grier Limestone Member, Lexington Limestone.
S.p. - 140 ft above the top of the Tyrone Limestone.
Q.n. - Ford, Ky.

C.n. - **5072-CO.**
G.l. - Immediately behind Old Crow Distillery, northern bluff of Glenns Creek Road, 0.7 mi west of intersection with Hanley Lane.
F. - Lower part of the Lexington Limestone; mostly basal Curdsville Limestone Member, but with some float blocks of Logana Member in the sample.
Q.n. - Frankfort East, Ky.
S.n. - Frankfort East B.

C.n. - **5073-CO.**
G.l. - Northwest of Old Crow Distillery on first sharp north bend of Glenns Creek Road, 0.9 mi west of intersection with Hanley Lane.
F. - Upper part of the Logana Member, Lexington Limestone.
S.p. - Just below 4-ft-thick dalmanellid coquina.
Q.n. - Frankfort East, Ky.
S.n. - Frankfort East B.

C.n. - **5092-CO.**
G.l. - From east side of Blackburn Memorial Bridge crossing Kentucky River, take first road north towards Kentucky Utilities Plant; section in railroad and road exposures.
F. - Logana Member, Lexington Limestone.
S.p. - 10–12 ft above base of Logana Member.
Q.n. - Tyrone, Ky.
S.n. - Tyrone C, or Ky. Utilities Plant.

C.n. - **5096-CO.**
G.l. - Just before ferry crossing Tates Creek Road and Kentucky River, take road west toward Daniel Boone YMCA Camp; collection from near top of northern bluff of Kentucky River, 0.4 mi southeast of YMCA Camp, in abandoned railroad bed.
F. - Grier Limestone Member, Lexington Limestone.
Q.n. - Valley View, Ky.

C.n. - **5100-CO.**
G.l. - Crisman Mill Road, 0.2 mi west of Hickman Creek crossing.
F. - Lower part of Curdsville Limestone Member, Lexington Limestone.

Q.n. - Little Hickman, Ky.
S.n. - Little Hickman B.

C.n. - **6034-CO.**
G.l. - Section on New Watts Mill Road, 0.1 mi southwest of intersection with Kentucky Route 39.
F. - Tyrone Limestone.
S.p. - 112 ft above the base of section.
Q.n. - Little Hickman, Ky.
S.n. - Little Hickman A.

C.n. - **6127-CO.**
G.l. - Southernmost tributary of Landing Run, 2,000 ft east of intersection of U.S. Route 31E and Kentucky Route 46, between Balltown and Culvertown, Ky.
F. - Rowland Member, Drakes Formation.
Q.n. - New Haven, Ky.

C.n. - **6419-CO.**
G.l. - 1.4 mi east of Wilmore, Ky., 500 ft north of main road.
F. - Lower part of the Logana Member, Lexington Limestone.
Q.n. - Wilmore, Ky.

C.n. - **7310-CO.**
G.l. - 1700 ft east of Kentucky Route 89, 1.5 mi southeast of Winchester, Ky.
F. - Strodes Creek Member, Lexington Limestone.
Q.n. - Winchester, Ky.

C.n. - **7328-CO.**
G.l. - Roadcut along Maple Street, Winchester, Ky.; 2.6 mi north of I-64.
F. - Strodes Creek Member, Lexington Limestone.
Q.n. - Austerlitz, Ky.

C.n. - **7353-CO.**
G.l. - 1 mi northeast of junction of Cook Road and Lair Road in stock pond.
F. - Lower tongue of Millersburg Member, Lexington Limestone.
Q.n. - Shawhan, Ky.

C.n. - **7471-CO.**
G.l. - Section along road to U.S. Route 25, 0.4 mi west of Sadieville, Ky.
F. - Clays Ferry Formation.
S.p. - 97 ft above base.
Q.n. - Sadieville, Ky.

C.n. - **7791-CO.**
G.l. - Road exposure on west side of Kentucky River at bridge crossing Central Kentucky Parkway on northern side of parkway.
F. - Lower part of the Logana Member, Lexington Limestone.

S.p. - 35 ft above Tyrone Limestone-Lexington Limestone contact.
Q.n. - Salvisa, Ky.
S.n. - Salvisa B.

Non-USGS Collections

C.n. - **KY-1.**
G.l. - Roadcut on north side of Orphanage Road (State Route 371), 0.25 mi west of intersection with Kentucky Route 17, just north of I-275E, Fort Mitchell, Kenton County, Ky.
F. - Kope Formation.
S.p. - Clay shale bed 54 ft above the road level.
Q.n. - Covington, Ky.

C.n. - **KY-2.**
G.l. - Road exposures at junction of Kentucky Routes 8 and 445, below and just west of I-275E bridge over Ohio River, Fort Thomas, Campbell County, Ky.
F. - Kope Formation.
S.p. - Talus blocks at base of exposure.
Q.n. - Newport, Ky.

C.n. - **KY-3.**
G.l. - 0.5 mi north of I-275E at intersection of Taylor Mill Pike (Kentucky Route 16) and Riedlin (Mason) Road, southeast roadcut at curve in road, Forest Hills, Kenton County, Ky.
F. - Upper part of the Kope Formation and basal Fairview Formation.
S.p. - Thin packstone bed 12 ft above Kope Formation-Fairview Formation contact.
Q.n. - Covington, Ky.

C.n. - **KY-4.**
G.l. - Belk Island in the Cumberland River, near Rowena, Russell County, Ky. (Shideler locality 3.51B; now flooded by Lake Cumberland).
F. - Leipers Limestone.
S.p. - *Tetradium* biostrome exposed along river bank.
Q.n. - Jamestown, Ky.

C.n. - **KY-5.**
G.l. - Road exposure south of I-64 at exit ramp to Kentucky Route 60, 1 mi southeast of Owingsville, Bath County, Ky.
F. - Upper part of Bull Fork Formation.
S.p. - 50 ft below contact with Preachersville Member of the Drakes Formation.
Q.n. - Colfax, Ky.

C.n. - **IND-1.**
G.l. - Streambank exposures along Harveys Branch of Salt Creek, just north of Oldenburg, Franklin County, Ind.
F. - Bull Fork Formation.
S.p. - Upper 25 ft of the "Waynesville biofacies" and lower 15 ft of the "Liberty biofacies."
Q.n. - Batesville, Ind.

C.n. - **IND-2.**
G.l. - Exposures on the west bank of the spillway for Brookville Lake, just west of Brookville Dam Road and north of Indiana Route 101, north of Brookville, Franklin County, Ind.
F. - Bull Fork Formation.
S.p. - "Waynesville and Liberty biofacies."
Q.n. - Whitcomb, Ind.

C.n. - **IND-3.**
G.l. - Road exposure on the west side of Indiana Route 101 at the north end of the valley at the crest of Garr Hill, 5.0 mi north of Brookville, Franklin County, Ind.
F. - Tongue of the Saluda Formation and overlying Whitewater Formation.
S.p. - *Tetradium* biostrome and overlying shales and micritic limestone beds.
Q.n. - Whitcomb, Ind.

C.n. - **IND-4.**
G.l. - Streambank exposures along Bull Fork of Salt Creek at the Sayers Farm, 2 mi northeast of Hamburg, Franklin County, Ind.
F. - Saluda Formation.
S.p. - "Concrete Layer," 18-in-thick resistant micritic limestone bed 15 ft above the *Tetradium* biostrome.
Q.n. - Clarksburg, Ind.

C.n. - **IND-5.**
G.l. - Road exposure on the south side of Indiana Route 50 at the east edge of Versailles, Ripley County, Ind.
F. - Saluda Formation.
S.p. - Slabby micritic limestone beds above the main *Tetradium* biostrome.
Q.n. - Milan, Ind.

C.n. - **IND-6.**
G.l. - Road exposures near the top of the hill along U.S. Route 421, 3.1 mi north of the intersection with Indiana Route 56 in Madison, Jefferson County, Ind.
F. - Saluda Formation and its Hitz Limestone Member.
Q.n. - Madison-West, Ind.

C.n. - **IND-7.**
G.l. - Road exposures on both sides of U.S. Route 27, 1.0 mi south of Richmond, Wayne County, Ind.
F. - Whitewater Formation.
S.p. - Upper 30 ft of the rubbly facies of the Whitewater Formation (*Rhynchotrema dentatum* zone).
Q.n. - Richmond, Ind.

C.n. - **OH-1.**
G.l. - Stream exposures along West Fork, west of Mill Creek, just above dairy and concrete retaining wall, near Cumminsville, western Cincinnati, Hamilton County, Ohio.
F. - Kope Formation.
Q.n. - Cincinnati-West, Ohio.

C.n. - **OH-2.**
G.l. - Road exposures on the east side of Ohio Route 125 in the Whiteoak Creek valley, 0.8 mi west of junction with U.S. Route 68 in Georgetown, Brown County, Ohio.
F. - Fairview Formation.
Q.n. - Higginsport, Ohio-Ky.

C.n. - **OH-3.**
G.l. - Streambank exposures along Taylors Creek, near Dent, Hamilton County, Ohio (Shideler locality 1.22A27).
F. - Fairview Formation.
S.p. - Basal 15 ft.
Q.n. - Addyston, Ohio.

C.n. - **OH-4.**
G.l. - Stonelick Creek, just south of Route 131 bridge, 2 mi west of Newtonsville, Clermont County, Ohio.
F. - Grant Lake Formation.
S.p. - Corryville Member of the Grant Lake Formation.
Q.n. - Newtonsville, Ohio.

C.n. - **OH-5.**
G.l. - North side of Conrail Railroad tracks, 500 ft south of I–75 and south of Maud, Butler County, Ohio.
F. - Grant Lake Formation.
S.p. - Mount Auburn Member of the Grant Lake Formation.
Q.n. - Glendale, Ohio.

C.n. - **OH-6.**
G.l. - Road exposure on the west side of U.S. Route 42, 2 mi northeast of Waynesville, Warren County, Ohio.
F. - "Waynesville biofacies" in the Bull Fork Formation.
S.p. - *Treptoceras duseri* shale unit.
Q.n. - Waynesville, Ohio.

C.n. - **OH-7.**
G.l. - Excavation for spillway (now under water) at junction of Caesars Creek and Flat Fork Creek, plus exposures along emergency spillway of Caesars Creek Lake, off Clarksville Road, Warren County, Ohio.
F. - "Waynesville and Liberty biofacies" in the Bull Fork Formation.
S.p. - Section from *Treptoceras duseri* shale unit to the upper *Thaerodonta clarksvillensis* zone.
Q.n. - Oregonia, Ohio.

C.n. - **OH-8.**
G.l. - Streambank exposures along the north and south forks of Harpers Run at the Whip-Poor-Will Girl Scout Camp, south of Stout Road and west of Middleboro Road, Warren County, Ohio.
F. - "Waynesville biofacies" in the Bull Fork Formation.
S.p. - *Treptoceras duseri* shale unit.
Q.n. - Oregonia, Ohio.

C.n. - **OH-9.**
G.l. - Streambank exposures along Stony Hollow, 1,000 ft north of Todds Fork, just north of Clarksville, Clinton County, Ohio.
F. - "Waynesville biofacies" in the Bull Fork Formation.
S.p. - *Treptoceras duseri* shale unit.
Q.n. - Clarksville, Ohio.

C.n. - **OH-10.**
G.l. - Exposures on the north side of spillway at Cowan Lake, just west of the Ohio Route 730 bridge, 5 mi southwest of Wilmington, Clinton County, Ohio.
F. - Bull Fork Formation.
S.p. - Upper 40 ft of the formation just below contact with Preachersville Member of the Drakes Formation.
Q.n. - Clarksville, Ohio.

C.n. - **OH-11.**
G.l. - Small road exposure at junction of U.S. Route 22 and Ohio Route 380, just west of Todds Fork, Clinton County, Ohio.
F. - "Liberty biofacies" in the Bull Fork Formation.
Q.n. - Clarksville, Ohio.

C.n. - **OH-12.**
G.l. - Railroad exposure on the north edge of Wright Brothers Memorial Park, just south of Ohio Route 444 and Huffman Dam, Greene County, Ohio.
F. - Bull Fork Formation.
S.p. - Upper part of the Bull Fork Formation 45 ft below contact with Preachersville Member of the Drakes Formation.
Q.n. - Fairborn, Ohio.

C.n. - **OH-13.**
G.l. - Small stream flowing southwest into the Stillwater River at Cricket Holler Boy Scout Camp, northwest of Dayton, Montgomery County, Ohio.
F. - Bull Fork Formation.
Q.n. - Dayton-North, Ohio.

C.n. - **OH-14.**
G.l. - Stream exposures along Collins Run, 0.25 mi upstream and north of Pffeifer Park, 1.5 mi south of the Oxford, just west of U.S. Route 27, Butler County, Ohio.
F. - "Waynesville and Liberty biofacies" in the Bull Fork Formation.
S.p. - "Waynesville biofacies" from the "Trilobite shale unit" to the contact with the overlying "Liberty biofacies."
Q.n. - Millville, Ohio.

C.n. - **OH-15.**
G.l. - Addisons Creek, southeast flowing tributary of Four-Mile Creek, just south of the spillway for Acton Lake, 3 mi north of Oxford, Butler County, Ohio.
F. - Bull Fork Formation.
S.p. - Upper 15 ft of the "Waynesville biofacies" and the lower 30 ft of the "Liberty biofacies."
Q.n. - Oxford, Ohio.

INDEX

A

Abstract ... 1
Acknowledgments ... 3
Actinoceras ... 81
   altopontense ... 82
   curdsvillense ... 84
   kentuckiense ... 82
Actinoceratidae ... 80
Actinoceratoidea ... 23, 80
Actinocerida ... 80
albersi, Isorthoceras ... 60
Allumettoceras cf. *A. tenerum* ... 103
altopontense, Actinoceras ... 82
amoenum, Beloitoceras ... 97
amplicameratum, Ordogeisonoceras ... 38, 39
Anaspyroceras cf. *A. cylindricum* ... 66
annularis, Monomuchites ... 56
apical angle ... 24
Arctic nautiloid fauna ... 10, 19
Ashlock Formation ... 9
Augustoceras ... 110
   commune ... 111
   shideleri ... 111

B

Baltoceratidae ... 27
Bardstown Member, Drakes Formation ... 9
Barrandeocerina ... 92
Beloitoceras ... 97
   amoenum ... 97
   bucheri ... 98
   huronense, cf. *B.* ... 99
Black River Group ... 12
Black River Formation ... 12
Blackriver nautiloid fauna ... 10, 12
Brannon Member, Lexington Limestone ... 8
Brassfield Limestone ... 20
brevicameratum, Richmondoceras ... 71
bucheri, Beloitoceras ... 98
Bull Fork Formation ... 5, 9
byrnesi, Treptoceras ... 51

C

Calloway Creek Limestone ... 9
cameral deposits ... 24
cameral length ... 24
Cameroceras ... 73
   rowenaense ... 74
   sp. ... 76
   trentonense ... 76
Camp Nelson Limestone, High Bridge Group ... 6
carletonense, Plectoceras cf. *P.* ... 93
Cartersoceras ... 28
   popei ... 30
   shideleri ... 28
Catheys Formation ... 9
central canal ... 81

Champlainian Provincial Series, Middle
   Ordovician ... 5, 6
Cincinnatian nautiloid fauna ... 10
Cincinnatian Provincial Series, Upper
   Ordovician ... 4, 8
Cincinnati arch ... 4
cincinnatiensis, Treptoceras ... 52
clarksvillensis, Pleurorthoceras ... 37
Clays Ferry Formation ... 5, 8
commune, Augustoceras ... 111
costalis, Monomuchites cf. *M.* ... 55
cross-sectional shape (shell) ... 24
Cumberland Formation ... 9
Cumberland sag ... 5, 9
Curdsville Limestone Member, Lexington
   Limestone ... 7
curdsvillense, Actinoceras ... 84
curdsvillense, Deiroceras ... 87
cylindricum, Anaspyroceras cf. *A.* ... 66
Cynthiana Limestone ... 16

D

davisi, Triendoceras(?) ... 79
Decorah Shale/Decorah Group ... 13
Deicke Bentonite Bed ... 12, 15
Deicke volcanic event ... 15
Deiroceras curdsvillense ... 86
Diestoceras ... 104
   eos ... 108
   indianense ... 105
   shideleri ... 107
Diestoceratidae ... 104
Dillsboro Formation ... 5, 9
Distribution and Faunal Affinities of
   Ordovician Nautiloids in the Cincinnati
   Arch Region ... 10
Drakes Formation ... 9
duseri, Treptoceras ... 46
dyeri, Orthonybyoceras ... 89

E

Edenian Nautiloid Fauna ... 16
Edenian Stage, Upper Ordovician ... 8
Ellesmerocerida ... 27
Endoceratidae ... 72
Endoceratoidea ... 23, 72
Endocerida ... 72
endosiphuncular deposits ... 24, 28
eos, Diestoceras ... 108
Euorthoceras ... 46

F

faberi, Trocholites ... 90
Fairview Formation ... 5, 9
ferecentricum, Ormoceras ... 86
floweri, Pojetoceras ... 35
fosteri, Treptoceras ... 48\

G

Galena shelf ... 4
Gamachian Stage, Upper Ordovician ... 20
Geisonoceras rivale ... 39
Geisonoceratidae ... 38
gorbyi, Gorbyoceras ... 63
Gorbyoceras ... 62
   gorbyi ... 63
   hammelli ... 64
   tetreauense ... 65
Grant Lake Formation/Grant Lake Limestone ... 5, 9
Grier Limestone Member, Lexington
   Limestone ... 7
gyroforme, Manitoulinoceras ... 112

H

halli, Plectoceras ... 94
hammelli, Gorbyoceras ... 64
High Bridge Group, Middle Ordovician ... 4, 6
Hitz Limestone Bed ... 5

I

huronense, Beloitoceras cf. *B.* ... 99
indianense, Diestoceras ... 105
Introduction ... 2
Isorthoceras ... 59
   albersi ... 60
   rogersensis ... 61
   sociale ... 61
   strangulatum ... 59
   strigatum ... 61

J

Jessamine dome ... 5

K

kentuckiense, Actinoceras ... 82
Kirkfieldian Stage, Middle Ordovician ... 5
Kope Formation ... 5, 8

L

Laphamoceras cf. *L. scofieldi* ... 101
Late Ordovician Extinction Event ... 20
Leipers Limestone ... 5, 9, 19
Leray-Rockland beds, Middle Ordovician ... 12
Lexington Limestone ... 4, 5, 6
Lexington platform ... 3, 4
Locality Register ... 120
Logana Member, Lexington Limestone ... 7, 99

M

Maelonoceras cf. *M. praematurum* ... 100
major, Oncoceras ... 95
Manitoulinoceras gyroforme ... 100

Entry	Page
Maysville Group, Upper Ordovician	3
Maysvillian and Early Richmondian Nautiloid Faunas	17
Maysvillian Stage, Upper Ordovician	8, 9
Michelinoceratidae	33
Middle Ordovician Champlainian Provincial Series Lithostratigraphy in Kentucky	6
Middle to Late Richmondian Nautiloid Fauna	19
Millersburg Member, Lexington Limestone	8
Millbrig Bentonite Bed	15
Millbrig volcanic event	15
Monomuchites	54
annularis	56
costalis cf.	55
obliquum	58
Mud Cave bentonite bed	6
murrayi, Murrayoceras cf. *M.*	31
Murrayoceras cf. *M. murrayi*	31

N

Nautiloid Biostratinomy	21
Nautiloid Classification	23
Nautiloidea	23, 90
Nautiloid Fauna of the Cincinnatian Provincial Series	16
Nautiloid Fauna of the Leipers Limestone	5, 19
Nautiloid Fauna of the Lexington Limestone	5, 15, 16
Nautiloid Fauna of the Tyrone Limestone	5, 10
Nautiloid Taphonomy	21
Non-silicified Nautiloids	22

O

obliquum, Monomuchites	58
Oncoceras	94
major	95
sp.	96
Oncoceratidae	94
Oncocerida	94
Ordogeisonoceras amplicameratum	38, 39
Ordovician Lithostratigraphy in the Cincinnati Arch Region	4
Oregon Formation, High Bridge Group	6
Ormoceras ferecentricum	86
Ormoceratidae	85
Orthoceras regularis	34
Orthoceratidae	33
Orthoceratoidea	23, 27
Orthocerida	23, 32
Orthonybyoceras dyeri	89

P

Entry	Page
Pencil Cave bentonite bed	6, 12
Platteville Formation/Group, Middle Ordovician	12
Plectoceras	92
carletonense, cf. *P.*	93
halli	94
Plectoceratidae	92
Pleurorthoceras	36
clarksvillense	37
selkirkense	37
Point Pleasant Tongue, Clays Ferry Formation	8, 33
Pojetoceras floweri	35, 38
Polygrammoceras	67
sp. A	68
sp. A, cf. *P.*(?)	68
sp. B	69
popei, Cartersoceras	30
praematurum, Maelonoceras cf. *M.*	100
Preachersville Member, Drakes Formation	9
Proteoceras tyronensis	43
Proteoceratidae	41
Pseudoceratidae	58

Q

Queenston delta	3

R

radial canal(s)	81
References Cited	113
Regional Ordovician Paleogeographic Setting	3
regularis, Orthoceras	34
Richmondian nautiloid fauna	10
Richmondian Stage, Upper Ordovician	8, 9
Richmondoceras brevicameratum	71
rivale, Geisonoceras	39
Rocklandian Nautiloid Extinction	12
Rocklandian Stage, Middle Ordovician	5
rogersensis, Isorthoceras	61
rowenaense, Cameroceras	74
Rowland Member, Drakes Formation	9

S

Saluda Dolomite Member, Drakes Formation	10
Saluda Formation	5
scofieldi, Laphamoceras cf. *L.*	101
Sebree trough	4, 7
segment compression ratio (SCR)	24
segment height	24
segment length	24
segment position ratio (SPR)	24
selkirkense, Pleurorthoceras	37
shell ornament	24
shideleri, Augustoceras	111
shideleri, Cartersoceras	28
shideleri, Diestoceras	107
Silicified Nautiloids	22
siphuncle segment measurements	24
sociale, Isorthoceras	61
Species Problem in Nautiloid Systematics	23
strangulatum, Isorthoceras	59
strigatum, Isorthoceras	61
Strodes Creek Member, Lexington Limestone	8
Systematic Descriptions	24
Systematic Paleontology	27

T

Taconic Landmass	3, 8
Tanglewood bank	8
Tanglewood Limestone Member, Lexington Limestone	8
Tarphycerida	90
Tarphycerina	90
tenerum, Allumettoceras cf. *A.*	103
tetreauense, Gorbyoceras cf. *G.*	65
transversum, Treptoceras	53
trentonense, Cameroceras	76
Trenton Group, Middle Ordovician	16
Trenton nautiloid fauna	10
Trenton shelf	4
Treptoceras	44
byrnesi	51
cincinnatiensis	52
duseri	46
fosteri	48
transversum	53
Triendoceras(?) *davisi*	79
Tripteroceratidae	102
Trocholites faberi	91
Trocholitidae	90
Tyrone Limestone, High Bridge Group	5, 6
tyronensis, Proteoceras	43

U

Upper Ordovician Cincinnatian Provincial Series Lithostratigraphy	8

V

Vaginoceras	77
sp. A	77
sp. B	78
Valcouroceratidae	109

W

Entry	Page
Watertown Limestone	6, 13
Whitewater Formation	10

PLATES 1–22

Contact photographs of the plates in this report are available, at cost, from U.S. Geological Survey Library, Federal Center, Denver, Colorado 80225.

PLATE 1

FIGURES 1–6, 11. *Cartersoceras shideleri* Flower, 1964a (p. 28).

    1, 2, 4, 5. Lateral, dorsal (×1), adoral, and adapical (×2) views of silicified phragmocone. Tyrone Limestone (Rocklandian), USGS 6034-CO. USNM 158631.

    3. Dorsal natural section exposing siphuncle rod (×1). Same collection as figure 1 above. USNM 158627.

    6. Dorsal natural section exposing siphuncle rod (×1). Same collection as figure 1 above. USNM 158633.

    11. Thin section (×3) of a partially silicified phragmocone. Same collection as figure 1 above. USNM 468656.

7–10. *Cartersoceras popei* new species (p. 30).

    7, 9. Dorsal views (×2, ×1) of silicified portion of phragmocone preserving siphuncular rod. Note dorsal groove. Same collection as figure 1 above. Holotype USNM 158625.

    8. Dorsal view of more adoral portion of phragmocone (×1). Same collection as figure 1 above. Paratype USNM 468664.

    10. Dorsal view showing siphuncular rod preserving expanded outline of siphuncle segments ×2. Note dorsal groove. Same collection as figure 1 above. USNM 158624.

12–17. *Murrayoceras* cf. *M. murrayi* (Billings), 1857 (p. 31).

    12, 13. Dorsal and ventral views of nearly complete silicified phragmocone (×1). Same collection as figure 1 above. USNM 468674.

    14, 15. Dorsal and adapical views (×1.5) of silicified specimen exposing siphuncular rod. Same collection as figure 1 above. USNM 468675.

    16. Dorsal view of silicified specimen exposing siphuncular rod (×1.5). Same collection as figure 1 above. USNM 468672.

    17. Dorsal view of weathered silicified specimen exposing siphuncle (×1.5). Same collection as figure 1 above. USNM 468673.

U.S. GEOLOGICAL SURVEY

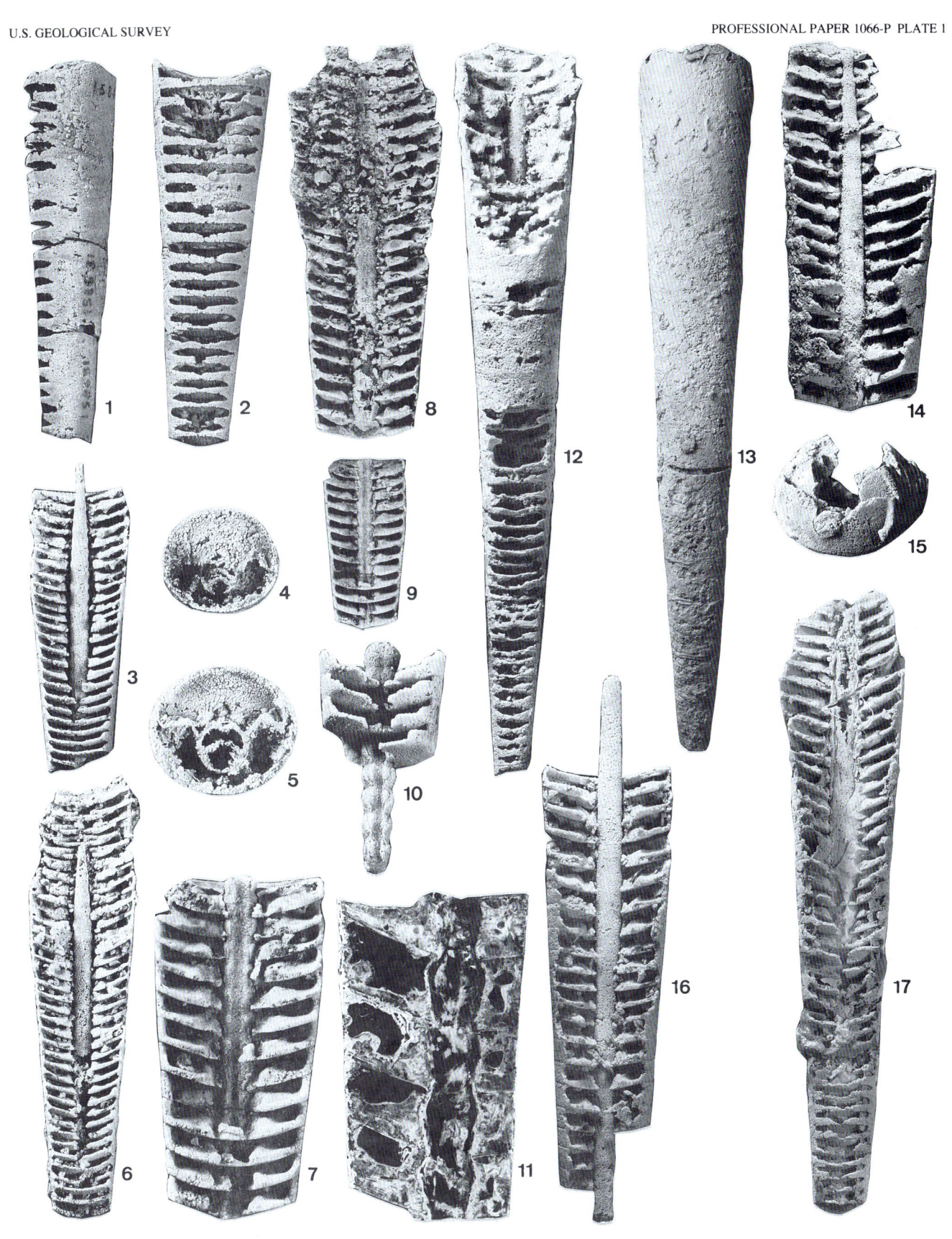

PROFESSIONAL PAPER 1066-P PLATE 1

CARTERSOCERAS AND MURRAYOCERAS

PLATE 2

FIGURES 1–9. *Pojetoceras floweri* new species (p. 35).
- 1, 9. Dorsal-ventral section of phragmocone, venter to right (×3.5, ×1). Note cameral deposits and thin ventral parietal deposits adapically. Tyrone Limestone (Rocklandian), USGS 6034-CO. Holotype USNM 468676.
- 2, 3. Adoral and exterior views of silicified paratype (×1). Same collection as figure 1 above. Paratype USNM 468677.
- 4, 6. External and adapical views of more adapical section of phragmocone (×1). Same collection as figure 1 above. Paratype USNM 468678.
- 5. External view of silicified adapical portion of phragmocone (×1). Same collection as figure 1 above. Paratype USNM 468679.
- 7. Incomplete fragment of phragmocone revealing cameral deposits (×1.5). Same collection as figure 1 above. Paratype USNM 468681.
- 8. Exterior of silicified adapical portion of phragmocone (×1). Same collection as figure 1 above. Paratype USNM 468683.

10–15. *Pleurorthoceras clarksvillense* (Foerste), 1924b (p. 37).
- 10. Ventral view of an internal mold showing impressions of cameral deposits (×1). Bull Fork Formation (Richmondian), Addisons Creek, near Oxford, Ohio (locality OH-15). MU 269T.
- 11. Nearly complete specimen holoperipherally encrusted by the bryozoan *Spatiopora* (×1). Bull Fork Formation (Richmondian), Harveys Branch, Oldenburg, Ind. (locality IND-1). MU 444T.
- 12. Naturally weathered internal mold of phragmocone exposing siphuncle (×1). Bull Fork Formation (Richmondian), near Blanchester, Ohio. Holotype USNM 48255.
- 13–15. Dorsal-ventral sections through adoral and adapical portions of a nearly complete phragmocone (×1), illustrating siphuncle segments and adapical development of cameral deposits. Same collection as figure 10 above. Same specimen illustrated by Flower (1962). MU 268T.

U.S. GEOLOGICAL SURVEY

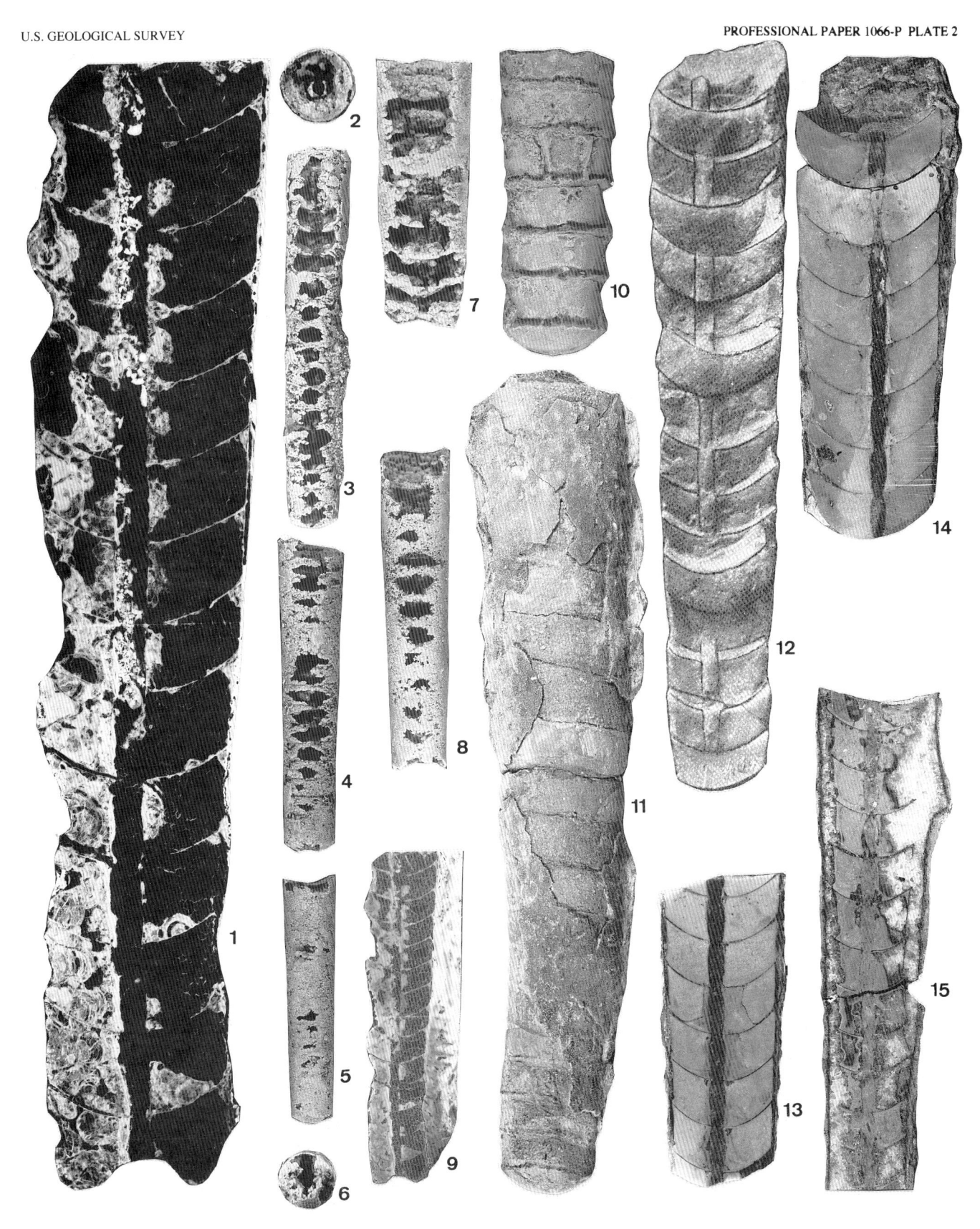

PROFESSIONAL PAPER 1066-P PLATE 2

POJETOCERAS AND *PLEURORTHOCERAS*

PLATE 3

FIGURES 1–8. *Ordogeisonoceras amplicameratum* (Hall), 1847 (p. 40).
    1–3. Dorsal-ventral sections exposing siphuncle (×1) and adoral view (×1), venter on right. "Trenton Limestone," Ludlow, Ky. Syntype of *Orthoceras ludlowense* Miller and Faber, UC 6457.
    4, 5, 8. Dorsal-ventral sections of a more adapical portion of phragmocone (×1) with magnified view of adapical piece showing cameral deposits and siphuncle segments (×2), venter to right. Kope Formation (Edenian), Cincinnati, Ohio. UC18229.
    6, 7. Exterior and adapical view of partially shelled internal mold (×1) showing faint longitudinal lirae and dorsal position of siphuncle. Trenton Limestone (Shermanian), Middleville, N.Y. Lectotype AMNH 29688.
9, 10. *Geisonoceras rivale* (Barrande), 1866 (p. 39).
Adoral view and dorsal-ventral section showing siphuncle and cameral deposits (×1), venter on left. Middle Silurian, Czechoslovakia. From Barrande (1866, pl. 209, figs. 2, 10).

U.S. GEOLOGICAL SURVEY

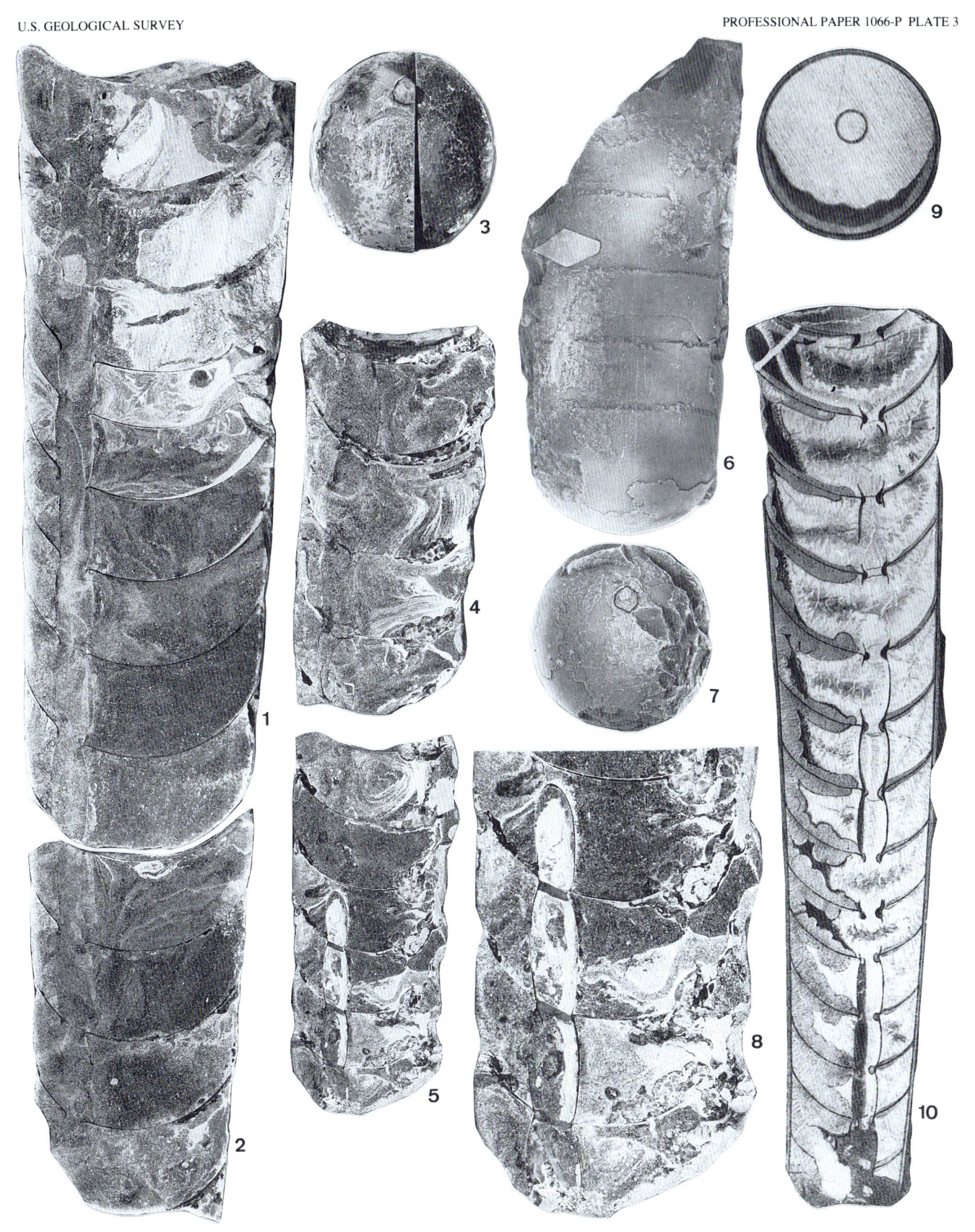

PROFESSIONAL PAPER 1066-P PLATE 3

ORDOGEISONOCERAS AND *GEISONOCERAS*

PLATE 4

FIGURES 1–7. *Proteoceras tyronensis* (Foerste), 1912 (p. 43).
    1–3. Dorsal and lateral (×1) and magnified ventral view exposing siphuncle (×2) of a silicified phragmocone. Tyrone Limestone (Rocklandian), USGS 6034-CO. USNM 158629.
    4. Adapical view of phragmocone showing cross section of endosiphuncular deposits (×1). Same collection as figure 1 above. USNM 468687.
    5, 7. Dorsal and adapical views (×1) of weathered silicified specimen exposing cameral deposits. Same collection as figure 1 above. USNM 468689.
    6. Dorsal view of silicified specimen exposing cameral deposits (×1). Same collection as figure 1 above. USNM 468688.

8–12. *Treptoceras duseri* (Hall and Whitfield), 1875 (p. 46).
    8–10. Dorsal and adapical views (×1.25) of adoral piece and dorsal-ventral section of adapical piece exposing siphuncle segments (×1.25), venter on right. "Hudson River Group," Waynesville, Ohio. Holotype UCal 34325.
    11. Dorsal-ventral section of adoral portion of phragmocone (×1), venter on left. Fairview Formation (Maysvillian), Dent, Ohio (locality OH-3). MU 29766.
    12. Dorsal-ventral section of a fragmentary specimen (×1), venter on right. Fairview Formation (Maysvillian), Taylor Mill, Ky. (locality KY-3). USMN 468727.

U.S. GEOLOGICAL SURVEY

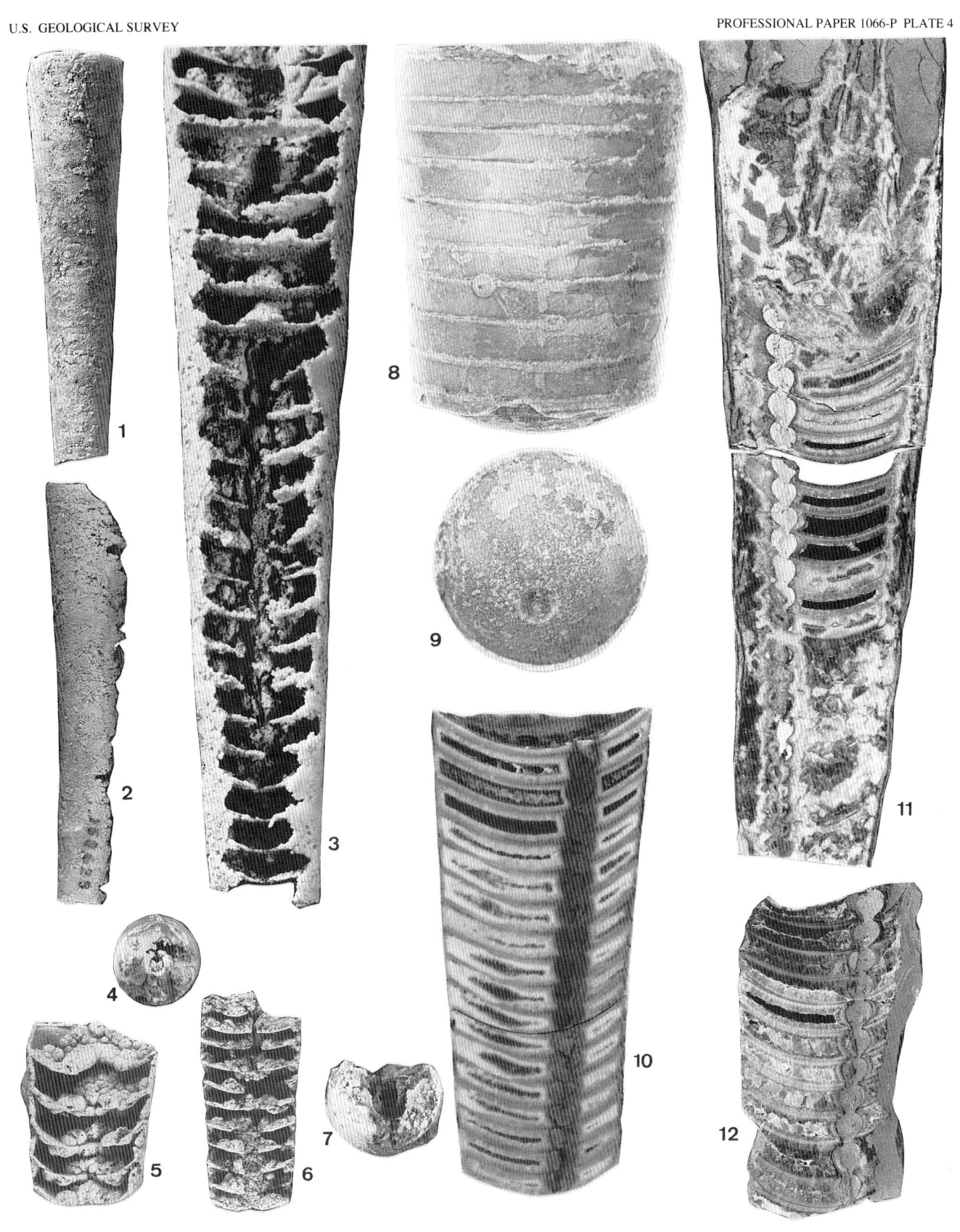

PROFESSIONAL PAPER 1066-P PLATE 4

PROTEOCERAS AND *TREPTOCERAS*

PLATE 5

FIGURES 1–6. *Treptoceras duseri* (Hall and Whitfield), 1875 (p. 46).
Dorsal-ventral sections through a nearly complete phragmocone (×1.25) plus a magnified view of adapical pieces showing endosiphuncular deposits (×2.5), venter on left. "*Treptoceras duseri* Shale," Bull Fork Formation (Richmondian), Warren County, Ohio (locality OH-8). MU 417T.

7–9. *Treptoceras fosteri* (Miller), 1875 (p. 48).
    7, 9. Dorsal-ventral section of a nearly complete phragmocone (×1) and magnified view of adapical end showing endosiphuncular deposits (×2), venter on left. Same collection as figure 1 above. MU 418T.
    8. Dorsal-ventral sections of less complete phragmocone (×1), venter on right. "Richmond Group," Clinton County, Ohio. Lectotype UC 356.

U.S. GEOLOGICAL SURVEY

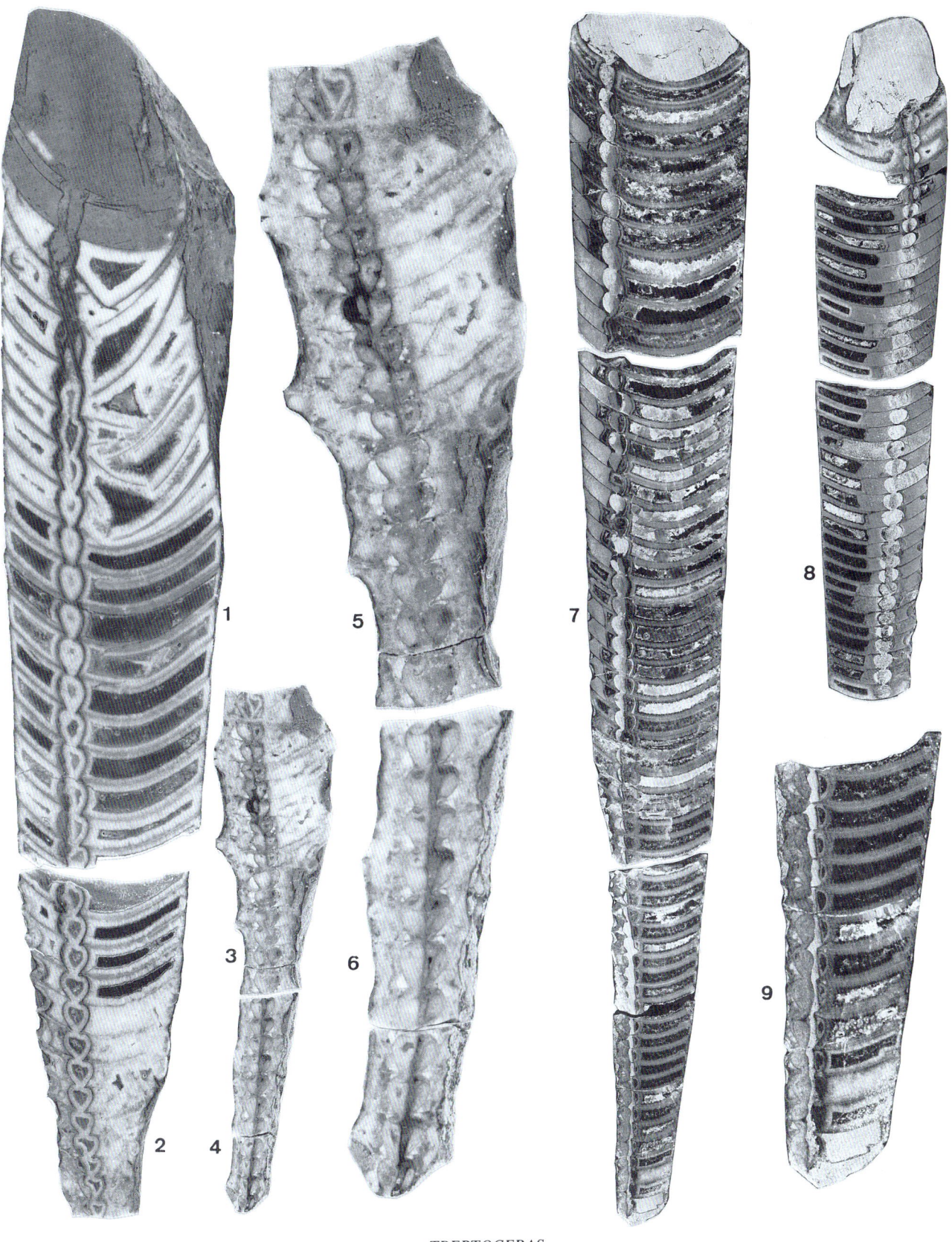

PROFESSIONAL PAPER 1066-P PLATE 5

TREPTOCERAS

PLATE 6

FIGURES 1–5, 7, 8. *Treptoceras fosteri* (Miller), 1875 (p. 48).
1. Exterior of specimen preserving much of the body chamber (×1). "*Treptoceras duseri* Shale," Bull Fork Formation (Richmondian), Waynesville, Ohio (locality OH-6). MU 419T.
2. Dorsal-ventral section of slightly crushed phragmocone (×1), venter on right. Bull Fork Formation (Richmondian), Oldenburg, Ind. (locality IND-1). UC 28389.
3. Dorsal view of shelled specimen preserving dorsal color bands (×1.5). Same collection as figure 1 above. MU 276T.
4. Internal mold preserving much of the body chamber (×1). "Liberty biofacies" in the Bull Fork Formation (Richmondian), Caesars Creek Lake, Ohio (locality OH-7). MU 420T.
5, 7. Adoral and adapical views of phragmocone showing position of siphuncle (×1). "*Treptoceras duseri* Shale," Bull Fork Formation (Richmondian), Clarksville, Ohio (locality OH-9). MU 421T.
8. Internal mold of phragmocone (×1). "Liberty biofacies" in Bull Fork Formation (Richmondian), Clinton County, Ohio (locality OH-11). MU 422T.
6. *Treptoceras byrnesi* (Miller), 1875 (p. 51).
Dorsal-ventral section through phragmocone revealing siphuncle segments (×1), venter on left. Grant Lake Formation (Maysvillian), Clermont County, Ohio (locality OH-4). MU 11035.

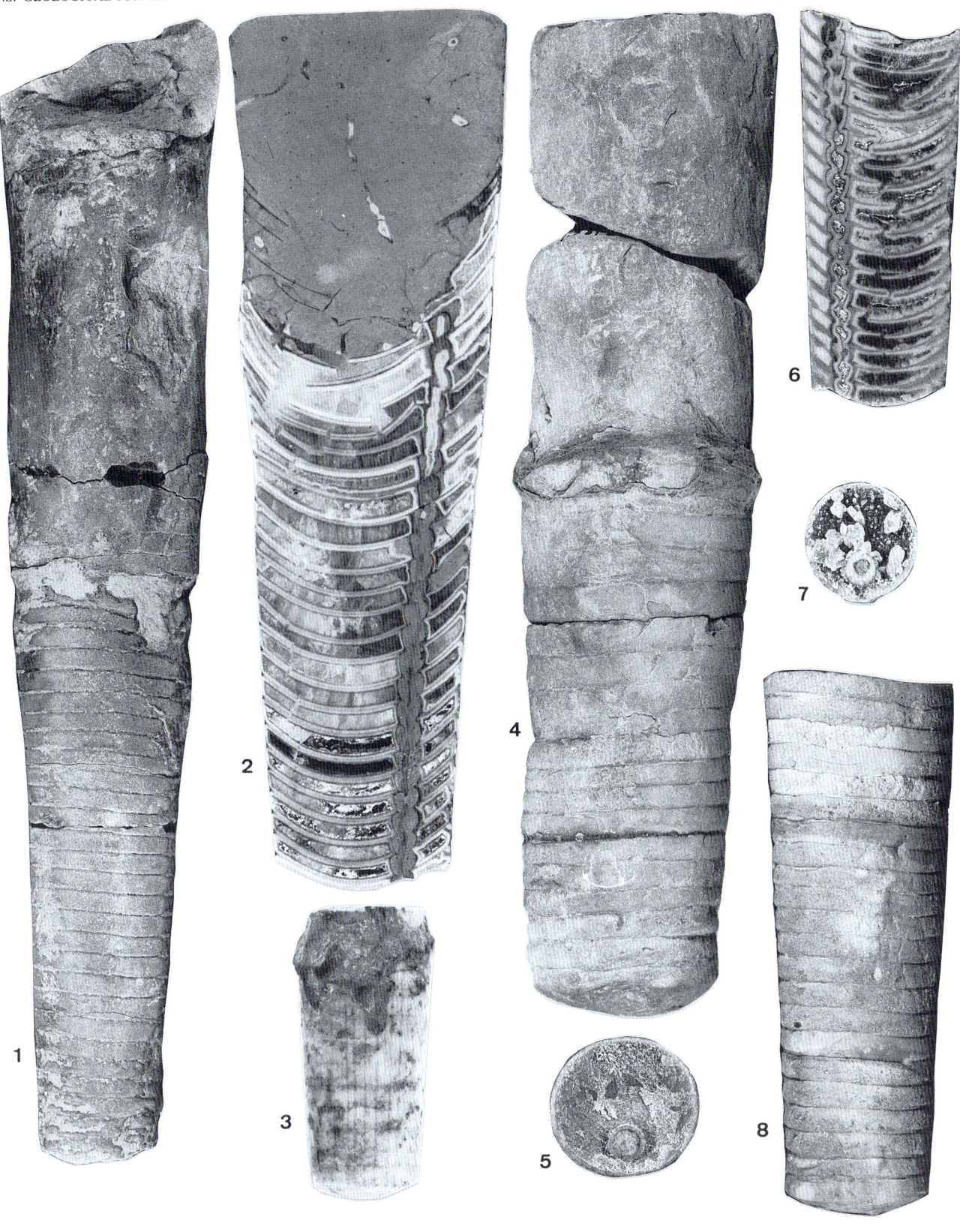

TREPTOCERAS

PLATE 7

FIGURES 1–8. *Treptoceras cincinnatiensis* (Miller), 1875 (p. 52).
- 1. Large specimen consisting of phragmocone and complete body chamber (×1). "*Treptoceras duseri* Shale," Bull Fork Formation (Richmondian), Clarksville, Ohio (locality OH-9). MU 423T.
- 2, 3, 5. Dorsal exterior, adoral view, and dorsal-ventral section revealing siphuncle segments (×1). Upper Ordovician, "hills behind Cincinnati, Ohio." Holotype MCZ 3404.
- 4, 6, 7. Dorsal-ventral section (×1), magnified view of apical end showing endosiphuncular deposits (×2), and adapical view showing position of siphuncle (×1). "*Treptoceras duseri* Shale," Bull Fork Formation (Richmondian), Warren County, Ohio (locality OH-8). MU 426T.
- 8. Dorsal-ventral section of a nearly complete phragmocone (×1), venter on left. Same collection as figure 4 above. MU425T.

U.S. GEOLOGICAL SURVEY

PROFESSIONAL PAPER 1066-P PLATE 7

1 2 3 4 5 6 7 8

TREPTOCERAS

PLATE 8

FIGURES 1–3. *Treptoceras cincinnatiensis* (Miller), 1875 (p. 52).
    1. Mold of body chamber holoperipherally encrusted by bryozoan *Spatiopora* (×1). "Liberty biofacies" in Bull Fork Formation (Richmondian), Brookville, Ind. (locality IND-2). MU 427T.
    2. Dorsal-ventral section through incomplete phragmocone exposing siphuncle (×1), venter on left. "Liberty biofacies" in Bull Fork Formation (Richmondian), Oxford, Ohio (locality OH-14). MU 425T.
    3. Internal mold of a large phragmocone weathered to expose siphuncle adapically (×1). "Waynesville biofacies" in Bull Fork Formation (Richmondian), Oxford, Ohio (locality OH-14). MU 443T.

4–9. *Treptoceras transversum* (Miller), 1875 (p. 53).
    4, 8. Dorsal-ventral section exposing siphuncle showing cameral and parietal deposits (×1) and a magnified view of the same (×2). Kope Formation (Edenian), Cincinnati, Ohio. Lectotype UC 1328C.
    5. Nearly complete shelled phragmocone showing transverse external ornament (×1). Kope Formation (Edenian), Ft. Thomas, Ky. (locality KY-2). USNM 468692.
    6. Dorsal-ventral section through phragmocone exposing siphuncle (×1). Same collection as figure 4 above. Paratype UC 1328B.
    7. Shelled fragment of phragmocone showing transverse ornament (×1). Same collection as figure 4 above. Paratype UC 1328A.
    9. Adorally crushed phragmocone holoperipherally encrusted by bryozoa (×1). Same collection as figure 5 above. USNM 468693.

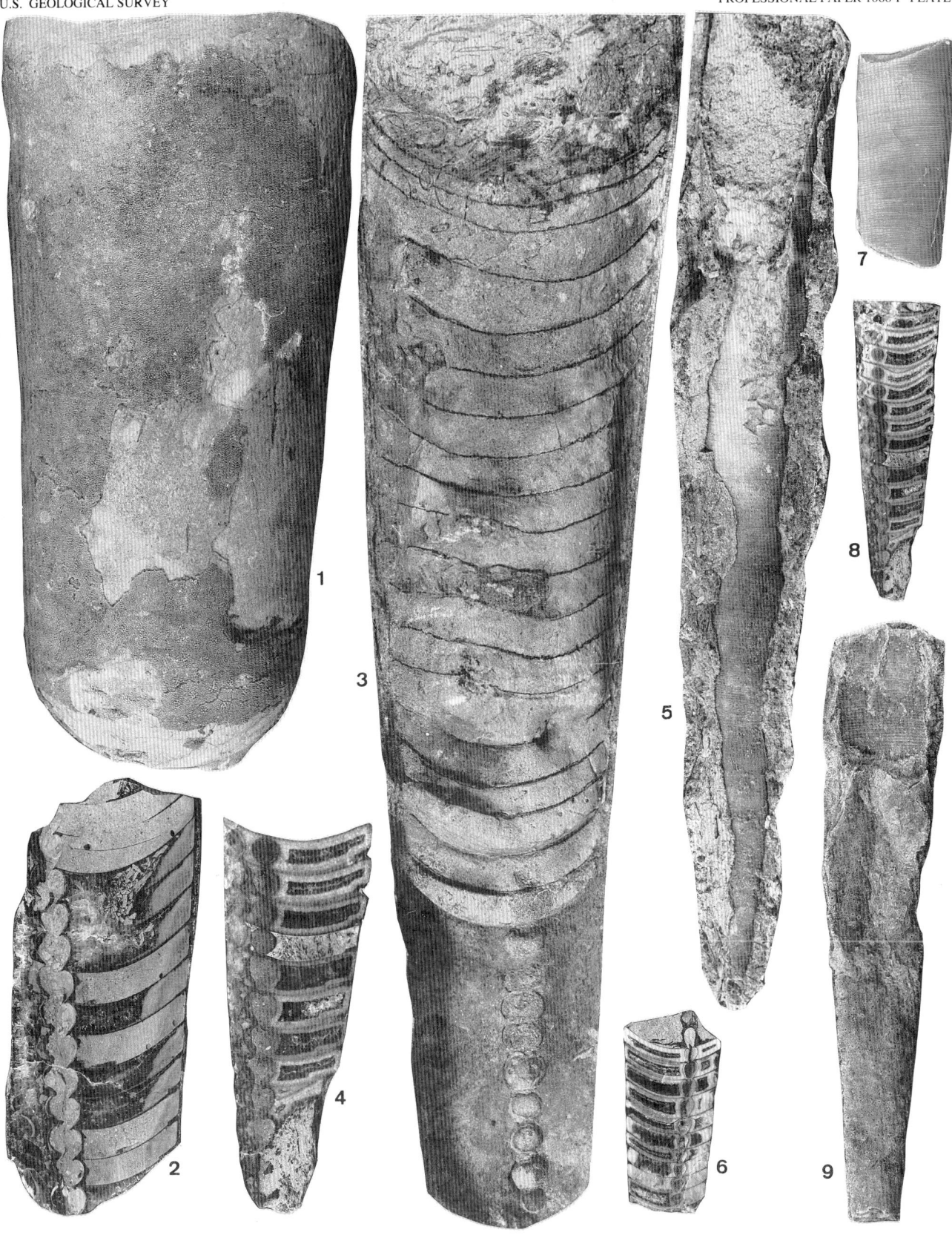

TREPTOCERAS

PLATE 9

FIGURES 1–3, 11–15. *Monomuchites annularis* new species (p. 56).

  1, 3. Exterior and adapical view of large silicified specimen, crushed adorally (×1). Tyrone Limestone (Rocklandian), USGS 6034-CO. Paratype USNM 158630.

  2. Dorsal view of naturally sectioned silicified specimen showing distinctive cameral deposits (×2). Same collection as figure 1 above. Paratype USNM 158671.

  11. Lateral view (×1) of exterior of silicified specimen showing slight curvature of shell, crushed adorally. Same collection as figure 1 above. Holotype USNM 158674.

  12, 13. Adoral and lateral views (×1) of internal mold of large specimen. Platteville Limestone (Rocklandian), Mineral Point, Wis. USNM 25275.

  14. Internal mold of a nearly complete specimen showing faint shell curvature (×1). Platteville Limestone (Rocklandian), Dixon, Ill. USNM 162068.

  15. Natural dorsal-ventral section through silicified specimen (×1) showing supracentral siphuncle and cameral deposits, venter to left. Same collection as figure 1 above. Paratype USNM 158676.

4–10. *Monomuchites obliquum* new species (p. 58).

  4. Adoral view (×1.5) of silicified specimen showing cameral deposits. Same collection as figure 1 above. Paratype USNM 158643.

  5–8. Dorsal, ventral, lateral, and adapical views (×1) of silicified specimen showing oblique annuli and central position of siphuncle. Same collection as figure 1 above. Holotype USNM 158653.

  9, 10. Ventral view of exterior (×1) and dorsal view exposing siphuncle and ventral parietal deposits (×1). Same collection as figure 1 above. Paratype USNM 158648.

16–19. *Monomuchites* cf. *M. costalis* Wilson, 1961 (p. 55).

  16, 17. Adoral and dorsal views of small silicified fragment (×1). Same collection as figure 1 above. USNM 158672.

  18, 19. Adoral and lateral views of small silicified fragment (×1). Same collection as figure 1 above. USNM 158667.

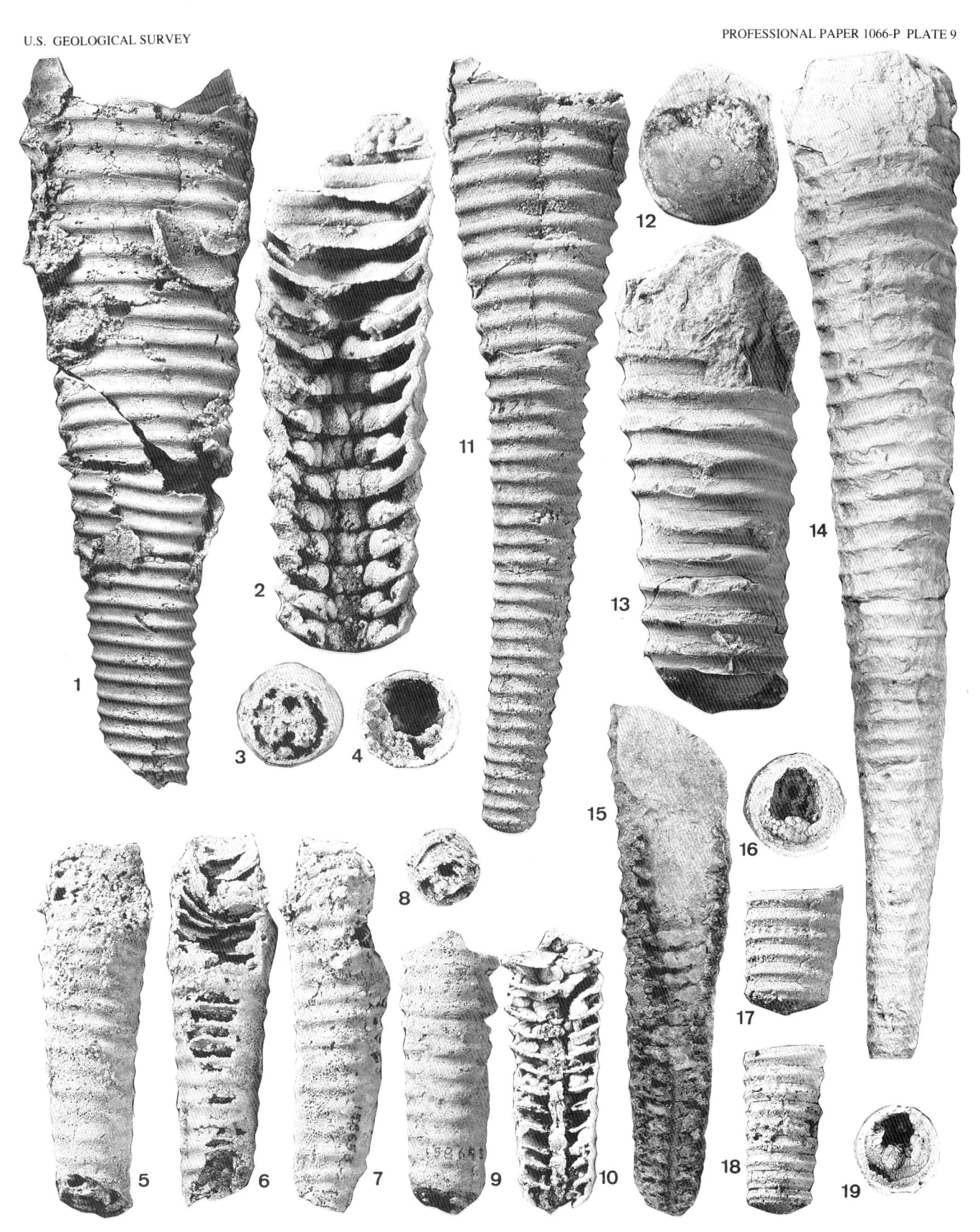

MONOMUCHITES

PLATE 10

Figures 1–9. *Isorthoceras albersi* (Miller and Faber), 1894b (p. 60).

    1. Shelled exterior showing longitudinal ornament (×1). "Trenton Limestone," Ludlow, Ky. Lectotype UC 360.

    2, 3. Dorsal-ventral section exposing siphuncle and cameral deposits (×1) and magnified view of same (×2), venter to left. Same collection as figure 1 above. Paralectotype UC 361.

    4. Tangential section through adapical portion of phragmocone revealing siphuncle and cameral deposits (×1.5). Point Pleasant Tongue of Clays Ferry Formation (Shermanian), Cincinnati, Ohio. MU 447T.

    5. Silicified exterior showing longitudinal ornament (×1). Logana Member, Lexington Limestone (Kirkfieldian), USGS 5092-CO. USNM 468697.

    6. Adoral view (×1.5) showing subcentral siphuncle. Grier Limestone Member, Lexington Limestone (Shermanian), USGS 5096-CO. USNM 468702.

    7. Silicified exterior showing longitudinal ornament (×1.5). Grier Limestone Member, Lexington Limestone (Shermanian), USGS 5067-CO. USNM 468695.

    8. Silicified exterior exposing part of siphuncle (×1.5). Same collection as figure 7 above. USNM 468696.

    9. Silicified exterior showing longitudinal ornament (×1.5). Same collection as figure 5 above. USNM 468698.

10–12. *Isorthoceras rogersensis* (Foerste), 1914 (p. 61).

    10, 11. Shelled exterior showing cancellate ornament (×1) and exfoliated venter showing cameral chambers (×1). "Rogers Gap Limestone" (Edenian?), Rogers Gap, Ky. Lectotype USNM 87193A.

    12. Naturally weathered section exposing siphuncle segments and cameral deposits (×1), venter on left. Same collection as figure 10 above. Paralectotype USNM 87193B.

13, 14. *Isorthoceras strangulatum* (Hall), 1847 (p. 59).

    13. Shelled internal mold of body chamber (×1). Trenton Limestone (Shermanian), Middleville, N.Y. Syntype AMNH 29714.

    14. Partially shelled internal mold with body chamber and much of phragmocone (×1). Same collection as figure 13 above. Syntype AMNH 29715.

15. *Isorthoceras sociale* (Hall), 1862 (p. 59).
Internal mold with body chamber and much of phragmocone (×1) and fragments of shell showing growth line. "*Orthoceras*" coquina in Elgin Member, Maquoketa Formation (Maysvillian), Graf, Iowa. MU 14566.

16. *Isorthoceras strigatum* (Hall), 1847 (p. 59).
Partially shelled internal mold of phragmocone and body chamber showing longitudinal ornament (×1). Compare with figures 1–9. Same collection as figure 13 above. Syntype AMNH 29692.

U.S. GEOLOGICAL SURVEY

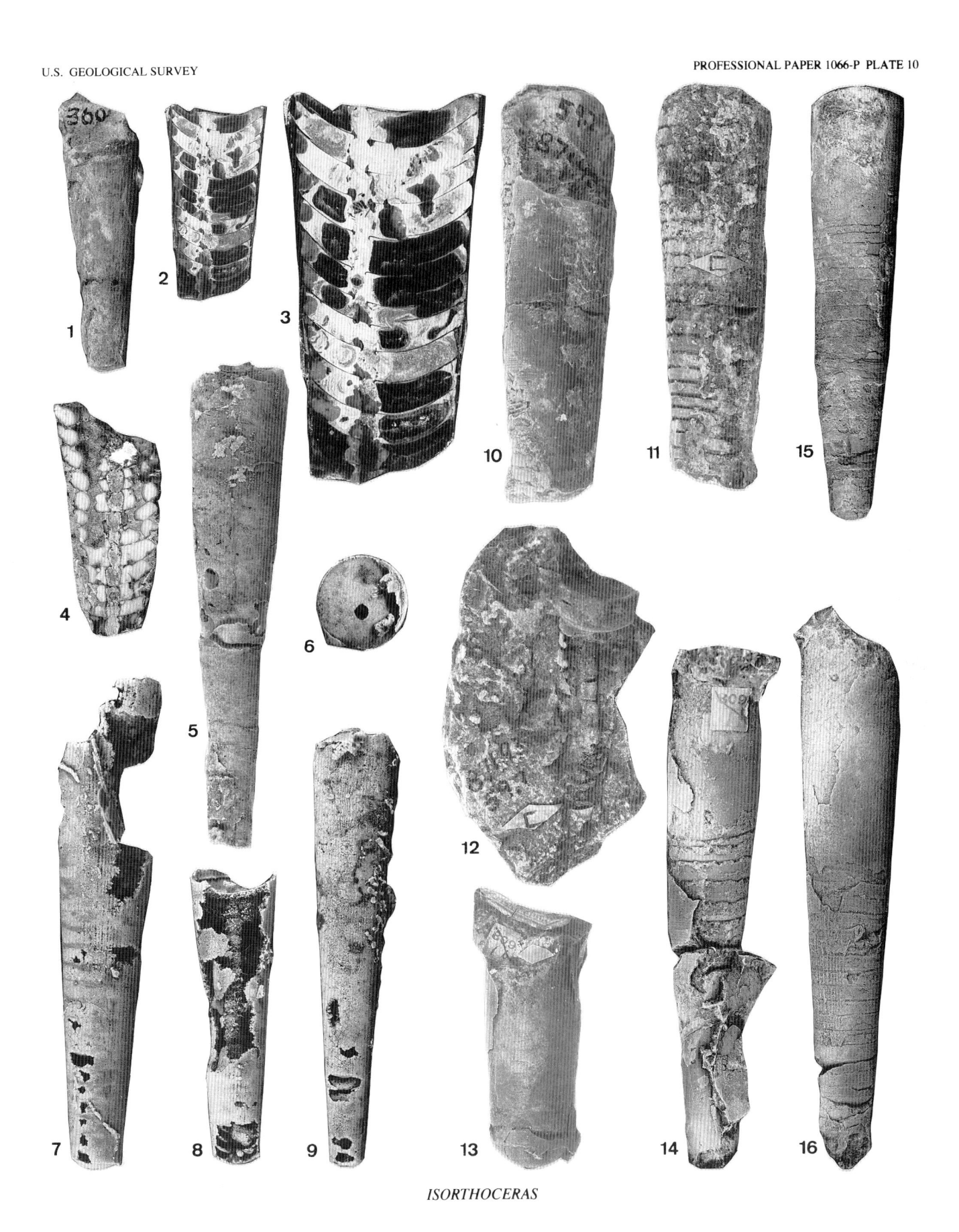

PROFESSIONAL PAPER 1066-P PLATE 10

ISORTHOCERAS

PLATE 11

FIGURES 1–3. *Polygrammoceras*? cf. *P.* sp A (p. 68).
Dorsal-ventral section revealing siphuncle, portion of lateral exterior, and adapical view showing siphuncle position in large calcite-replaced specimen (×1). Strodes Creek Member, Lexington Limestone (Shermanian), USGS 7310-CO. USNM 468694.

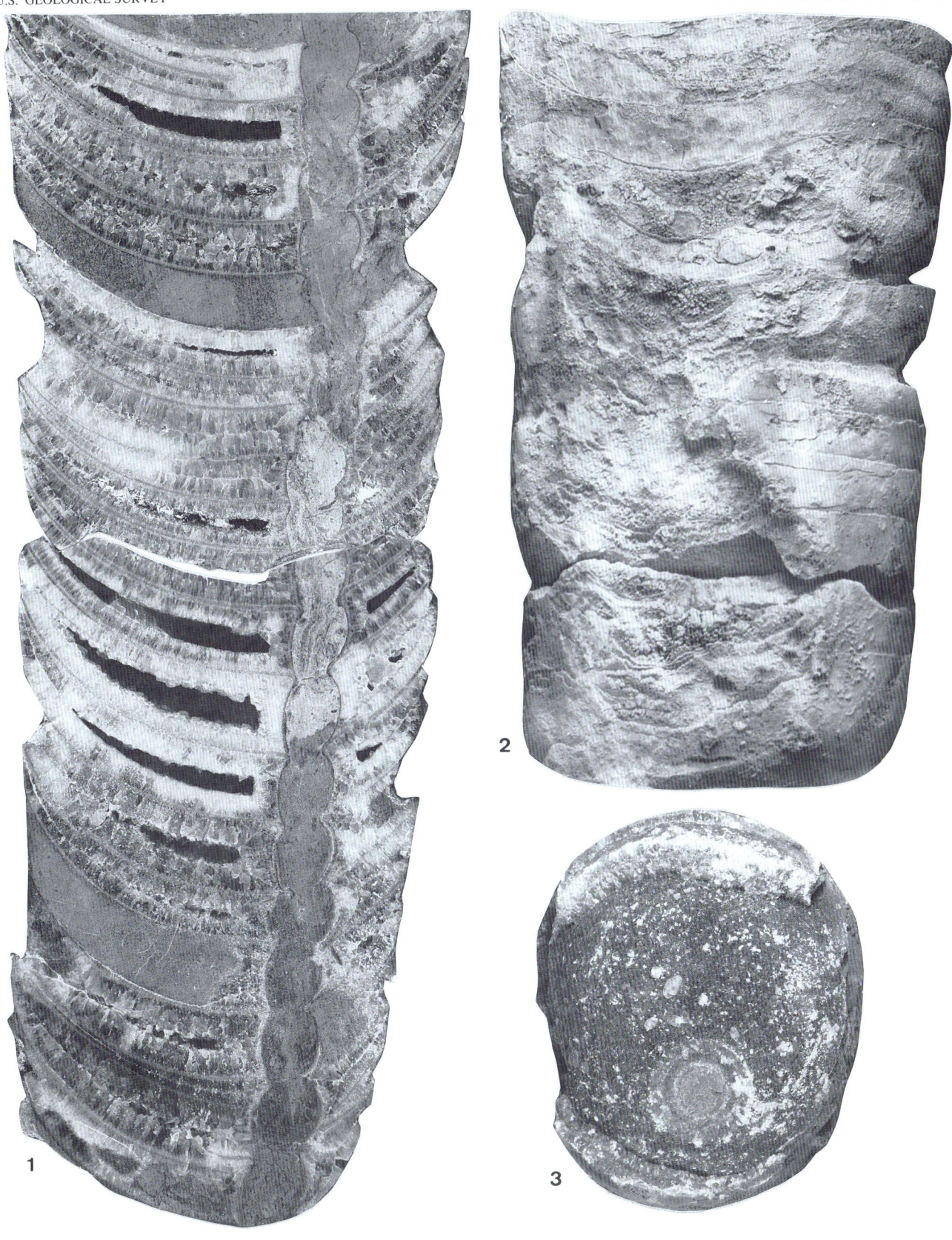

U.S. GEOLOGICAL SURVEY   PROFESSIONAL PAPER 1066-P PLATE 11

POLYGRAMMOCERAS ?

PLATE 12

FIGURES 1, 2. *Polygrammoceras* sp. A (p. 68)
Dorsal and ventral views of silicified specimen (×1) showing longitudinal ornament and globular siphuncle segments. Grier Limestone Member, Lexington Limestone (Shermanian), USGS 5067-CO. USNM 468710.

3. *Gorbyoceras* cf. *G. tetreauense* Wilson, 1961 (p. 65)
Lateral view of silicified specimen (×1.5) showing slight curvature of shell and external ornament. Logana Member, Lexington Limestone (Kirkfieldian), USGS 5092-CO. USNM 468708.

4, 6, 7, 10, 11. *Gorbyoceras hammelli* (Foerste), 1910 (p. 64).

    4, 7. Exterior of adoral portion of shelled phragmocone (×1) and tangential section through adapical piece of the same specimen showing siphuncle segments (×2). Saluda Formation (Richmondian), Versailles, Ind. (locality IND-5). MU 408T.

    6. Section through calcite-replaced phragmocone showing siphuncle segments (×1). Hitz Limestone Member of the Saluda Formation (Richmondian), Madison, Ind. (locality IND-6). MU 446T.

    10, 11. Calcite-replaced phragmocone and adapical view (×1). Same collection as figure 6 above. MU 409T.

5. *Anaspyroceras* cf. *A. cylindricum* (Foerste), 1932 (p. 66).
External mold showing oblique annuli and slender shell form (×1). Curdsville Limestone Member, Lexington Limestone (Kirkfieldian), Curdsville, Ky. USNM 48290.

8, 9. *Polygrammoceras*? sp. B (p. 69).
Lateral exterior showing longitudinal furrows and adapical view showing central position of siphuncle (×1.5). Tyrone Limestone (Rocklandian), USGS 6034-CO. USNM 468711.

12, 13. *Gorbyoceras gorbyi* (Miller), 1894 (p. 63).
Lateral and ventral views of a nearly complete internal mold of a large phragmocone showing oblique annuli and slender, slightly curved shell (×0.8). Saluda Formation (Richmondian), Hamburg, Ind. (locality IND-4). MU 402T.

POLYGRAMMOCERAS, ANASPYROCERAS, AND GORBYOCERAS

PLATE 13

FIGURES 1–8. *Richmondoceras brevicameratum* new species (p. 71).

    1, 3, 4. Internal mold showing most of body chamber and adoral portions of phragmocone (×1), adapical view showing subcentral position of siphuncle (×1), and dorsal-ventral section through attached phragmocone revealing siphuncle segments (×1). Saluda Formation (Richmondian), Laugherty Creek, near Versailles, Ind. Holotype MU 398T.

    2. Internal mold of adoral portion of large phragmocone showing approximate septa (×1). Saluda Formation (Richmondian), Oxford, Ohio. Paratype MU 399T.

    5. Tangential section of adoral portion of phragmocone revealing siphuncle segments (×1). Upper part of the Bull Fork Formation (Richmondian), Clinton County, Ohio (locality OH-10). MU 400T(A).

    6. Dorsal-ventral section exposing siphuncle segments (×1). Same collection as figure 5 above. MU 400T(B).

    7, 8. Dorsal-ventral section through adapical portion of phragmocone showing siphuncle segments (×2) and a magnified view of the same showing annulosiphonate deposits and cameral deposits. Same collection as figure 5 above. MU 400T(C).

U.S. GEOLOGICAL SURVEY

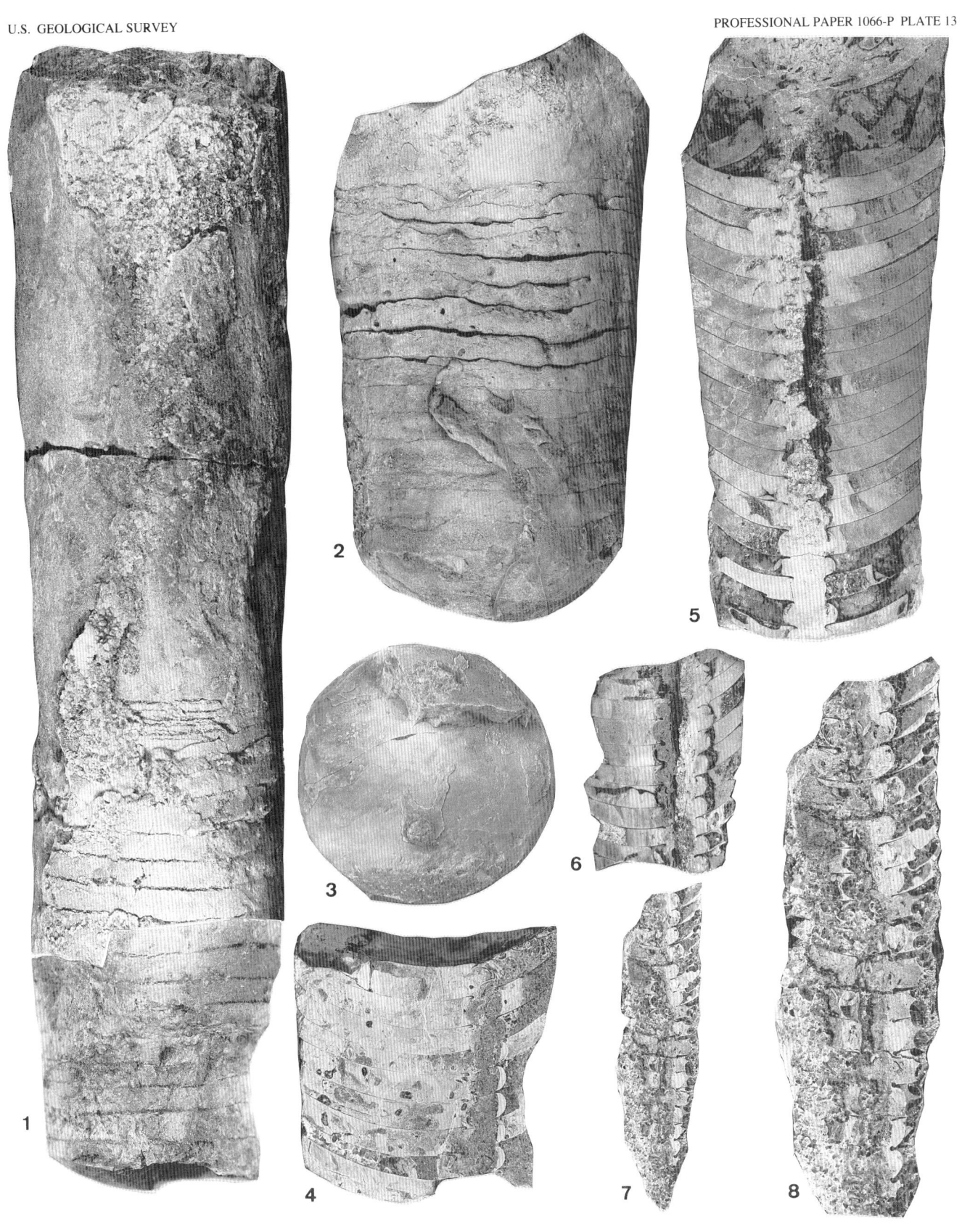

PROFESSIONAL PAPER 1066-P PLATE 13

RICHMONDOCERAS

PLATE 14

FIGURES 1, 2. *Cameroceras rowenaense* new species (p. 74).
Dorsal-ventral and adapical view of calcite-replaced specimen showing siphuncle structure and position (×1). Leipers Limestone (Maysvillian), Rowena, Ky. (locality KY-4). Holotype MU 266T.

3, 4. *Cameroceras* cf. *C. trentonense* Conrad, 1842 (p. 75).
Adoral and lateral views of silicified siphuncle showing location of endosiphuncular tube and rate of expansion (×1). Grier Limestone Member, Lexington Limestone (Shermanian), USGS 4073-CO. USNM 468712.

5, 6. *Cameroceras*? sp. (p. 76).
Adoral view showing depressed cross section of siphuncle (×1) and tangential section revealing large-diameter siphuncle and macroholochoanitic septal necks (×1). Upper part of the Bull Fork Formation (Richmondian), Owingsville, Ky. (locality KY-5). USNM 468713.

U.S. GEOLOGICAL SURVEY

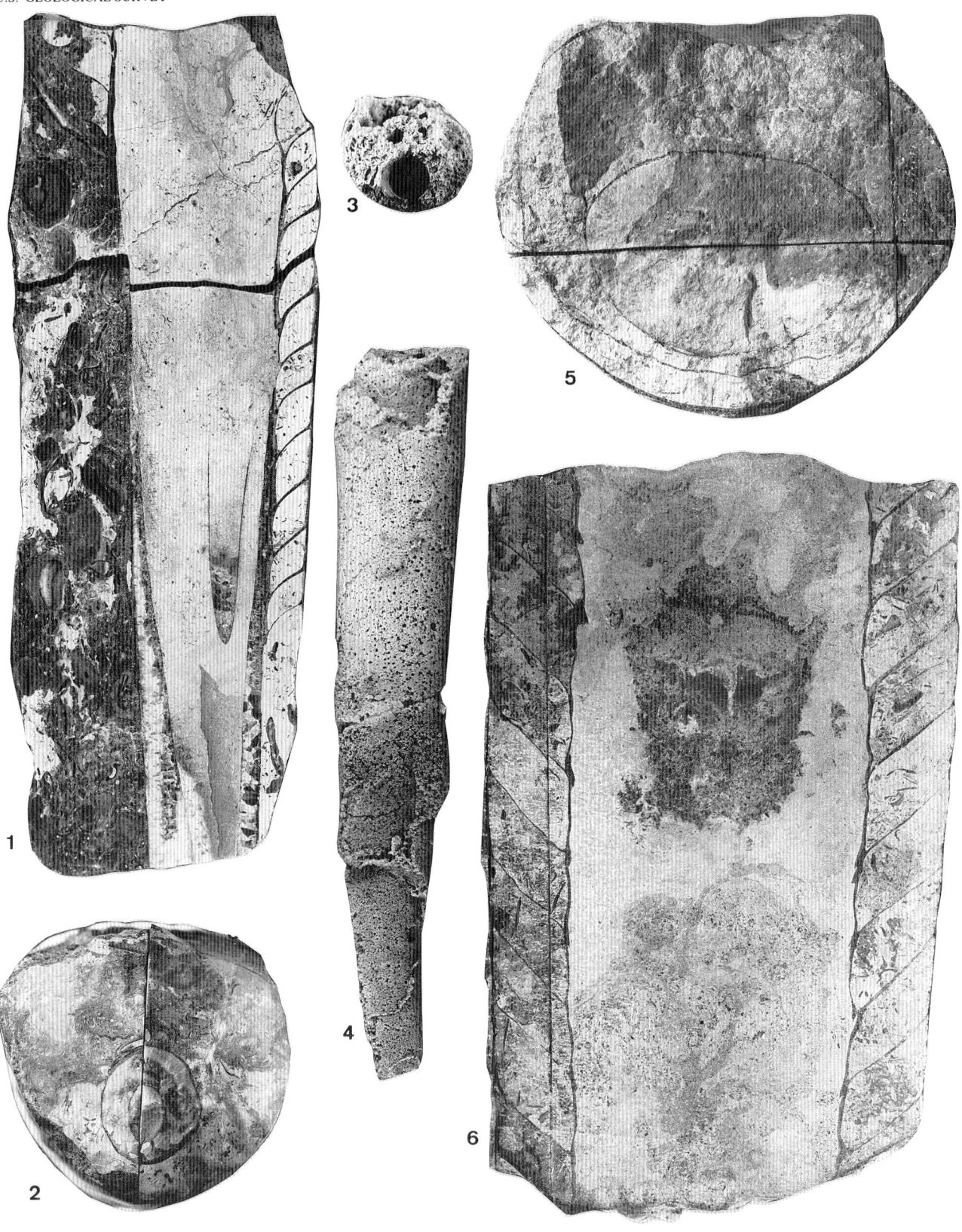

PROFESSIONAL PAPER 1066-P PLATE 14

CAMEROCERAS

PLATE 15

FIGURES 1–4, 8–12. *Vaginoceras* sp. A (p. 77).
- 1–4. Lateral exterior, dorsal-ventral section, adoral, and adapical views of a large silicified specimen (×1). Note faint, raised transverse ornament on exterior and siphuncle diameter and position. Tyrone Limestone (Rocklandian), USGS 6034-CO. USNM 158668.
- 8–10. Lateral (×1), dorsal (×1), and adoral (×1.5) views of the apical end of a silicified siphuncle revealing endosiphuncular blades typical of genus. Same collection as figure 1 above. USNM 158652.
- 11, 12. Dorsal and lateral views (×1) of apical end of a silicified siphuncle. Same collection as figure 1 above. USNM 468714.

5–7. *Vaginoceras* sp. B (p. 78).

Lateral, adoral, and adapical views (×1) of silicified portion of siphuncle revealing endosiphuncular blades typical of genus. Curdsville Limestone Member, Lexington Limestone (Kirkfieldian), USGS 5072-CO. USNM 468718.

U.S. GEOLOGICAL SURVEY

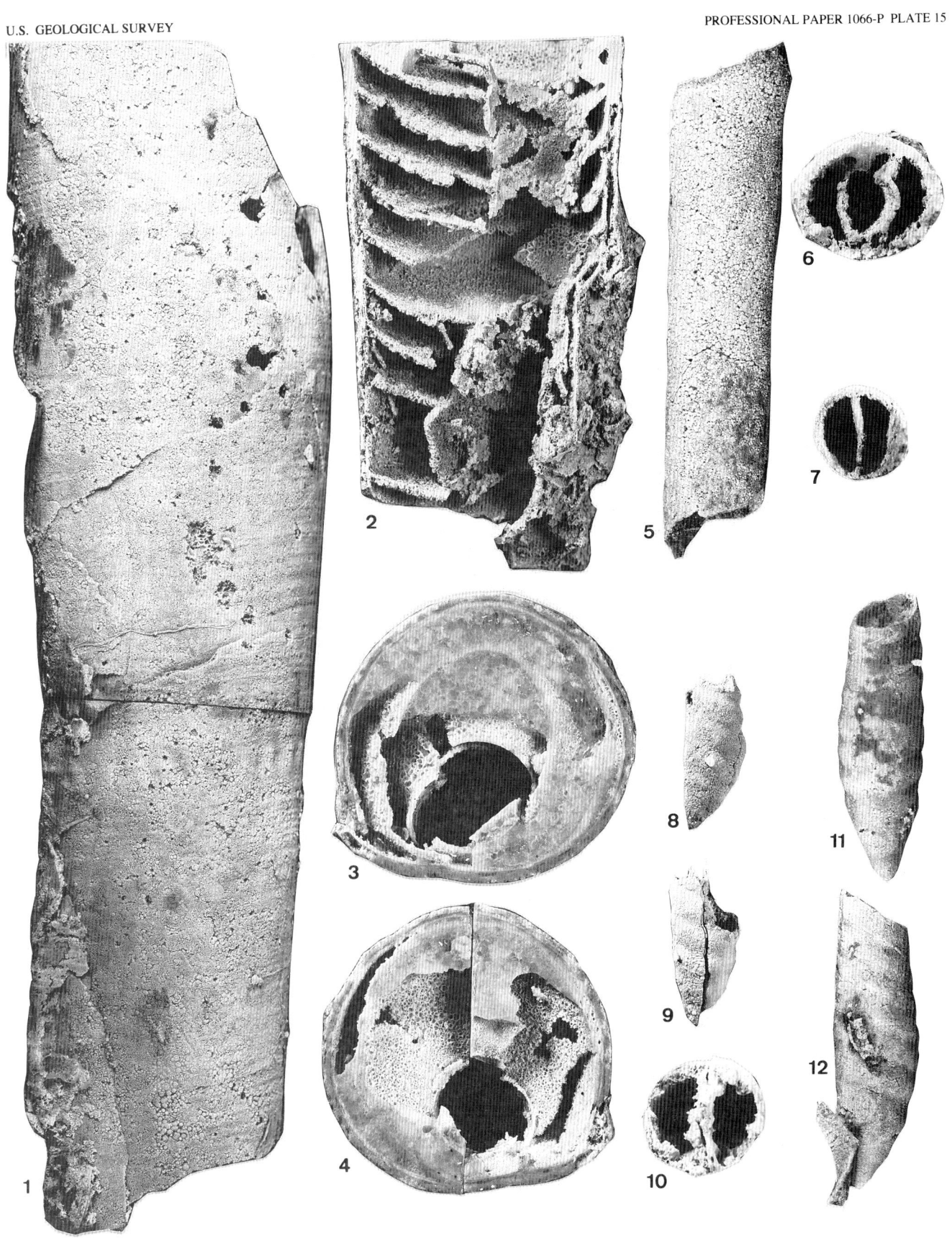

PROFESSIONAL PAPER 1066-P PLATE 15

VAGINOCERAS

PLATE 16

FIGURES 1–5. *Triendoceras? davisi* new species (p. 79).
  1. Ventral view of infilling of large siphuncle showing impressions of narrow connecting rings and initially rapid expansion of tube (×0.8). Kope Formation (Edenian), Ft. Mitchell, Ky. (locality KY-1). Holotype MU 410T.
  2. Ventral view of adapical infilling of siphuncle showing rapid expansion of endosiphuncular tube (×1). Kope Formation (Edenian), Cincinnati, Ohio (locality OH-1). Paratype MU 411T.
  3. Cross section through infilling of siphuncle tube showing subtriangular shape, venter up (×1). "Fulton Shale" (Edenian), Moscow, Ohio. Paratype MU 414T.
  4. Portion of phragmocone preserving several cameral chambers (×1). Same collection as figure 1 above. Paratype MU 412T.
  5. Lateral view of infilling of siphuncle tube (×1). Kope Formation (Edenian), Cincinnati, Ohio. Paratype MU 413T.
  6. *Actinoceras altopontense* Foerste and Teichert, 1930 (p. 82).
  Lateral view of a large silicified phragmocone revealing cameral chambers and rapid rate of shell expansion (×1). USGS 6034-CO. USNM 158640.

U.S. GEOLOGICAL SURVEY

PROFESSIONAL PAPER 1066-P PLATE 16

TRIENDOCERAS (?) AND *ACTINOCERAS*

PLATE 17

FIGURES 1–5. *Actinoceras kentuckiense* Foerste and Teichert, 1930 (p. 82).

    1, 2, 4, 5. Adoral (×1), dorsal (×0.8), dorsal-ventral section (×1), and adapical views (×1) of a large silicified specimen. Note septal necks and the vascular canal system evident in dorsal-ventral adapical section. Tyrone Limestone (Rocklandian), USGS 6034-CO. USNM 158670.

    3. Large silicified portion of phragmocone exposing siphuncle segments and camerae (×1). Same collection as figure 1 above. USNM 158647.

6–10. *Actinoceras altopontense* Foerste and Teichert, 1930 (p. 82).

    6. Adapical view of silicified siphuncle segment showing central canal and impressions of radial canals (×1). Same collection as figure 1 above. USNM 158658.

    7. Adapical view of more adoral siphuncle segment showing decreased diameter of central canal and impressions of radial canals (×1). Same collection as figure 1 above. USNM 158657.

    8, 9. Adapical and dorsal views of silicified specimen exposing siphuncle (×1). Same collection as figure 1 above. USNM 158641.

    10. Large silicified specimen exposing siphuncle (×1). Same collection as figure 1 above. USNM 158642.

11. *Actinoceras curdsvillense* Foerste and Teichert, 1930 (p. 84).
Dorsal view of natural section of silicified mold (×1). Note highly recurved, short septal necks. Curdsville Limestone Member, Lexington Limestone (Kirkfieldian), Curdsville, Ky. Holotype USNM 48425.

U.S. GEOLOGICAL SURVEY

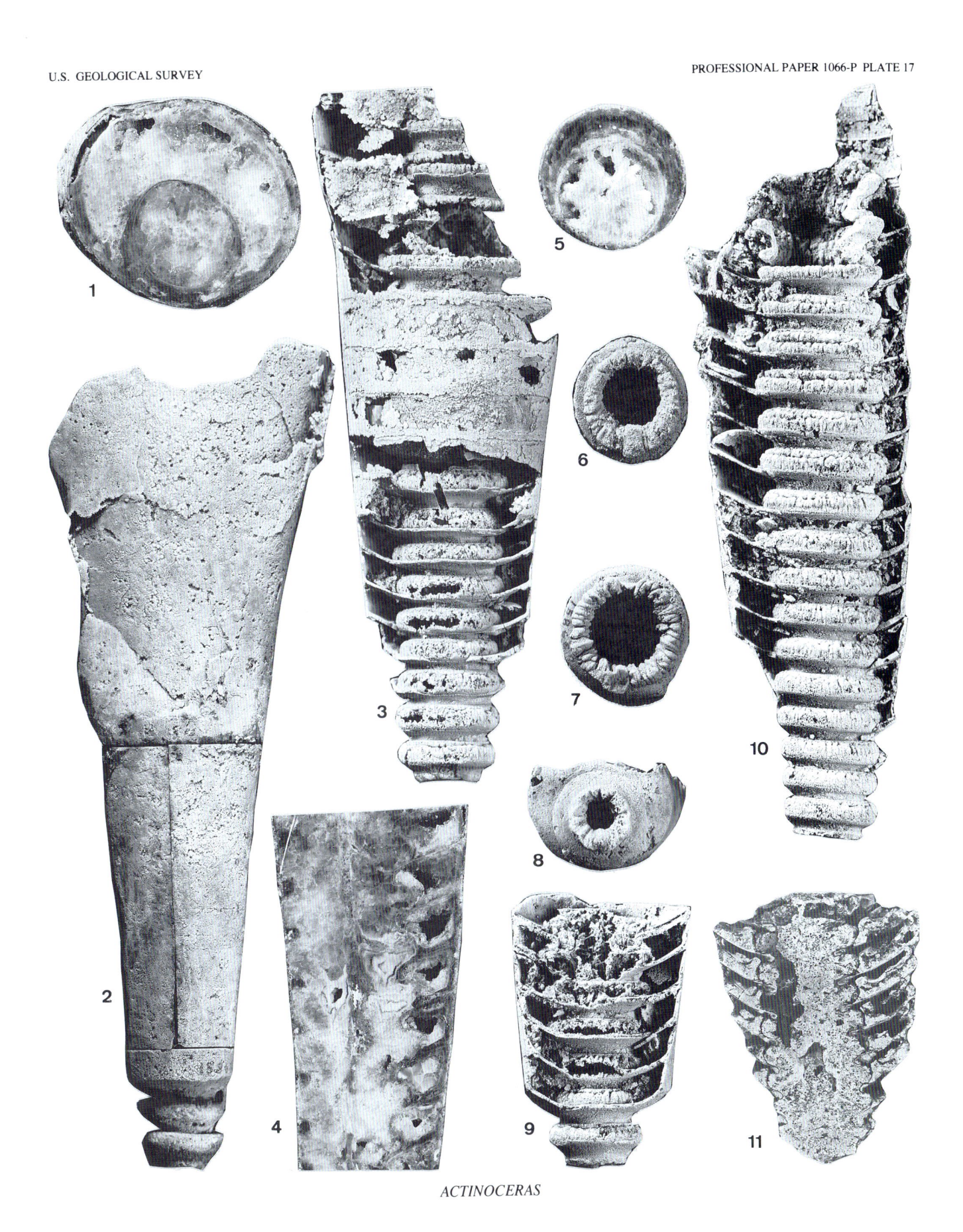

PROFESSIONAL PAPER 1066-P PLATE 17

ACTINOCERAS

PLATE 18

FIGURES 1–7. *Ormoceras ferecentricum* Foerste and Teichert, 1930 (p. 86).
    1–3. Adoral, ventral, and adapical views (×1) of silicified specimen exposing siphuncle. Tyrone Limestone (Rocklandian), High Bridge, Ky. Holotype USNM 82214.
    4–6. Ventral, adoral, and lateral views (×1) of adapical portion of silicified phragmocone exposing siphuncle. Tyrone Limestone (Rocklandian), USGS 6034-CO. USNM 158639.
    7. Nearly complete silicified phragmocone (×1). Same collection as figure 4 above. USNM 468721.

8, 9. *Deiroceras curdsvillense* Foerste and Teichert, 1930 (p. 87).
    8. Dorsal-ventral section of silicified specimen exposing siphuncle (×1), venter on left. Curdsville Limestone Member, Lexington Limestone (Kirkfieldian), Curdsville, Ky. Holotype USNM 48221.
    9. Dorsal-ventral section of silicified specimen exposing siphuncle (×1), venter on left. Curdsville Limestone Member, Lexington Limestone (Kirkfieldian), USGS 5100-CO. USNM 468723.

10–17. *Orthonybyoceras dyeri* (Miller), 1875 (p. 89).
    10. Dorsal-ventral section exposing siphuncle (×1.25), venter on left. "Waynesville biofacies" in Bull Fork Formation (Richmondian), Weisburg, Ind. UC 31380.
    11. Tangential section exposing siphuncle and annulosiphonate deposits (×1). Fairview Formation (Maysvillian), Cincinnati, Ohio. UC 352A.
    12. Dorsal-ventral section exposing siphuncle and annulosiphonate deposits (×1). Fairview Formation (Maysvillian), Cincinnati, Ohio. UC 1330A.
    13, 15, 16. Adapical, adoral, and dorsal-ventral section (×1) of adoral portion of phragmocone. Upper Ordovician, "hills behind Cincinnati, Ohio." Holotype MCZ 3403.
    14, 17. Lateral exterior of internal mold and dorsal-ventral section (×1) of the same, exposing siphuncle segments and deposits (×1). "Maysville Group" (Maysvillian), Covington, Ky. Holotype of *Ormoceras? covingtonense* Foerste and Teichert, 1930. USNM 48258.

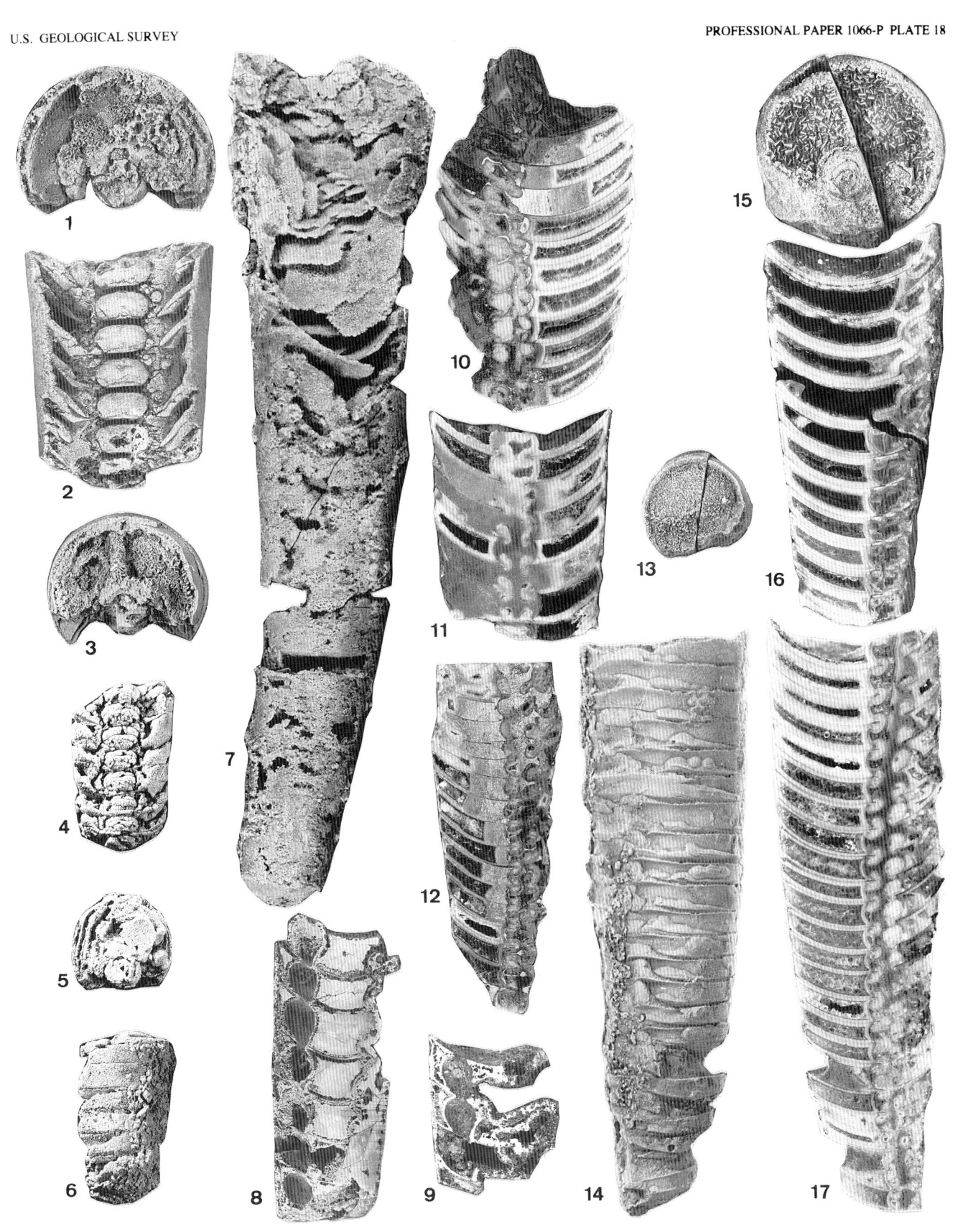

ORMOCERAS, DEIROCERAS, AND ORTHONYBYOCERAS

PLATE 19

FIGURES 1–3. *Plectoceras* cf. *P. carletonense* Foerste, 1933 (p. 93).
Lateral, ventral, and dorsal views of silicified inner whorl showing ventral sinus and camerae (×1). Tyrone Limestone (Rocklandian), USGS 6034-CO. USNM 158651.

4, 5. *Plectoceras halli* (Foord), 1891 (p. 94).
4. Apertural view showing depressed subquadrate whorl profile (×1). Label states "Trenton, near Lorette, Quebec." Flower collection, uncataloged.
5. Lateral view of a large specimen showing development of lateral costae and fine oblique growth lines (×1). Same collection as figure 4 above.

6–10. *Trocholites faberi* Foerste, 1929c (p. 91).
6, 7. Lateral and ventral views (×1) of a large specimen with complete body chamber. Note U-shaped hyponomic sinus. Kope Formation (Edenian), Ft. Mitchell, Ky. (locality KY-1). USNM 468724.
8, 9. Fragmentary specimen preserving part of body whorl and inner whorl (×1). Adoral view of outer whorl (×1). Same collection as figure 6 above. USNM 468725.
10. Small, slightly deformed internal mold with body chamber and phragmocone (×1). Same collection as figure 6 above. USNM 468726.

11–15. *Oncoceras major* new species (p. 95).
Dorsal, ventral, lateral, apertural, and adapical views (×1) of a silicified specimen preserving complete body chamber and most of phragmocone. Note distinctive transverse costae. Tyrone Limestone (Rocklandian), USGS 6034-CO. Holotype USNM 158665.

U.S. GEOLOGICAL SURVEY

PROFESSIONAL PAPER 1066-P PLATE 19

PLECTOCERAS, *TROCHOLITES*, AND *ONCOCERAS*

PLATE 20

FIGURES 1, 2. *Oncoceras* sp. (p. 96)

    1. Adapical view of silicified specimen (×1) showing cross-sectional shape and position of siphuncle. Grier Limestone Member, Lexington Limestone (Shermanian), USGS 4879-CO. USNM 468732.

    2. Dorsal view of adapical portion of phragmocone (×1) showing external ornament of fine transverse growth lines. Grier Limestone Member, Lexington Limestone (Shermanian), USGS 4883-CO. USNM 468731.

3–6. *Beloitoceras* cf. *B. huronense* (Billings), 1865 (p. 99).

    3, 4, 6. Dorsal, lateral, and apertural views of small silicified specimen (×1) preserving most of body chamber. Tyrone Limestone (Rocklandian), USGS 6034-CO. USNM 158666.

    5. Lateral view of part of phragmocone (×1.5). Same collection as figure 3 above. USNM 468733.

7–12. *Maelonoceras* cf. *M. praematurum* (Billings), 1860 (p. 100).

    7, 8, 11, 12. Apertural (venter down), adapical, dorsal, and lateral views (×1) of silicified specimen with incomplete aperture. Same collection as figure 3 above. USNM 158649.

    9, 10. Apertural (venter down) and lateral views of complete body chamber (×1). Same collection as figure 3 above. USNM 468735.

13–18. *Laphamoceras* cf. *L. scofieldi* Foerste, 1932 (p. 101).

    13–15. Dorsal, lateral, and ventral views of a small silicified specimen (×2) showing external ornament and sponge(?) encrustations. Same collection as figure 3 above. USNM 158663.

    16–18. Lateral and dorsal views of silicified specimen (×1) showing strong curvature of shell. Same collection as figure 3 above. USNM 158654.

19–22. *Allumettoceras* cf. *A. tenerum* (Billings), 1860 (p. 103).

    19. Adapical view (×1) showing depressed cross section and ventral siphuncle. Logana Member, Lexington Limestone (Kirkfieldian), USGS 6419-CO. USNM 468738.

    20. Adoral view of a smaller specimen (×1) showing depressed cross section and expanded siphuncle segments. Logana Member, Lexington Limestone (Kirkfieldian), USGS 4865-CO. USNM 468739.

    21, 22. Lateral and dorsal views (×1) of most complete phragmocone observed. Same collection as figure 19 above. USNM 468640.

23–27. *Augustoceras shideleri* Flower, 1946 (p. 111).
Ventral, dorsal, adoral, adapical, and lateral views of a nearly complete specimen preserving body chamber (×1). Leipers Limestone (Maysvillian), Rowena, Ky. (locality KY-4). MU 428T.

28, 29. *Manitoulinoceras gyroforme* Flower, 1946 (p. 112).
Ventral and lateral views of an internal mold of phragmocone (×1) showing short camerae, marginal expanded siphuncle segments, and strong shell curvature. Upper part of the Bull Fork Formation (Richmondian), Fairborn, Ohio (locality OH-12). USNM 468719.

U.S. GEOLOGICAL SURVEY

PROFESSIONAL PAPER 1066-P PLATE 20

ONCOCERAS, BELOITOCERAS, MAELONOCERAS, LAPHAMOCERAS ALLUMETTOCERAS, AUGUSTOCERAS, AND *MANITOULINOCERAS*

PLATE 21

FIGURES 1, 2, 5. *Diestoceras shideleri* Foerste, 1924b (p. 107).
        1, 2. Ventral and lateral views of a large, nearly complete specimen preserving all of the body chamber and much of the phragmocone (×1). Note slight endogastric curvature of shell adapically. Saluda Formation (Richmondian), north of Brookville, Ind. (locality IND-3). MU 429T.
        5. Ventral view of internal mold of body chamber showing well-developed hyponomic sinus (×1). Saluda Formation (Richmondian), Hamburg, Ind. (locality IND-4). MU 430T.
    3. *Beloitoceras amoenum* (Miller), 1879 (p. 97).
Lateral view of nearly complete internal mold preserving body chamber (×1). Bull Fork Formation (Richmondian), Dayton, Ohio (locality OH-13). USNM 468668.
    4. *Diestoceras eos* (Hall and Whitfield), 1875 (p. 108).
Lateral view of internal mold of body chamber (×1), venter to right. Upper part of the Bull Fork Formation (Richmondian), Clinton County, Ohio (locality OH-10). USNM 468730.

U.S. GEOLOGICAL SURVEY

PROFESSIONAL PAPER 1066-P PLATE 21

DIESTOCERAS AND *BELOITOCERAS*

PLATE 22

FIGURES 1, 3. *Diestoceras shideleri* Foerste, 1924b (p. 107).
Lateral and apertural views of body chamber showing general dimensions and subtriangular aperture shape (×1). Saluda Formation (Richmondian), Hamburg, Ind. (locality IND-4). MU 430T.

2. *Beloitoceras amoenum* (Miller), 1879 (p. 97).
Lateral view of internal mold of nearly complete phragmocone with attached body chamber (×1). Whitewater Formation (Richmondian), Richmond, Ind. (locality IND-7). USNM 468669.

4, 5. *Beloitoceras bucheri* Flower 1946 (p. 98).
Ventral and lateral views of a nearly complete, shelled specimen (×1). Note slender phragmocone and fine transverse ornament. Same collection as figure 1 above. MU 433T.

6–9. *Diestoceras indianense* (Miller and Faber), 1894 (p. 105).

6, 7, 9. Apertural, lateral, and ventral views of a nearly complete internal mold, including body chamber (×1). Note triangular aperture. Same collection as figure 1 above. MU 432T.

8. Ventral view of a partially shelled specimen showing ventral hyponomic sinus and external growth lines (×1). Same collection as figure 1 above. MU 431T.

U.S. GEOLOGICAL SURVEY

PROFESSIONAL PAPER 1066-P PLATE 22

DIESTOCERAS AND *BELOITOCERAS*